AF786089

The Helmholtz Legacy in Physiological Acoustics

Archimedes

NEW STUDIES IN THE HISTORY AND PHILOSOPHY OF SCIENCE AND TECHNOLOGY

VOLUME 39

EDITOR

JED Z. BUCHWALD, *Dreyfuss Professor of History, California Institute of Technology, Pasadena, CA, USA.*

ASSOCIATE EDITORS FOR MATHEMATICS AND PHYSICAL SCIENCES

JEREMY GRAY, *The Faculty of Mathematics and Computing, The Open University, Buckinghamshire, UK.*
TILMAN SAUER, *California Institute of Technology*

ASSOCIATE EDITORS FOR BIOLOGICAL SCIENCES

SHARON KINGSLAND, *Department of History of Science and Technology, Johns Hopkins University, Baltimore, MD, USA.*
MANFRED LAUBICHLER, *Arizona State University*

ADVISORY BOARD FOR MATHEMATICS, PHYSICAL SCIENCES AND TECHNOLOGY

HENK BOS, *University of Utrecht*
MORDECHAI FEINGOLD, *California Institute of Technology*
ALLAN D. FRANKLIN, *University of Colorado at Boulder*
KOSTAS GAVROGLU, *National Technical University of Athens*
PAUL HOYNINGEN-HUENE, *Leibniz University in Hannover*
TREVOR LEVERE, *University of Toronto*
JESPER LÜTZEN, *Copenhagen University*
WILLIAM NEWMAN, *Indiana University, Bloomington*
LAWRENCE PRINCIPE, *The Johns Hopkins University*
JÜRGEN RENN, *Max-Planck-Institut für Wissenschaftsgeschichte*
ALEX ROLAND, *Duke University*
ALAN SHAPIRO, *University of Minnesota*
NOEL SWERDLOW, *California Institute of Technology, USA*

ADVISORY BOARD FOR BIOLOGY

MICHAEL DIETRICH, *Dartmouth College, USA*
MICHEL MORANGE, *Centre Cavaillès, Ecole Normale Supérieure, Paris*
HANS-JÖRG RHEINBERGER, *Max Planck Institute for the History of Science, Berlin*
NANCY SIRAISI, *Hunter College of the City University of New York*

Archimedes has three fundamental goals; to further the integration of the histories of science and technology with one another: to investigate the technical, social and practical histories of specific developments in science and technology; and finally, where possible and desirable, to bring the histories of science and technology into closer contact with the philosophy of science. To these ends, each volume will have its own theme and title and will be planned by one or more members of the Advisory Board in consultation with the editor. Although the volumes have specific themes, the series itself will not be limited to one or even to a few particular areas. Its subjects include any of the sciences, ranging from biology through physics, all aspects of technology, broadly construed, as well as historically-engaged philosophy of science or technology. Taken as a whole, *Archimedes* will be of interest to historians, philosophers, and scientists, as well as to those in business and industry who seek to understand how science and industry have come to be so strongly linked.

For further volumes:
http://www.springer.com/series/5644

Erwin Hiebert
1919–2012

The Helmholtz Legacy in Physiological Acoustics

Springer

Erwin Hiebert (Deceased)

ISSN 1385-0180 ISSN 2215-0064 (electronic)
ISBN 978-3-319-06601-1 ISBN 978-3-319-06602-8 (eBook)
DOI 10.1007/978-3-319-06602-8
Springer Cham Heidelberg New York Dordrecht London

Library of Congress Control Number: 2014941950

© Springer International Publishing Switzerland 2014
This work is subject to copyright. All rights are reserved by the Publisher, whether the whole or part of the material is concerned, specifically the rights of translation, reprinting, reuse of illustrations, recitation, broadcasting, reproduction on microfilms or in any other physical way, and transmission or information storage and retrieval, electronic adaptation, computer software, or by similar or dissimilar methodology now known or hereafter developed. Exempted from this legal reservation are brief excerpts in connection with reviews or scholarly analysis or material supplied specifically for the purpose of being entered and executed on a computer system, for exclusive use by the purchaser of the work. Duplication of this publication or parts thereof is permitted only under the provisions of the Copyright Law of the Publisher's location, in its current version, and permission for use must always be obtained from Springer. Permissions for use may be obtained through RightsLink at the Copyright Clearance Center. Violations are liable to prosecution under the respective Copyright Law.
The use of general descriptive names, registered names, trademarks, service marks, etc. in this publication does not imply, even in the absence of a specific statement, that such names are exempt from the relevant protective laws and regulations and therefore free for general use.
While the advice and information in this book are believed to be true and accurate at the date of publication, neither the authors nor the editors nor the publisher can accept any legal responsibility for any errors or omissions that may be made. The publisher makes no warranty, express or implied, with respect to the material contained herein.

Printed on acid-free paper

Springer is part of Springer Science+Business Media (www.springer.com)

Envoi

In July 2012, I spent a short time with Erwin at his home in Belmont. His son, Tom, was there to help, since Erwin's beloved wife, Elfrieda, was in hospital and passed away 2 months later. We had first met in the fall of 1971 when I entered Harvard as a graduate student in the history of science. Throughout the nearly half century since then Erwin and Elfrieda remained, and will remain in the memories of their students and friends, paragons of friendship, warmth, intellectual engagement, and, most of all, decency, true morality, and concern.

Erwin worried about this last product of his scholarship, for he knew that the end of life was fast approaching. Having seen the work as it evolved over the years, I knew it to be the fruits of decades of thought and research, stimulated and assisted by the insights of Elfrieda, and so we told him that it would be published in the series "Archimedes."

Erwin and Elfrieda were the very best of people and the very best of academics. There are, and certainly will always be, few like them. They will be sorely missed by their children, family, friends, colleagues, and students.

California Institute of Technology
Pasadena, CA, USA

Jed Buchwald

Eloge

Erwin Hiebert died on November 28, 2012, less than 3 months after his wife Elfrieda died, on September 2. For the dozens of graduate students, colleagues, and friends who tasted the intellectual excitement of discussions in the warmth of their home, these deaths mark the passing of an age. For more than three decades, as a Professor of the History of Science first at the University of Wisconsin, Madison, then at Harvard, Erwin passionately engaged his students and colleagues in explorations that ranged from nuclear physics to experimental physiology, thermodynamics to Cantorian set theory, quantum mechanics to comets. Acoustics was always a favorite subject, and the best evenings were those that Elfrieda closed by playing on her beloved piano.

Erwin was the third of seven children of a Mennonite Bretheren minister, who raised his children in an urban Russian Mennonite community in Winnipeg, Manitoba. In later years, he loved to point to his early education, first in Faraday Grade School and then in Sir Isaac Newton High School, as indicators of his subsequent career as a Historian of Science. However, that career was far in the future for a young man who spent his summers following the wheat harvest from Oklahoma to the Dakotas to pay for his postsecondary education at Tabor College in Hillsboro, Kansas. After 2 years, he transferred to Bethel College, where in 1941 he earned a bachelor's degree in Mathematics and Chemistry. In 1943 he received a master's degree in Chemistry and Physics from the University of Kansas in Lawrence.

Also in 1943, Erwin married Elfrieda, née Franz. Elfrieda was already a highly accomplished pianist, who in 1938 had received the highest award in the National Music Competition in Colorado Springs, Colorado; when the two met, she was studying music at Tabor College. Immediately after their marriage, the young couple moved to Chicago, where in 1945 Elfrieda earned a bachelor's degree and in 1946 a master's degree in Music from the University of Chicago. Erwin was enlisted as a Research Chemist at Standard Oil Company of Indiana in those years, and Elfrieda was Assistant Music Librarian for the University of Chicago.

Erwin's work with Standard Oil was under the jurisdiction of the Chicago Metallurgical Labs of the Manhattan Project; "within months of the Japanese surrender in August 1945" he and other scientists were coming together "to discuss the political and social responsibility of the scientist, civilian control of atomic energy, the economics of atomic power, the freedom of scientific information, relations with the Soviet Union, etc."[1] From 1947 to 1948, Erwin carried these concerns to Washington, where he served as Assistant to the Chief of the Scientific Branch, War Department General Staff in Washington, DC, and Elfrieda worked as Copyright Cataloguer at the Library of Congress. Their first child, Catherine Anne, was born there in 1948.

Soon thereafter, the Hieberts returned to Chicago, where Erwin worked as a Research Chemist at the Institute for the Study of Metals while pursuing his M.Sc. in Physical Chemistry at the University of Chicago. There the Hieberts reveled in the company of "an international constellation of scientists, mathematicians, composers, philosophers, *Allewissers*, humanistically-minded foreign visiting professors, and loquacious political emigres," who had found "haven" in the university there.[2] Most memorable among these was Alexandre Koyré, who captivated Erwin "with his French charm, the boundless depths of his learning, and the scientific, historiographic, and philosophical expertise with which he approached and executed the history of science."[3]

Fortified by this "exposure to large doses of cultural history sandwiched in between expansive schemes for world government and global internationalism,"[4] in 1950 the Hieberts moved to Madison, where Erwin began working toward a joint Ph.D. degree in the History of Science and Physical Chemistry. Erwin's decision to study the History of Science "had a great deal to do with reflections about the war, the events leading up to it, its outcome, and prospects for the future,"[5] and his interests lay in the development of modern science, but the strongest influence on him at the University of Wisconsin was the medievalist, Marshall Clagett. As Erwin worked with Clagett, he began to see "that basic techniques and methods for study of the medieval period were not as far removed from the study of more recent periods as one might have assumed." Clagett moved him beyond a "myopic vision" of science, to "an appreciation for the immense differences that have served to identify so-called 'science' and views on 'nature' in disparate times and places: alternative time-bound customs for formulating and resolving seminal questions; the establishment of acceptable criteria for presenting logically unassailable arguments; and the degree of importance given at different times in history to the role of experimental verification and theoretical reasoning."[6]

[1] Erwin N Hiebert, (1993) "On Demarcations between science in context and the context of science" in Kostas Gavroglu, Jean Christianidis, Efthymios Nicolaidis eds. *Trends in the Historiography of Science,* (Boston Studies in the Philosophy of Science, v. 151) pp. 87–106 on 87–88.

[2] Ibid. 99.

[3] Ibid. 102.

[4] Ibid. 100.

[5] Ibid. 87.

[6] Ibid. 96–97.

The year 1954 was a banner year for the Hieberts: Erwin got his Ph.D.; their second daughter, Margaret Helen, was born; and the whole family moved to Göttingen. In Göttingen, Erwin was Fulbright Lecturer at the Max Planck Institute for Physics and Elfrieda took up a Fulbright Scholarship to study musicology at the University of Göttingen. Their third child, Thomas Nels, arrived in 1955.

Erwin's career as an academic Historian of Science began immediately after this German year, when he served as an Instructor in the Harvard History of Science Department from 1955 to 1957. In 1957, he joined the History of Science Department at the University of Wisconsin in Madison, Wisconsin, in 1957 as Assistant Professor, in 1960 as Associate Professor, and in 1963 as Professor. He joined the Harvard History of Science Department in 1970.

Erwin brought all of himself – his Mennonite convictions, his scientific expertise, and his wartime ruminations – to the study of the History of Science. For him, science constituted "one of the few common languages of all mankind. It is a language which can provide a most important basis for the communication of ideas between people of different political and ideological convictions. In their work, scientists the world over place the highest premium on intellectual honesty, personal integrity, hard work, tenacity, concentration, imagination, insight and curiosity." These values were always paramount for both Erwin and Elfrieda. If people "everywhere would throw themselves wholeheartedly into the building of a world community without regard to national interests, their actions would go a long way toward the creation of a world free from war,"[7] Erwin declared. For decades, he and Elfrieda threw themselves wholeheartedly into the work of the Social Concerns Committee of the Mennonite Congregation of Boston.

Within the History of Science, Erwin pursued his internationalist vision as a member of a number of organizations including the British Society for the History of Science and the Canadian Society for the History and Philosophy of Science. He was an elected Fellow of the Académie International d'Histoire des Sciences, an Overseas Fellow of Churchill College, Cambridge, an Honorability Sodalis of the Czechoslovak Society for the History of Science and Technology, and Auswärtiges Mitglied for the Sächsiche Adademie der Wissenschaft zu Leipzig. In addition, he served for 8 years as Vice President, 5 years as President, and an additional 3 years as ex officio member of the Council of the Division of the History of Science of the International Union of the History and Philosophy of Science of the International Council of Scientific Unions (ICSU).

Erwin developed warm relations with a large constellation of international scholars through the performance of these duties, as well as the many visiting appointments he accepted over the years. One of the residues of growing up in an émigré community was a joyful ease with Central and East European languages that supported Erwin's warm relations with people from those regions. Always sensitive

---

[7] Erwin N. Hiebert, (1961) *The Impact of Atomic Energy* (Faith and Life Press: Newton, Kansas,) p. 291.

to the difficulties of the political, intellectual, religious, ethnic, and scholarly conditions under which they worked, Erwin stood ready to respond generously in any way he could.

Erwin was equally devoted to building academic community at home. In addition to membership in the American Chemical Society and Sigma Xi, he was a member of the Association of Members of the Institute for Advanced Study in Princeton and an elected Fellow of the American Academy of Arts and Sciences.

Within the History of Science, Erwin worked to strengthen the community in any number of ways. On the academic side, from 1970 to 1980 he devoted himself to editing the chemistry articles for the *Dictionary of Scientific Biography*. On the administrative side, he was Chairman of the Wisconsin History of Science Department from 1960 to 1965 and of the Harvard History of Science Department from 1977 to 1984. In addition, he served as President of the Midwest History of Science Society from 1967 to 1968, as Vice President and then President of the American History of Science Society from 1971 to 1974, and as Chairman of Section L of the American Association for the Advancement of Science in 1982.

While Erwin was an effective and well-organized administrator, his real passion lay in the kinds of textual engagement he first encountered in discussions with Koyré and seminars with Clagett. His major publications, *Historical Roots of the Principle of Conservation of Energy* (1962) and *The Conception of Thermodynamics in the Scientific Thought of Mach and Planck* (1968), were well-defined problem studies rooted in his deep knowledge of the scientific texts. However, the tight focus of these published efforts belies the breadth of vision that he brought to all of his work.

That breadth of vision came out in Erwin's teaching, which can best be described as inspired. In the classroom, Erwin moved briskly and self-confidently across the fields of physics, chemistry, psychology, religion, philosophy, and whatever else might illuminate the material. His lectures seemed to emerge as the product of some kind of epic battle. Nothing was glib, no point was ever pat or even fixed; all of the issues were confronted and examined almost physically, as Erwin paced back and forth, occasionally knocking his glasses from his nose with the force of his gestures. "You know how it is when you try to start a tractor?" he would blurt to Harvard classes filled with students who most certainly did not know. They nonetheless listened and watched transfixed as he explained thermodynamic issues while jerking out the choke of a – for him – viscerally familiar tractor.

Erwin brought the same intensity to the graduate seminars that met weekly in the cozy confines of his living room. There he introduced generations of students to an ever-changing smorgasbord of texts by Mach, Duhem, Helmholtz, Durkheim, Planck, Cassirer, James, Pearson, Poincaré, Einstein, Bohr, or whoever else he was reading at the moment. Erwin engaged these materials as passionately in his living room as he did in his classes, and insisted that everyone else do the same. Evening after evening, the Hieberts' living room rang out with the sounds of sharp debate, with everyone involved defending, clarifying, and arguing their positions.

Erwin's embrace of diversity included his students as well as their ideas. One of the striking aspects of his legacy is the number of women who worked with him. In a time of transition for women in academe, questioning either their intellectual or their personal competence seems never to have occurred to him. As a result, Joan Bromberg, Caroline Merchant Iltis, Mary Jo Nye, Gisela Kutzbach, Susan Wright, Barbara Buck, Lorraine Daston, Joan Richards, Maila Walter, Sara Genuth, and Diana Barkan all earned their Ph.D.s under Erwin's tutelage and shared seminars as equals with his other students, including but not limited to Bernard Finn, Michael Crowe, Ed Daub, Thomas Hawkins, Roger Stuewer, Joe Dauben, Fred Gregory, Jed Buchwald, Keith Nier, Peter Galison, Richard Kremer, and Skuli Siggurdson.

Erwin's interests were very broad, but he did have favorites; year after year, his seminars would return to the writings of the brilliantly prolific, nineteenth-century polymath, Hermann von Helmholz. Various students emerged from these forays with insights into non-Euclidean geometry, experimental physiology, and the interaction between experiment and theory; what Erwin found in Helmholtz was an interaction between music and physics that was also evidenced in his marriage. Throughout the 1950s and 1960s, Elfrieda had devoted much of her time to raising their family, but she never abandoned the music for which, in 1970, she earned a Ph.D. from the University of Wisconsin. Her musicality permeated the seminars in her house, where she would illustrate acoustical points on the piano, sometimes accompanied by Erwin on the clarinet. Best of all were the seminars she closed by playing an entire piece to the exhilarated group.

The unique combination of Erwin and Elfrieda as physicist and musician flowered for undergraduates in the context of the Harvard House system. This saga began in the fall of 1975, when a group of students in Mather House gathered to construct a harpsichord under the tutelage of Frank Hubbard. When Hubbard took ill, just as the term was starting, Erwin and Elfrieda stepped into the breach. Week after week they considered with the students both the physical and the musical properties of the instrument they were constructing together. At the end of this term, Mather House had not only a harpsichord, but also a chamber music program, which thrived for the next 30 years under Elfrieda's devoted tutelage. Erwin remained ever true to his original commitment to Dunster House, which is next door to Mather; over the years, the influence of each of the Hieberts spilled liberally onto both of their adjoining Houses, bringing warmth, thought, and music into the everyday lives of their students.

Erwin formally retired from Harvard in 1989, but for many friends, colleagues, and a beloved group of children and grandchildren, the tradition of evenings at his house continued unabated. When they were not traveling, teaching, and studying in Europe, Erwin spent almost every day in his Widener Study, while Elfrieda taught and played music with the students in Mather House. At the time of his death, Erwin had just completed the manuscript of this book. From one point of view, it can be characterized as neatly fitting into a growing literature on the interactions between science and music in the German nineteenth century. From a larger perspective, though, its exploration of the interface between science and art, rationality and emotionality, brings together major themes and ideas that Erwin and Elfrieda pursued over the course of 69 years of marriage. Even as they are sorely missed, the science and art that permeated Erwin's and Elfrieda's world endures in the lives and work of the many people they touched.

Brown University
Providence, RI, USA

Joan Richards

Acknowledgments

Lin Garber has complete command of what can be achieved, at many levels, with the computer – especially in terms of its outreach to retrieve documents and events that are not readily or normally accessible. I have benefited enormously from Lin's linguistic skills, his organizational assistance, and his knowledge of the best way to plumb libraries for where documents may be located, including interlibrary loan books, films, and graphics. His knowledge of German and French is more than adequate for tracing books and documents.

Andy Wilson (Andrew Wilson, Access Services Librarian at Harvard's Loeb Music Library) deserves special thanks for his assistance in securing books and other documents.

In addition to members of my family (listed below), persons who have shown special interest in this project and have given me encouragement and spiritual support (in the German sense of *geistlich*) during the writing of these essays include Lorenz Krüger, Hans Jorg Rheinberger, Julia Kursell, Dieter Hoffmann, Lorraine Daston, and Skuli Sigurdssohn at the Max Planck Institut für Wissenschaftsgeschichte, Berlin.

Works in German and Dutch have been translated by the author when not otherwise specified. The author's knowledge of Low German (*Plautdietsch*) has been of great help in translating works from the Dutch.

Works in Japanese have been translated by Dong-Won Kim (Harvard, Ph.D., History of Science, 1991). Dong-Won is a trusted and cherished friend of all members of the Hiebert family, including the children, Catherine, Margaret, and Thomas, and in a special way of the grandchildren: David and Anitha Kerst; Sarah, Benjamin, and Daniel Hiebert; and Jonathan and Rebecca Beissinger.

Finally, I mention with love, affection, and sincere admiration my wife Elfrieda, Ph.D. musicologist and pianist, who has given me companionship and constructive musical criticisms throughout the more than 5 years during which I was engaged in working on the project that forms the corpus of this monograph.

Erwin Hiebert

Contents

Part II Shohé Tanaka, Just Intonation and the Enharmonium

Part III Max Planck

Part IV Adriaan Fokker. Theoretical Physics and Just Intonation Keyboards

Introduction

Over the past decade, there have been numerous studies in the history of science and musicology, which proffer accounts of the intertwined histories of music and physics in nineteenth-century Europe, particularly Germany.[8] Erwin Hiebert's *The Helmholtz Legacy in Physiological Acoustics, Just Intonation Theory, and Fixed-Tone Keyboards* fits squarely in that genre. Hiebert's work reminds us of the importance of music to the nineteenth-century *Bildungsbürgertum* from which both Hermann von Helmholtz and Max Planck so proudly hailed. Hiebert's material, cultural, and intellectual history charts out a territory, which includes the role of the sciences to aesthetics. As Tim Lenoir and Robert Brain have demonstrated, physiologists such as Helmholtz and Ernst Brücke were committed to using physiology to help elucidate theories of art, and a number of artists drew upon these scientists' work, which served as inspirations for some of their creations.[9] Lenoir correctly refers to the era as one of "educating the senses."[10] Similarly, German

[8] See, for example, Myles W. Jackson, *Harmonious Triads: Physicists, Musicians, and Instrument Makers in Nineteenth-Century Germany* (Cambridge, MA: MIT Press, 2006); David Pantalony, *Altered Sensations: Rudolph Koenig's Acoustical Workshop in Nineteenth-Century Paris* (Dordrecht, Heidelberg, London, N.Y.: Springer Verlag, 2009), Alexandra Hui, *The Psychophysical Ear: Musical Experiments, Experimental Sounds, 1840–1910* (Cambridge, MA: MIT Press, 2012); Benjamin Steege, *Helmholtz and the Modern Listener* (N.Y.: Cambridge University Press, 2012); Julia Kursell, "Hermann von Helmholtz und Carl Stumpf über Konsonanz und Dissonanz," in *Berichte zur Wissenschaftsgeschichte*, 31(2008): 130–143; idem, ed. *Physiologie des Klaviers. Vorträge und Konzerte zur Wissenschaftsgeschichte der Musik* (Berlin: Max-Planck-Institut für Wissenschaftsgeschichte, 2009) preprint 366, available at http://www.mpiwg-berlin.mpg.de/Preprints/P366.PDF, last accessed on 22 December 2012; and Alexandra Hui, Julia Kursell, and Myles W. Jackson, eds., *Music Sound and the Laboratory* (Chicago, IL: University of Chicago Press, 2013).

[9] Timothy Lenoir, *Instituting Science: the Cultural Production of Scientific Disciplines* (Palo Alto: Stanford University Press, 1997), pp. 132–178 and Robert Brain, "The Pulse of Modernism: Experimental Physiology and Aesthetic Avant-gardes circa 1900," in *Studies in the History and Philosophy of Science A* **39** (2008): 393–417.

[10] Lenoir, *Instituting Science* (1997), p. 151.

physicists and physiologists were committed to understanding crucial aesthetic components of the art of music, including the standardization of pitch and the implementation of various types of intonations. In addition, musical instruments such as reed pipes, pianos, and harmoniums were used to investigate scientific properties such as the speed of propagation of a sound wave, the ratio of the increase in density to the increase in pressure of sound waves, and the determination of specific heats.

Many scientists and historians of science know Helmholtz as a leading nineteenth-century physicist and physiologist who made important contributions to theories of vision, perception, acoustics, optics, electrodynamics, and thermodynamics. Few realize that he was also a very talented amateur pianist who spent decades studying the relationship among physics, physiology, and music. He enjoyed playing works by Mozart and Beethoven and was very familiar with the works of Wagner. His *Die Lehre von den Tonempfindungen* of 1863 is the seminal text of the period on physiological acoustics. Specifically, Helmholtz researched combination tones, the vibration of strings, the physics of organ pipes, and intonation. The piano for Helmholtz, who served as scientific advisor to the renowned piano manufacturers Steinway and Sons, was simultaneously a musical and scientific instrument. And similarly, Helmholtz used the harmonium, a popular nineteenth-century organ used for entertainment in small churches and wealthy homes, to study intonation. Comprised of reed pipes and powered by bellows, harmoniums could sustain tones for longer periods of time than an organ; therefore, they were particularly amenable to the study of pitch. They were also employed by music instructors to help singers regulate the dynamic shading of their voices.

Hiebert's section on Helmholtz underscores the importance of intonation to the scientist's work on physiological and physical acoustics. Indeed, intonation, or musical/tuning temperament, is the theme tying together all the chapters of this work. There are an infinite number of ways to divide up an octave, which represents two pitches, one twice as high as the other. The semitone, or half tone, is the smallest musical interval in Western tonal music. The octave is divided into 12 semitones, namely A, B flat, B, C, C sharp, D, E flat, E, F, F sharp, G, and G sharp. In just or pure intonation, musical intervals remain in their mathematical ratios. A major third, for example from C to E, represents two pitches with an integer ratio 5:4, while a minor third from A to C represents the integer ratio of 6:5. A perfect fifth, from C to G for example, is a musical interval of 3:2, and a perfect fourth, or from C to F, corresponds to a ratio of 4:3. Western music considers these ratios to be consonant. The problem is that all the concords of triadic music (octaves, fifths, and thirds) are incongruous in their pure forms of just intonation. For example, three pure major thirds are flatter in pitch of a pure octave by one fifth of a whole tone, while four pure minor thirds are slightly sharper than the pitch of a pure octave. Hence, some ratios must be altered, or tempered. As Hiebert reminds us, history has provided us with numerous versions of tuning temperament, which is critical to fix-tone instruments such as pianos, organs, and harmoniums, as performers on wind and bowed instruments can easily adjust pitches. Pythagorean temperament has pure octaves, fourths, and fifths and by and large avoids thirds, as they could not be tempered in a sufficiently consonant way. This became a problem with the rise of keyboard instruments such as the harpsichord during the early fifteenth century. What does one do with major thirds, which were generally absent in ancient Greek harmonies? Pythagorean temperament now needed to be replaced in order to produce a consonant third. Mean-tone temperament, an invention of the Middle Ages, strove for consonant fourths, fifths, and thirds. The pure third was cut in half (half mean-tone).

During the mid-nineteenth century, a new temperament arose, which is prevalent today, namely equal temperament, in which each semitone of an octave is the twelfth root of two sharper than the preceding pitch. Hence, a major third is no longer 5:4, or 1.25, but is now tempered to 1.2599, or $1{:}2^{(4/12)}$. A perfect fourth is now 1.3349, or $1{:}2^{(5/12)}$, rather than 1.3333, while a perfect fifth is 1.4983, or $1{:}2^{(7/12)}$, rather than 1.5000. Only octaves remain pure in the ratio of 2:1. Such a temperament enables modulations of keys in a piece; however, it comes with a price, as thirds, fourths, and fifths are all tempered. For some the sacrifice was too great, as skilled ears could easily hear the difference. Despite appreciating the advantages of equal temperament to the development of nineteenth-century music, Helmholtz bemoaned the loss of pure intervals. In particular, Helmholtz recognized that justly tuned harmoniums were crucial to voice training so that the vocalists could hear pure intervals. He turned to musicians and instrument makers who could produce keyboard instruments, which would have more semitones in the octave than the standard 12. He was particularly fascinated by the work of the English musician Robert H. M. Bosanquet who published *An Elementary Treatise on Musical Intervals and Temperament* in 1876, which featured an enharmonic harmonium comprised of octaves divided into 53 equal intervals. By the 1880s, harmoniums with more than 12 semitones per octave were piquing the curiosity of physicists and musicians alike.

In 1884, a Japanese music theory student, Shohé Tanaka, traveled to Berlin to study acoustics and electromagnetism with Helmholtz and his colleagues. Tanaka was particularly interested in designing a harmonium that was not based on equal temperament, but rather on just intonation. His *Enharmonium* featured 20 keys and 26 pitches per octave. In July of 1892, the Philip J. Trayser & Cie Harmonium Factory of Stuttgart announced their construction and marketing of four- and five-octave harmoniums based on the model of Tanaka's pure-tempered *Enharmonium.* The instruments won praise from prominent musicians of the period, including Franz Schulz, Professor of Music Theory and Composition at Berlin's *Königliche Academische Hochschule für Musik*; the Austro-Hungarian violinist Joseph Joachim; Martin Blumner, Director of the Berlin *Singakademie*; and Hans von Bülow, Chair of the Piano Department of Berlin's *Konservatorium für Musik.* While certainly not endorsed as an instrument to be played by virtuosi, the key attributes of Tanaka's instruments were its ability to assist cappella singing in pure intonation and its accentuation of classical music playing.

Max Planck, renowned theoretical physicist and founder of the quantum theory, was very much a part of the tradition to which Helmholtz had belonged a generation earlier. Being a graduate of a late nineteenth-century German gymnasium and a member of the *Bildungsbürgertum*, music played a critical role in the Nobel laureate's life. He was an active member of the *Akademische Gesangverein* in his student days in Munich, serving as second choirmaster and tenor soloist and leading the group while a *Privatdozent* at the University of Munich. Joseph Joachim was among his circle of innermost friends. Planck even studied composition with the renowned organist Joseph Rheinberger. Indeed, at one point in his life, he needed to decide whether to choose between a career in music or physics. Later in life, he was a prominent member of Grunewald's musical soirees both as a singer and piano accompanist. Berlin scientists and musicians alike congregated in his house to perform various forms of music.

Planck reviewed Leopold Zellner's *Vorträge über Akustik* of 1892, which introduced the young physicist to musical acoustics. Shortly thereafter, Planck became fascinated with musical temperament thanks to a just-tone harmonium designed by Carl Eitz and built by the highly reputable piano company Schiedmayer of Stuttgart. The instrument had been housed in Berlin's Institute for Theoretical Physics when Planck accepted his appointment. He quickly used the instrument for his analysis of vocal music and just intonation. In particular, Planck was interested in how the human ear could accommodate pitch and pitch variation. He argued that in time the listener could accommodate to the ear's deception when listening to tempered music.

Hiebert concludes with the acoustical work on just-intonation studies of the Dutch physicist Adriaan Daniel Fokker, whose major contribution to physics was the Fokker–Planck equation, which describes the change of state over time of a probability density function of a particle's velocity. He also collaborated with Albert Einstein on gravitational theory. During the 1920s, Fokker designed parabolic and hyperbolic reflecting surfaces in Haarlem's St. Bavo cathedral with a view to study and increase speech intelligibility. Not only was he interested in architectural acoustics, throughout the 1930s he was in contact with acousticians working on the standardization of acoustical methods, nomenclature, equipment, and pitch. Of particular interest to Hiebert is Fokker's study of the renowned seventeenth-century Dutch natural philosopher Christiaan Huygen's work on music theory. Huygens had proposed a new musical temperament, a 31-tone system per octave.

Unlike Planck, who never wished to influence musicians to use pure intonation, Fokker arranged for organs to be built based on 31 tones per octave, orchestrated concerts for these new instruments, and even attempted to compose microtonal music, or music whose tonality is based on intervals smaller than the typical 12 semitones of Western music. Indeed, Fokker became known as one of the "Dutch five" composers who wrote for 31-tone instruments as well as other microtonal instruments. Toward the end of his life, Fokker spent much of his time and energy supporting institutions committed to the microtonal tradition.

In short, Hiebert's work clearly demonstrates the active engagement of some of the leading physicists of the past century and a half with musical intonation. Whereas many natural philosophers of the early modern period waxed poetic about the harmony of the spheres, physicists of the nineteenth and early twentieth centuries were committed to experimenting with and improving the material world of music. Instruments such as the harmonium could be used to generate precise pitches necessary to train the ears of vocalists. Reed pipes were critical not only to organs but to the study of adiabatic phenomena, the precursor to thermodynamics. And Fokker was interested in experimenting in music itself by proffering compositions based on microtonality and designing microtonal keyboard instruments. Let us hope that this work will inspire young historians of physics and musicologists to seek out and analyze other cultures and historical periods with a view to illustrate the complex, intertwined, fruitful, and historically contingent relationship between music and physics.

Department of History and Nyu-Gallatin
New York University
New York, NY, USA

Myles Jackson

Part I
Helmholtz

Chapter 1
The Helmholtz Setting in the Johannes Müller Circle in Berlin

When Hermann von Helmholtz died over a 100 years ago German science lost its preeminent physicist and physiologist. Over a time span of 50 years, as teacher, researcher, and science mediator in Königsberg, Bonn, Heidelberg, and Berlin, Helmholtz initiated substantial advances in the science of medicine, physiological optics, physiological acoustics, thermodynamics, hydrodynamics, electrodynamics, and mathematics. His reflections on science and its philosophical dimensions belong to an important genre of epistemological writings on the natural sciences. They represent a precedent-setting example in which the scientist becomes the intermediary who engages in reflections on the philosophy of science from the perspective not of philosopher but of scientist.

This essay is devoted to exploring those landmarks in Helmholtz's scientific career that are associated with his contributions to various branches of musical acoustics. What that entails is an examination of his various writings on the sensations of tone, his experimental investigations in musical acoustics, his experiences in music-listening and music-making, his scientific training as physiologist and physicist, the impact of those experiences on his own musical career, and a discussion of the historical context in which his musical theories were conceived and evaluated by music colleagues. Beyond the various areas of musical acoustics that our analysis purports to examine and to evaluate, a peripheral and less structured intention is to follow, wherever possible, the Helmholtzian trends of thought that become inherently more complex by reaching beyond the sciences to perform a bridge with aesthetics and the diverse ways in which the human mind interprets or is taught, in different cultures, to interpret and understand music. Central to the overall focus of the Helmholtz legacy in the final analysis are his contributions to music theory. In this analysis they are examined principally in conjunction with the practical and culture-conditioned directives of the problem of temperament and intonation.

Working within a music-listening, music-making, and music-composing environment of music theorists who gradually were becoming conditioned to music written for and performed in equal temperament on fixed-tone instruments such as the piano, Helmholtz became involved in exploring the theoretical and practical

E. Hiebert, *The Helmholtz Legacy in Physiological Acoustics*, Archimedes 39,
DOI 10.1007/978-3-319-06602-8_1, © Springer International Publishing Switzerland 2014

feasibility of music-listening and music-performing in tempered intonation. Although he was keenly aware of the enormous technical difficulties involved in constructing instruments that are able to play in tempered intonation, not to speak of the cultural barriers that this option entailed, he gave his guarded support to the tempered position recognizing full well that there would be occasions on which equal temperament or, indeed, a temperament other than tempered intonation, would be preferred. The situation that this choice created for Helmholtz, and the arguments advanced to support his position in the midst of science colleagues and musician compatriots who were steeped in equal temperament, is presented in considerable detail in the section on just intonation.

The introductory sections of our analysis deal with Helmholtz's early work on physiology and more specifically with alignments and orientations taken in what became the "new physiology" in the Johannes Müller circle in Berlin during the 1830s and 1840s. Hermann Helmholtz (1821–1894) is introduced in this context as a student of medicine and the life sciences who lived and worked in the circle of like-minded and "new physiology"-oriented fellow students and colleagues associated with Müller at the University of Berlin. The expositional phase in the development of Helmholtz's formative training and scientific interests stretches from his 16th to his 25th year. It encompasses, as salient experiences and accomplishments, several years of formal military training in Berlin, and the completion of a medical degree as surgeon. As it turned out, Helmholtz completed his entrance examinations to medical school and was admitted in 1837 to one of the much-cherished positions at the royal school of medicine and surgery at the Friedrich-Wilhelm Institute – an institution also referred to as the *Pepinière*.

Neither Helmholtz's commitment to working in the Müller physiology circle nor his stint in the military were matters of his own choice. They were thrust upon him by circumstances connected with what seemed to lie ahead for him as a financially remunerative professional career. That professional career, as a point of departure into medicine and the life sciences, took the form of experimental investigations into the nature of putrefaction and fermentation, the consumption of matter associated with muscular action, and the role and theory of thermal phenomena in physiology. While at the royal regiment of the *Garde du Corps* in Potsdam, Helmholtz apparently had ample time to explore works by well-known mathematical physicists such as Denis Poisson (1781–1840) and Carl Friedrich Gauss (1777–1885). In 1847 Helmholtz completed his treatise on the conservation of energy. The conservation principle, which runs like an Ariadne thread through the length and breadth of the physiological and physical sciences, was published as a pamphlet in Berlin when Helmholtz was only 26 years of age. Initially it had been rejected for publication in *Poggendorff's Annalen* for being too theoretical. That was a charge that no one honestly should have levelled against him at the time. Although Helmholtz later in life was recognized as a theoretical physicist, his image in 1847 was quite other, for conspicuous in all his efforts in the sciences was a thrust in the physiology-oriented direction of experimentation.

In the main section of this work an effort has been made to examine the physiology of music and its formulations within the ambience of Helmholtz's work and in relation to the efforts of others that Helmholtz was able to build upon. It treats the aesthetics of music as seen within Helmholtz's life and work and brings into focus his contributions to the physiological aspects of musical acoustics in relation to combination tones, the theory of temperament, the problems of intonation, and the practical implementation of his views on these matters in the construction of an harmonium designed to deliver approximate just intonation.

In general I have chosen to take Helmholtz's deliberations, experiments, and interpretations on the science of music as the point of departure for appraising not only his own contributions to the theory of music but also to search out links to the works of others and to examine the bearing his work had on scientists who in turn examined, enhanced, modernized, criticized, and on occasion rejected his analyses of the theory of music and its aesthetic correlates. My analysis rests on a tripod of discipline-connected and discipline-overlapping domains that are Helmholtzian to the core: (1) Physics, (2) Physiology, and (3) Aesthetics.

Concurrent with study of the problems of temperament and intonation from a mathematical and scientific perspective come questions that inevitably direct the attention of the historian of music, and in this case the music-attentive historian of science, to questions linked with the aesthetic aspects of music and the study of music as an essentially humanistic discipline. Exploring problems from the point of view of aesthetics entails, *ab initio*, an alertness to searching out the common frontiers of the exact sciences and the humanities in a manner that accords even-handedness and historical sensitivity to both the scientific and the humanistic landscapes. To probe what lies at the borderland, and beyond, in the domain of the aesthetics of music connotes an effort to look *beyond* the discipline of music and not merely *within* the domain of the science of music itself. For example, the aesthetic problems that surface at the level of musical perception in the listener are uniquely circumscribed by the culture-conditioning of the listener. A given musical work may elicit contrary appraisals and emotional responses not only in two different cultures but in two individuals in the same culture. The responses may actually have little or nothing in common with what was in the mind of composer or performer. The contexts in which the evaluations, preferences, and responses of different individuals in different cultures fluctuate so conspicuously in the perception of music, constitutes a problem that is allied with the entanglement of the objective and subjective components of intonation, the perception of pitch intervals, and their so-called consonance or harmonic purity. Although music can be conceived and analyzed either as an external and objective aural sensation, or as an internal mental listening perception, it nevertheless, however conceived, should enjoy as valid a disciplinary status as any scientific or humanistic discipline, at least in principle.

Of all the arts, music has had most to rely on the scientific and mathematical analysis of its materials; and while music need not mirror any specific external reality, it is closer to mathematics, and has more in common with natural phenomena, than any of the other fine arts. Among the arts music stands out for having first most clearly revealed the intrinsic significance of number, ratio, and harmony – a significance that in one form or another is mirrored in the way that the mathematics of proportion and the idea of order have been applied to all the arts.[1] The essence of music, as contained in numbers, embodies a perspective on the nature of things that has prevailed throughout most of written history. Mathematicians and mathematically-minded philosophers have written a great deal about music in relation to mathematics. Leibniz felt that "music is a hidden exercise in arithmetic, of a mind unconscious of dealing with numbers."[2] This Leibnizian sentiment links the history of scientific thought with the history of Western musical thought at the levels of mathematics, music, ear, and brain.

The central problem we want to address relates to the manner in which just intonation came to intersect with the history of keyboard instruments and exert an influence on the development of Western music. Before proceeding with the main, it is essential to lay out and define a number of the basic musical terms that enter into what follows. To this end I have consulted standard books of musical literature such as different editions of the *Harvard Dictionary of Music* and other multivolume dictionaries of music such as the *New Grove Dictionary of Music* and *Musik in Geschichte und Gegenwart.*

By the early fourteenth century, the introduction of polyphony had given rise to a widening of the compass of the keyboard and to a fully chromatic instrument with seven natural and five chromatic keys per octave.[3] With keyboard instruments taking the lead in eighteenth and nineteenth century European music circles, the art of adjusting sounding frequencies by tuning became a matter of utmost practical and theoretical concern among composers, performers, and music theoreticians. Conspicuously on display in the science-sensitive musical environment of the 1860s were various theoretical endeavours to establish the magnitude of pitch intervals for different musical contexts. Concurrently, on the practical side, came the demand for keyboard instruments designed to furnish acoustically pure intervals thus to undergird a theoretical system of tuning that realizes, or at minimum approximates, just intonation. The pursuit of acoustically pure intervals entails the recognition and analysis of the notion of the harmonic series.

In principle it is possible to describe the vibration of an idealized string or elastic body that delivers what is called a pure tone consisting of a single

---

[1] For a wide range of historical studies on music and mathematics: Günter Schnitzler, ed., *Musik und Zahl. Interdisziplinäre Beiträge zum Grenzbereich zwischen Musik und Mathematik,* Bonn-Bad Godesberg, 1976. See esp. Martin Vogel's essay "Reine Stimmung und Temperierung."

[2] "Musica est exercitium arithmeticae occultum nescientis se numerari anime." This statement is made by Leibniz in a letter of 17 April 1712 addressed to the Russian mathematician Christian Goldbach. R. C. Archibald, "Mathematics and Music," *American Mathematical Monthly, 31* (1924), 1–2.

[3] The family of instruments with keyboard mechanisms, each of which has its own checkered history, includes: organs as the oldest keyboards that are sounded by air under pressure; the clavichord and harpsichord that are activated by struck or plucked mechanisms and that were in use from the fifteenth to the eighteenth centuries and revived at the end of the nineteenth; the pianoforte (piano) as an instrument central to musical life since the 3rd quarter of the eighteenth century; and the harmonium which is a pedal-operated organ with freely vibrating metal reeds – an instrument that developed during the first half of the nineteenth century.

frequency. In practice, however, no vibrating body produces a pure tone. All musical instruments produce composite tones called harmonics. They are produced simultaneously and belong to the harmonic series. The harmonic of lowest frequency is called the fundamental because it is more intense than the other frequencies and determines the pitch of the composite tone. The frequencies of the other harmonics are exact multiples of the frequency of the fundamental so that if x is the frequency of the fundamental the other harmonics in the series have frequencies 2x, 3x, 4x,. . .

The "harmonic series," also referred to as the harmonic overtone series, is made up of the individual pure sounds normally present as part of an ordinary musical tone. For example, a string or column of air vibrates simultaneously not only as a whole but as two halves, three thirds, etc. When a musical interval can be expressed by adjacent "partials" in the harmonic series it has the form $(x+1):x$. A partial is taken to be a component vibration of a particular frequency in a compound tone and need not, as in the case of a bell or plate, be harmonic. When the components belong to the harmonic series they are referred to as "overtones." In the Helmholtz terminology they are referred to as *Obertöne*, a contraction of *Oberpartialtöne*, as distinguished from the fundamental tone or *Grundton.* The seven simplest harmonics are represented by the ratios of frequencies of vibration $(x+1):x$, for whole number fractions of an idealized string: octave 2/1, fifth 3/2, fourth 4/3, major third 5/4, minor third 6/5, major sixth 5/3, minor sixth 8/5. Closely related to the theoretical problem of just intonation is the effort to build musical instruments capable of furnishing microtonal intervals; namely, intervals distinctly smaller than a semitone, such as a third of a semitone, a quarter of a semitone, etc. Microtones will be dealt with in what follows only insofar as their history impinges on and illuminates the discussion on the construction of fixed-tone instruments designed for playing in just intonation.

The expression "fixed-tone keyboard" is used to characterize musical instruments such as the piano, harpsichord, clavichord, pipe organ, and reed organ (harmonium), the pitch of whose tones cannot be altered during the act of playing. Unlike the human voice or the violin or the slide trombone, where the pitches can be adjusted to a large extent by the singer or player, the pitches on keyboards are fixed. A similar problem of pitch inflexibility maintains for other fixed-tone orchestral instruments among the woodwinds and brasses.[4] In contrast to the pitch inflexibility of fixed-tone keyboards the instrument whose pitch adjustment is near ideal is the human voice – this, despite the recognition that it not only is a much more complicated musical instrument than

[4] To a limited extent the woodwinds can adjust their pitches by lengthening or shortening an instrument's vibrating air column at its upper extremity, by alternative fingerings, and by embouchure control. The same problem of pitch control maintains for the brasses. The French horn presents a unique case since a skillful player is able to sound the entire chromatic scale on the valveless natural horn by means of embouchure control and hand-in-bell techniques. Some of the tones on the natural horn will take on a unique muffled character – a phenomenon that on occasion has been said to elicit its own artistic appeal.

any man-made device but that its anatomical structure is inordinately resistant to scientific investigation.[5]

Since antiquity the human voice has occupied a central place in much of mankind's music-making. Practically speaking all music prior to 1500 is vocal, as is nine-tenths of the music of the sixteenth century. During the so-called *Baroque* period (1600–1750) the performance of vocal and instrumental music were about equal, whereas after 1750 instrumental music gained the upper hand. Nevertheless, throughout the nineteenth and twentieth centuries, and into more recent times, the human voice has served as a point of departure for music-theoretic discussions and analyses that relate to problems connected with the control of pitch and the ideal of just intonation. This relationship essentially maintains for all musical instruments that belong to the keyboard family, the woodwind family, and most of the brass family. Their pitches can only be controlled, and this to a limited extent, by the performing artist. The human voice has been referred to in connection with the literature on intonation as the musical instrument best able, by natural endowment, to sound in just intonation. This however is only the case provided the vocalist has not, by hardened habituation, been conditioned to adjust his or her pitches to the temperament of an accompanying fixed-tone instrument like the piano. The tuning problem here touched upon is one that carries over into all musical ensembles that include a keyboard instrument, since it is standard practice to expect all musical instruments to tune to the keyboard prior to performance.

Conventional 12 key per octave keyboard instruments such as the piano and the harmonium usually are tuned to equal temperament more or less. The principle of equal temperament is based on the division of the octave into 12 equal semitones. In this tuning system, no interval other than the octave is acoustically correct or pure. The deviation of the fifth normally is too small to be readily perceived; with thirds the deviation is considerably greater. Given the situation, it commonly has been asserted that the modern ear has become accustomed to the "errors" in pitch that are intrinsic to all fixed-tone instruments. The reason for persevering with equal temperament in the case of keyboard instruments such as the piano is that it would take somewhere between 19 and 31 keys per octave to approximate just intonation in all modulations – a difficult assignment for both keyboard builder and keyboard performer. For string instruments such as the violin and blown instruments such as the trumpet or the clarinet, pitch adjustments are possible to some degree by means of fingerboard adjustments and embouchure control. On the other hand, for fixed-tone keyboards such as the piano the solution to the intonation problem, while feasible from both theoretical and mathematical points of view, becomes technologically formidable for the keyboard designer and builder and manually challenging for the performer.

---

[5] Among orchestral instruments only the slide trombone and slide trumpet can claim true pitch flexibility. Except for its open strings the violin and the family of string instruments can claim maximum tuning independence.

During Mozart's lifetime in the second half of the eighteenth century, the rapidity with which keyboards were developing led to the demand for an instrument more complex in its mechanism than any other known machine at the time. By 1791, when Mozart died and Beethoven was 21, the pianoforte was beginning to eclipse the clavichord and harpsichord. By the middle of the nineteenth century, and against the background of five centuries of keyboard construction, the pianoforte had become the dominant musical instrument in Western music. The pianoforte was so thoroughly entrenched in society that the idea of redesigning keyboards that might be able to approximate just intonation was inconceivable except in the minds of a small number of music theorists. Unable to convert their theoretical ideas into concrete form for playable instruments theorists nevertheless continued to develop elaborate and imaginative schemes designed to achieve just intonation. After the introduction of electro-acoustic music and the development of microtonal instruments in the middle of the twentieth century it became plausible to anticipate keyboards able to function in just-tone intonation. It was, however, an option that altered the fundamental equation about what constitutes and defines desirable music.

In spite of all built-in, historically frozen, and technically formidable resistances to change, it is readily demonstrated from the history of music that from the time that fixed-tone keyboards came to be integrated into Western music there always have been persistent and resolute efforts to design, construct, and perform music on keyboards in just intonation. Most of the enthusiasm for such activities has been generated by mathematicians and physicists whose calculations and thought experiments – scientific *Gedankenexperimente* – have been provoked not only by the drive toward realizing the aesthetic subtleties of "pure" intonation but by the incentive and gratification rooted in the techno-theoretic inventiveness of instrument craftsmanship. The essential point of leverage for just intonation *aficionados* has been that although technically demanding for instrument designers and builders and manually next to unmanageable for the ordinary keyboard player, the perennial lure in the direction of reaching the highest possible degree of purity in musical intonation has been sufficiently challenging to entice a small number of persons to explore the goal of just intonation on fixed-tone keyboards. Helmholtz was a pioneer in the investigation of fixed-tone keyboards. His student, the Japanese physicist Shohé Tanaka (1868–1945), Max Planck (1858–1943), and Adriaan Fokker (1887–1972) followed in Helmholtz's footsteps with their own passionate curiosity and inventiveness.

The history of science makes manifest that significant insights on how to proceed in a developing field of scientific endeavor have been brought about by a new way of thinking about, and perhaps reformulating, a problem of long standing that has had its own prior history of negative or limited resolution. Such problems conceivably are worthy of re-examination either from a hitherto unformulated theoretical perspective or from a point of view that brings different or newly discovered information to the situation. Hermann Helmholtz has been regarded by scholars from various disciplines as one who, more than others in his generation, identified and hunted down problems of longstanding puzzlement.

Helmholtz's career was launched in physiology – a field in which his own research convinced him early on that the problems of greatest interest to him suggested that he learn physics for the sake of advancing his physiology, and then mathematics for the sake of advancing his physics; the result being that he mastered all three disciplines. Eventually his investigations spilled over into the theory of cognition, psychology, and aesthetics. His personal involvement in music-making and his interest in the longstanding problems of just intonation led him to formulate a set of questions whose implementation and solution made extensive use of music theory, music history, his own proficiency with experimental investigations involving scientific and musical instruments, and his acumen for recognizing and probing aesthetic dimensions of music that lay hidden in the mind of the music theoretician.

We have stressed the point that Helmholtz's approach to the study of music was tempered to a large extent by his experiences as a physiologist, but that he nevertheless was prepared to apply mathematics and the full theoretical and experimental regalia of physics to a theory of aural reception and interpretation that encompasses pitch perception and the harmoniousness of musical intervals. It is also evident that he approached his studies in music within a cultural environment in which high-minded and discriminatory music-listening and music-making were fostered enthusiastically at the level of family and local community. The situation was one in which at one level the sub-culture of music or the so-called "material culture" of music, as practice, set the stage, and at another level made its way into music-theoretic concerns that were supportive.

References

Goldbach, Christian, and R.C. Archibald. 1924. Mathematics and music. *American Mathematical Monthly* 31: 1–2.

Schnitzler, Günter (ed.). 1976. *Musik und Zahl. Interdisziplinäre Beiträge zum Grenzbereich zwischen Musik und Mathematik*. Bonn: Verlag für Systematische Musikwissenschaft.

Chapter 2
New Directions in Physiology in the Johannes Müller Circle in Berlin

Helmholtz's academic career began with the study of chemistry, clinical medicine, and physiology. The research with Johannes Müller, which led to a doctoral dissertation in 1842 at the University of Berlin, put Helmholtz in contact with the ideas of Müller and Müller's students, and convinced him at the very start of his studies of the importance of experimental medicine and the urgency of launching a programme within the discipline of physiology based on observation and experimentation using the techniques of chemistry and physics.

Johannes Müller (1801–1858), Professor of Anatomy and Physiology and Director of the Museum of Comparative Anatomy at the University of Berlin, more than any other individual of his time, came by the end of the nineteenth century to be recognized as Germany's prime mover for an experiment-inspired programme in biological and medical research. In large part the robustness of Müller's programme was owing to the fact that his students had achieved considerable eminence in the life sciences by the end of the century. As will be shown, Müller's own exemplary experimental investigations were realized within the context of the doctrine of *Naturphilosophie* – a philosophy of nature most closely linked with German idealism and the romanticism of the philosopher Schelling.

In a work of 1826 on the comparative physiology of the sense of sight Müller discovered that the sensory systems of the body are able to respond to various stimuli in a fixed and characteristic way with an energy (the term Müller used) specific not to the nature of the stimuli but to the particular organ of sense that receives the stimulus. According to Müller's law of specific nerve energies [*die Lehre von den specifischen Sinnesenergien*] which encapsulates the law, the eye registers the sense of sight, the ear registers the sense of sound, and so forth. This discovery turned out to embody far-reaching scientific and epistemological implications and demonstrates that what can be known about the so-called objective world of nature is registered in the mental apparatus of the body subjectively, and depends more on the physiological structure of human sensors such as the eye or the ear than on the objective physical agents of sensation. These implications go a long way toward explaining why physicists such as Helmholtz and Ernst Mach became so deeply involved in the principles and the terminology of sensory physiology and

E. Hiebert, *The Helmholtz Legacy in Physiological Acoustics*, Archimedes 39,

DOI 10.1007/978-3-319-06602-8_2, © Springer International Publishing Switzerland 2014

psychology in the course of working on problems whose origins spring from physics. Müller's two-volume *Handbuch der Physiologie des Menschen* (1834/1840) was translated into several languages and represents a landmark in physiological research and teaching texts. Throughout the second half of the nineteenth century, Müller's views on sensation provided strong support for the mechanistic or physicalistic conception of life.

A pioneer in the study of comparative and pathological anatomy, embryology, and physiology using methods drawn from the physical sciences, Müller nevertheless held fast to the notion of a non-physical vital force, a *Lebenskraft* that exists apart from the forces that are at work in the domain of physics and chemistry. The concept of *Lebenskraft* was regarded by Müller, in a general sense, as an essence akin to the soul, or as a reality living in the body next to the soul. In some contexts the *Lebenskraft* concept was placed alongside the so-called "imponderables of nature" – immaterial essences such as the nerve principle, animal heat, and galvanic electricity. For Müller, *Lebenskraft* was taken to be the unitary cause and steward of all living phenomena. Fundamentally different from inorganic forces, the *Lebenskraft* was in conflict with inorganic forces. As Müller expressed it, the secrets of the laws of physics and chemistry were known to *Lebenskraft* and had developed alongside the sensory and motor organs of the breathing and digestive apparatus. They had developed as a part of the organization of the body. *Lebenskraft* had no specific locus in the body, was present everywhere, and acted at no specific point. It possessed the potential not only of being able to control nutrient material and invigorate matter capable of becoming living, but also of rejecting matter after it had served its function in the living process. In regard to the basic tenets of *Lebenskraft* Müller had the good company of the physiologist Rudolph Wagner (1805–1864) and the chemists Justus Liebig (1803–1873) and Friedrich Wöhler (1800–1882).[1] On the other hand Müller's own students rejected *Lebenskraft* outright. For them it would have been antithetical to postulate a philosophy that condones behind-the-scene explanations or metaphysical principles that appeal to non-physical vital properties or forces.

The close-knit group of four physiologists in the Müller circle – a group that for an entire generation came to be looked upon as the founders of the new German physiology – included Hermann Helmholtz, Ernst Brücke, Emil Du Bois-Reymond, and Karl Ludwig. All except Ludwig had studied personally with Müller in Berlin in the late 1830s and the 1840s. Helmholtz, in particular, went out of his way to recognize Müller as the one person who had convinced him that he would not have to abandon his first love, physics, even if he dug himself into problems traditionally considered to belong to physiology. He remarked that his acquaintance with Müller had "definitely altered his intellectual standards."[2]

---

[1] Emil Du Bois-Reymond, "Festrede zur Feier des Leibnizschen Jahrestages." *Sitzungsberichte der königlichen Preussischen Akademie der Wissenschaften zu Berlin*, Berlin, 1894, 625–628.

[2] Johannes Steudel, "Johannes Müller," *Dictionary of Scientific Biography* (referred to hereafter as the *DSB*) 9 (1974) 567–574. Quotation on 568. See also R. Steven Turner, "Hermann Helmholtz," *DSB*, 6 (1972) 241–253.

Karl Ludwig (1816–1895), an enthusiastic Müllerite, was a medical student at the University of Marburg, and, in fact, was the senior physiologist in the Müller circle. He became the most outspoken proponent for the explanation of living phenomena in terms of mechanics. He had invented the kymograph and the blood pump, introduced graphic methods into physiology, and made major contributions to knowledge about blood circulation. A visit to Berlin in 1847 brought him in contact with Helmholtz, Brücke, and Du Bois-Reymond and led to a lifelong collaboration in the pursuit of problems in experimental physiology based on physical principles. Ludwig's mastery of physicochemical principles, and his ingenious experimental use of custom-built instruments, enabled him to become one of the most prominent experimental physiologists of the nineteenth century. As chair of physiology at the University of Leipzig, and as a phenomenally successful teacher (an accolade that Helmholtz never received), Ludwig in 1865 established an institute for physiology in Leipzig that acquired worldwide recognition.[3]

Ernst Brücke (1819–1892), assistant to Müller, graduated from the University of Berlin with a doctorate in medicine and surgery. One of the most versatile physiologists of his day, Brücke exhibited a lifelong fascination with the theory of art. His programme for the new physiology led to observations on optical images, after-images, stereoscopic vision, and reflections from the retina of the eye – foundational studies that Helmholtz was able to draw on in his three classic volumes on physiological optics. In 1849 Brücke was appointed Professor of Physiology in Vienna. In this city, with its flourishing scientific and artistic cultures, Brücke not only established an internationally recognized school for physiology but was able to pursue his persuasion that it is meaningful to engage in the rational study of the humanities and the arts using scientific methods. He wrote a work on the characteristic features of the physiology and taxonomy of linguistics based on the transcription of sounds of one language into the alphabetic characters of another language. Using a lip-measuring device he studied the lengths of strongly and weakly accented syllables in verse. Highly acclaimed not only as a man of enormous learning and prodigious scholarly output but as an outstanding teacher, Brücke was known among his students for his modesty. It was said that "[He] was interested only in explaining the events of nature with a view to their objective regularity."[4]

Emil Du Bois-Reymond (1818–1896) is counted among the founders of a physiology that is based predominantly on physics, chemistry, and the mechanistic interpretation of living phenomena. An outspoken mouthpiece for antivitalistic philosophy, he provided the essential inspiration and laid out well-defined methods of research that led to the establishment of the new field of electrophysiology.[5] As an assistant to Johannes Müller in the 1840s, Du Bois-Reymond remained in close contact with Brücke, Helmholtz, and Ludwig. When Müller died in 1858

[3] George Rosen, "Karl Ludwig," *DSB*, *8* (1973) 540–542.

[4] Erna Lesky, Ernst Brücke, *DSB*, *2* (1970) 530–532. Quotation on 532.

[5] K. E. Rothschuh, "Emil Heinrich Du Bois-Reymond," *DSB*, *4* (1971) 200–205.

Du Bois-Reymond, 40 years old at the time, inherited the chair in physiology in Berlin. From that time on his contributions to the development of modern physiology display a strong preference for the explanation of processes in living organisms based not so much on mechanics as on molecular and atomic mechanics – a unique position even among the Müllerites. His physics-directed research ventures into physiology led to significant contributions in animal electricity, electrophysiology, and the physics of muscles and nerves. After much agitation in Vienna, and with support from the Prussian government, he was able to establish a physiological institute in Berlin that apart from Ludwig's institute in Leipzig and Brücke's institute in Vienna became the largest and most handsomely equipped physiology institute in all of Germany. When the original Müller foursome disbanded in 1858 it was Du Bois-Reymond who would become the most outspoken voice for the physicalistic and mechanistic physiology that had received its initial stimulus in the Berlin of the Johannes Müller circle.[6]

The compilation of Du Bois-Reymond's many public lectures and published essays, the *Reden*, feature the relation of the natural sciences to philosophy, politics, history, and theology. Presented in a tone best characterized as a blend of Francophilic Bravado and Germanophilic *Gründlichkeit*, the *Reden* elicited discussions and controversy that lasted for several decades. In essence the *Reden* demonstrate Du Bois-Reymond's keen ability to identify essential and explore-worthy links between the natural sciences, the humanistic disciplines, and cultural history. He also drew attention to acknowledging that the history of science was the most neglected area of cultural history.[7]

In his scientific papers, so too in his *Reden*, Du Bois-Reymond sought to distance himself at all costs from the hypothetical vital forces that for him and most of his fellow physiologists in the Müller circle had become the *bête noir* of the discipline; this in stark contrast to the views of their own teacher who had never been able to rid himself entirely of the *Naturphilosophie* that already had dominated his own medical training at the University of Bonn in the 1820s. When Müller died in 1858 Du Bois-Reymond penned a memorial address of 182 pages for his teacher and colleague.[8] In this address he provided a well-documented and historically informative account of Müller's scientific outlook and contributions He characterized Müller's role as a leader and reformer in the development of physiology, and touched significantly on the changing ways of thinking that were taking place in the discipline of physiology during the Müller years. In the history of biology Johannes Müller has been referred to as the Albrecht Haller of the nineteenth century. In the discipline of comparative anatomy the allusion is to Müller as the

[6] The emergence of the modern physiology laboratory in Berlin in the age of electricity and the machine is examined in the career of Du Bois-Reymond by Sven Dierig, in *Wissenschaft in der Maschinenstadt: Emil Du Bois-Reymond und seine Laboratorien in Berlin*, Göttingen, 2006.

[7] Estelle Du Bois-Reymond, ed. *Reden von Emil Du Bois-Reymond*, I–XXII, 2 vols., Leipzig, 1912.

[8] Emil Du Bois-Reymond, "Gedächtnisrede auf Johannes Müller," Gehalten an der Leibniz-Sitzung der Akademie der Wissenschaften, 1858, *Reden* #VI, Leipzig, 1912, 135–317.

George Cuvier of Germany. Du Bois-Reymond intimates that Müller's death in 1858 was the greatest loss to the University of Berlin since the death in 1851 of Carl Jacobi, who apart from Karl Friedrich Gauss at Göttingen was the leading German mathematician of his time.[9]

An examination of Du Bois-Reymond's comprehensive mid-century evaluation of the Müller period in physiology convinced its author – who was well aware of the enormous changes that had taken place in physiology – that on behalf of his other closest colleagues in the Müller circle (Ludwig, Brücke, and Helmholtz) it was his duty to put on record the full story of how the discipline of physiology had come to be recognized and certified as a modern *scientific* discipline alongside physics and chemistry. With unreserved high praise for Müller's physiological and medical research, Du Bois-Reymond acknowledged, in particular, the effectiveness of Müller's rapport with students and the widely acclaimed efficacy of his teaching and publications – singling out especially the 2-volume *Handbuch der Physiologie des Menschen* (1840). He nevertheless recognized that it was imperative to come to terms with an ostensibly inconsistent collusion of ideas in Müller's scientific outlook. On the one hand Müller had championed the mechanistic conception of life. In his own physiological research, and in the research of his students, he had promoted the principles and the use of experimental methods that draw on physics and chemistry. On the other hand Müller had condoned the notion of *Lebenskraft* throughout his entire career. In spite of this fundamental concept as anchor point in Müller's perspective, his students, to a person, had rejected *Lebenskraft* as an empty metaphysical hangover from the *Naturphilosophie* of earlier times. By the middle of the nineteenth century, a decade before Müller died, the *Lebenskraft* idea virtually had vanished from the writings of physiologists – a state of affairs that was taken proudly to be the result of the implementation of the mechanistic philosophy and physicalistic methods set in motion almost exclusively by Müller and the Müllerites.

Du Bois-Reymond's reconstruction of Müller's reasoning on this score is compelling and complicated and need not be drawn out here. The gist of the argument is given by the following excerpt from one of Du Bois-Reymond's *Reden*.

> A description of Müller as physiologist would not be complete if his relationship to the principle question of biology and the essence of life processes and their active forces were left untouched. Everyone knows that from the start and until the end of his life Müller was a resolute vitalist. . . .
>
> Müller adopted the idea of an integrated *Lebenskraft* which, although distinguished from physical and chemical forces, comes into conflict with these forces and acts in organisms as cause and uppermost guardian according to a determined plan. All the puzzles of physics and chemistry bow down before this force. In death it vanishes without a corresponding action in its place. . . .[10]

[9] Estelle Du Bois-Reymond, *Reden*, #VI, 135–136.

[10] Estelle Du Bois-Reymond, *Reden* #VI, 205–206. "Ueber die Lebenskraft," *Reden* #I (1848) (1–26) provides an earlier exposé of the *Lebenskraft* doctrine.

Du Bois-Reymond's mid-century evaluation of Johannes Müller's role in the development of modern biology and physiology is one that historians of the life sciences have continued to belabor. It rests upon the assumption that Müller, presumably, was the last individual of his generation to have been in command of the whole of biological knowledge – this in addition to being regarded as a pioneer in various sub-branches of the life sciences that were the purview of his and his student's experimental research. Many years later, in a lecture presented to the Physical Society in Berlin and commemorating the 50th anniversary of Helmholtz's doctorate, Du Bois-Reymond referred to physiology as the "queen of the sciences."[11] In the estimation of persons such as Du Bois-Reymond and Helmholtz, who counted themselves as among the earliest pioneers of the new science of physiology, it was the dominance of the physical and the mechanical-mathematical approach that by the end of the century had given physiology its essential modern character. A prominent historian of biology has remarked, in reference to the mechanistic approach and accomplishments of the Müller circle scientists, that from the 1840s onward German physiology travelled a path that took the discipline from being "a playground for dilettantism to a scientific physiology."[12]

The mechanistically conceived theoretical ideas and physical methods that were being put in practice led not only to an abundance of new discoveries in medicine and physiology but gave the discipline and its practitioners a new profile. The Müllerites had made no attempt to define life or clarify what meaning might be given to life's vital forces. Rather, the functioning of living processes was expressed and explained solely in the terms of physics and chemistry. By the end of the century the new physiology was firmly established in Vienna, Leipzig, Heidelberg, and Berlin. The school of physiology and its laboratory in Vienna had become world renowned. Brücke and his students had demonstrated, in experiments on the electric stimulus of muscles, that the magnitude of the stimulus effect was caused not solely by the amount of current applied, as Du Bois-Reymond had claimed, but was governed as well by a time factor. In Leipzig Carl Ludwig had established a research programme to clarify physiological problems by correlating study of the anatomy of a particular organ with the physiochemical changes that occur in its functioning. To achieve this end he had created physical, chemical, and anatomical divisions in the laboratory and set his sights on the design and refinement of instruments that provide the information he was looking for. Together with his students he devised a mercurial blood pump to study circulation, opened up the new field of investigations on the process of diffusion and osmosis through body membranes, and set up experimental methods to study salivary and renal secretions and respiration.

---

[11] Estelle Du Bois-Reymond, *Reden*, No. XVII (1892), 643–648.

[12] K. E. Rothschuh, *Physiologie im Werden*, Stuttgart, 1969, p. 167. "So wird aus einem Tummelplatz des Dilettantismus eine naturwissenschaftliche Physiologie." See, in particular, Rothschuh's last chapter entitled: "Ursprünge und Wandlungen der physiologischen Denkweise im 19. Jahrhundert."

In a lecture to the Berlin Academy of Sciences on the occasion of a Leibniz Anniversary celebration Du Bois-Reymond seized the opportunity to reflect once more on matters he had dealt with almost 40 years earlier in the memorial address for Johannes Müller.[13] The interim years had given Du Bois-Reymond the retrospective credentials to revisit and reevaluate the determinative factors that had paved the way for the accomplishments of the Müller group during the Berlin years and beyond. How had it come about that Müller, proponent of the *Lebenskraft* idea, had provided so fertile an intellectual and academic haven to pursue the mechanistic line of thought in physiology? How had Müller's students become such enthusiastic supporters and successful practitioners of the mechanistic outlook on living phenomena while rejecting their scientific mentor's *Lebenskraft* doctrine? Du Bois-Reymond sought, these many years later, to shed new light on such questions. He felt that it was important to recognize that Müller not only was the leading physiologist of Germany at the time but that he also was a skillful and influential expositor and writer. Du Bois-Reymond inferred that when Müller wrote on the subject of *Lebenskraft* he had laid out its essence and implications with such clarity, and so forthrightly and transparently, that it became relatively straightforward, especially for persons who were knowledgeable about physiology and adept at using analytical tools and sharp arguments, to show that the *Lebenskraft* doctrine rested on an illusion (*Trugbild*) that readily could be exposed. Accordingly, Müllers's vitalism was openly challenged. He had listened to his students' arguments but stood his ground. They had listened to his arguments and stood their ground. Mutual tolerance and respect were taken for granted.[14]

A related issue that Du Bois-Reymond chose to dwell on in the 1894 *Festrede* was that the task of demonstrating the sterility of the *Lebenskraft* doctrine in the biological sciences – a task approached resolutely out of respect for Müller – had led to raising questions and invoking analytical reasoning that, as by-product, opened up new trajectories for strengthening the physicalist point of view. The cardinal fault of the vitalists, as Du Bois-Reymond emphasized, was embodied in positing an erroneous conception of force (*Kraft*), but force is not, as the vitalist had conceived, associated with matter as an essence that exists separate from matter. Rather, force is a conceptual notion invoked to explain observed changes in matter. As Newton had shown, force is a mathematical concept. When associated with living matter it becomes a conceptually empty notion that serves no function in the life sciences.[15]

The focus of this study is Helmholtz (1821–1894). He was the youngest among the four members of the Johannes Müller (1801–1858) circle whose new directions in physiology developed into a scientific discipline whose early history we are exploring. He began his academic studies at the *Pepinière*, an institute for the

[13] Emil Du Bois-Reymond, "Festrede zur Feier des Leibnizschen Jahrestages," *Sitzungsberichte der königlichen Preussischen Akademie der Wissenschaften zu Berlin*, 1894, 623–641.

[14] Emil Du Bois-Reymond, 1894, "Festrede..." 626–627.

[15] Emil Du Bois-Reymond, "Festrede..." 1894, 628–629.

training of medical doctors. At the same time he attended lectures on physiology at the University of Berlin. Many years later Helmholtz recalled:

> With this study I came directly under the influence of a sensitive teacher [*eines tiefsinnigen Lehrers*], the physiologist Johannes Müller, who at the same time had led Du Bois-Reymond, Brücke, Ludwig and Virchow into physiology and anatomy. Johannes Müller still was struggling with puzzling questions concerning the nature of life – questions that essentially were metaphysical, and also questions concerning the newly developing scientific outlook on the nature of life. However, the conviction that the knowledge of facts cannot be replaced by anything else was developing in him with steady firmness. That he himself still was struggling [in this way] probably made his influence on students all the greater.[16]

Working with Müller at the University of Berlin, Helmholtz completed his doctoral dissertation on the structure of the nervous system in invertebrates in 1842.[17] To the extent that freedom from medical studies and other assignments permitted, Helmholtz simultaneously immersed himself in the classic works of eighteenth-century mathematician-physicists such as Leonhard Euler (1707–1783), Daniel Bernoulli (1700–1782), Jean le Rond d'Alembert (1717–1783), and Joseph Louis Lagrange (1736–1813). His secret love [*Lieblingsthema*] from the days of his youth had been physics. When Du Bois-Reymond first met Helmholtz in 1845 he wrote to his friend Eduard Hallmann (1813–1855), who at the time was Müller's assistant,

> In the meantime an acquaintance with Helmholtz has given me much pleasure. This, according to Brücke and little me [*sauf la modestie*], is the third organic physicist in our league. He is a fellow who has devoured [*gefressen*] chemistry, physics, and mathematics with spoons; he stands entirely with us on our *Weltanschauung*, and is rich with ideas and new ways of looking at things.[18]

References

Du Bois-Reymond, Emil. 1894a. Festrede zur Feier des Leibnizschen Jahrestages. In *Sitzungsberichte der königlichen Preussischen Akademie der Wissenschaften zu Berlin*, 625–628. Berlin: The Academy.

Du Bois-Reymond, Emil. 1894b. Festrede zur Feier des Leibnizschen Jahrestages. In *Sitzungsberichte der königlichen Preussischen Akademie der Wissenschaften zu Berlin*, 623–641. Berlin: The Academy.

Du Bois-Reymond, Estelle, ed. 1912a. *Reden von Emil Du Bois-Reymond*, I–XXII, 2 vols. Leipzig: Verlag Von Veit.

---

[16] Helmholtz, "Erinnerungen," Tischrede gehalten bei der Feier des 70. Geburtstages, Berlin, 1891, *Vorträge und Reden*, 5th ed. 1903, vol. 1, 9.

[17] Arminius Helmholtz, *De fabrica systematis nervosi evertebratorum*, Berlin, 1842.

[18] Estelle Du Bois-Reymond (ed.), *Jugendbriefe von Emil Du Bois-Reymond und Eduard Hallmann*, Berlin, 1918, 122–123.

Du Bois-Reymond, Emil. 1912b. Gedächtnisrede auf Johannes Müller. In *Gehalten an der Leibniz-Sitzung der Akademie der Wissenschaften*, 1858, Reden #VI, 135–317. Leipzig: Veit & Co.

Du Bois-Reymond, Estelle (ed.). 1918. *Jugendbriefe von Emil Du Bois-Reymond und Eduard Hallmann*, 122–123. Berlin: Reimer.

Helmholtz, Arminius. 1842. *De fabrica systematis nervosi evertebratorum*. Berlin: Berolini, Nietackian.

Helmholtz. 1891. Erinnerungen. In *Tischrede gehalten bei der Feier des 70*. Geburtstages, Berlin, 1891, *Vorträge und Reden*, 5th ed. 1903, vol. 1, 9.

Lesky, Erna. 1970. Ernst Brücke. *DSB* 2: 530–532.

Rosen, George. 1973. Karl Ludwig. *DSB* 8: 540–542.

Rothschuh, K.E. 1969. *Physiologie im Werden*, 167. Stuttgart: Fischer.

Rothschuh, K.E. 1971. Emil Heinrich Du Bois-Reymond. *DSB* 4: 200–205.

Steudel, Johannes. 1974. Johannes Müller. *DSB* 9: 567–574.

Steven Turner, R. 1972. Hermann Helmholtz. *DSB* 6: 241–253.

Chapter 3
From Physiology to Energy Conservation

Neither Helmholtz's commitment to working in the Müller physiology circle nor his stint in the military were matters of his own choice. They were thrust upon him by circumstances connected with what seemed to lie ahead for him as a financially remunerative professional career. Many years after his training in Berlin, Helmholtz told an audience that he had always wanted to become a physicist. it was in 1886 on the occasion of his acceptance of the Graefe-Medallion in Heidelberg that he remarked: "From early youth my inclination and my interest was directed toward physics. My father explained to me that as much as it pained him he had to tell me that physics was not a science that could guarantee a livelihood…., but that if I wanted to study medicine I also would be able to practice natural science [*Naturwissenschaft*]".[1]

The physiologists who had worked alongside Helmholtz in the Müller circle in Berlin knew from the start that their colleague had set his sights on becoming a physicist – this in spite of his keen interest in physiology and his commitment to pursuing physiological problems from a point of view that makes full use of the intellectual and experimental tools of the physical sciences. Helmholtz's colleagues also were convinced that he matched their image of an enthusiast for cross-disciplinary investigations; they knew that he would do what it takes to steer physiology in the new mechanistic direction that had been developed within the Müller circle. For Helmholtz it was evident that since he had begun his career in the fields of medicine and physiology – because of limited financial resources in the family – he would set his mind in the direction of transforming physiology into an exact science along physicalistic guidelines. In this he had the enthusiastic support of every member of the Müller group. First, however, he was faced with the task of fulfilling his military service in the Prussian army – an assignment that he had contracted for in 1838 and that was to terminate 10 years later. On completion of his tours in internal medicine, pediatrics, and obstetrics Helmholtz found time in Müller's laboratory at

[1] Hermann von Helmholtz, Antwortrede gehalten beim Empfang der Graefe-medaille zu Heidelberg 1886 *Vorträge und Reden*, 3rd ed., vol. 2, Braunschweig, 1896, 314.

E. Hiebert, *The Helmholtz Legacy in Physiological Acoustics*, Archimedes 39, DOI 10.1007/978-3-319-06602-8_3, © Springer International Publishing Switzerland 2014

the Friedrich-Wilhelms-Institute at the University of Berlin to undertake research on putrefaction and fermentation. On promotion to staff surgeon in the cavalry division at Potsdam, Helmholtz was privileged for 5 years to spend a substantial fraction of his time in research at the army barracks laboratory. His studies on the production of animal heat resulting from chemical transformations in muscle contraction were published in 1845. These studies turned Helmholtz's mind in the direction of reflecting in a more general way on the connection between mechanical, thermal, and chemical energy.

On the days Helmholtz was assigned to duty as military hospital guard (*Lazarettwache*) in Potsdam he made use of his leisure time to organize his evolving thoughts on energy conservation (*Erhaltung der Kraft*). The experiments he had undertaken on muscle contraction in the laboratory at the army barracks had convinced him that there was no need to invoke non-physical forces to explain living phenomena. By the beginning of 1847 Helmholtz's formulation of the energy conservation idea had reached a stage of development that prompted him to draw up a preliminary version of an article for publication. He decided to send it to his closest friend Du Bois-Reymond.

> Enclosed I send you my effort at an introduction to conservation of force [*Konstanz der Kraft*], not because I believe to be finished with it. . ..but because I do not yet foresee how often I will have to rework it before it is completed, and because I would like to know whether you take the exposition to be such as will find access to physicists. In my last reworking of it I have pulled myself together, and in so far as it is not urgently necessary, have thrown overboard all that smelled of philosophy [*alles was nach Philosophie roch*]. In that case a number of mental lacunae [*Gedankenlücken*] may remain.[2]

In July 1847, Helmholtz, then a 26-year-old physiology student at the University of Berlin, presented the members of the *Physikalische Gesellschaft* in Berlin a lecture entitled *Ueber die Erhaltung der Kraft*. It was a topic that none of the society's 50-some members seems to have given serious thought. The lecture was rejected for publication in Poggendorff's *Annalen der Physik* allegedly for being both too theoretical and too speculative. The claim that it was too theoretical was a charge that no one honestly might have levelled against Helmholtz at the time or in later life. Although he was recognized as a theoretical physicist most of his life, his characteristic ways of thinking and working were decisively experiment directed. Conspicuous in all his efforts in the sciences was a thrust in the direction of experimentation that was closely coupled with an inventive drive for designing and refining instruments that would set the pace for experimental and observational novelty. This certainly was the case for his venture into the field of musical acoustics – the central theme in this paper.

On the recommendation of Helmholtz's somewhat older colleague C.G. Jacobi (1804–1851), who at the very beginning of his career became known for his

[2] Helmholtz to Du Bois-Reymond, Potsdam, 12 Feb. 1847, C. Kirsten et al. (eds.), *Dokumente einer Freundschaft. Briefwechsel zwischen Hermann von Helmholtz und Emil du Bois-Reymond, 1846–1896*, Berlin, 1986, 78–79.

creative contributions to the mathematics of elliptic functions, Helmholtz's lecture was accepted for publication. As a clearly copied draft with minor alterations it was delivered in Helmholtz's handwritten version to publisher Georg Reimer in August 1847. The lecture was published as a 68-page pamphlet that year with title *Ueber die Erhaltung der Kraft*.[3]

The principle of conservation of energy was presented by Helmholtz in the form of a mathematically formulated universal guideline valid for all processes that occur in nature. It was a theoretical claim that over a lifetime would be put to test by its author in experimental research programmes that encompass not only mechanics, acoustics (as a branch of mechanics), hydrodynamics, thermodynamics, and chemistry, but biology, physiology, and the life sciences in general. The analytical section of the memoir is given over to an elaborate mathematical explanation and demonstration of the principle of conservation of mechanical energy and its provision that the forces between the smallest parts of matter act along the lines joining the parts. The enunciated principle is then applied to a number of mechanical theorems and to the force-equivalents of heat and electrical processes.

Helmholtz notes at the outset that his memoir is addressed mainly to physicists; that it is presented "purely in the form of a physical premise independent of any philosophical foundation [*rein in der Form einer physikalischen Voraussetzung...unabhängig von einer philosophischen Begründung*]." In expressing so bold a disavowal of metaphysical and philosophical underpinnings and so far-reaching a claim, Helmholtz nevertheless latched on to the law of causality as anchor point in formulation of the principle. He remarked, "We are justified and indeed impelled in this proceeding by the conviction that every change in nature *must* have a sufficient cause [*eine zureichende Ursache*]." This "conviction impels us to seek for causes to account for the change; and thus we proceed until we at length arrive at final causes which are unchangeable, and which therefore must in all cases, where the exterior conditions are the same, produce the same invariable effects."[4]

For Helmholtz the final aim of the theoretic sciences is "to discover the ultimate and unchangeable causes of natural phenomena," but he goes out of his way to mention that perhaps there are "processes in nature [*Vorgänge in der Natur*]" that arise from "spontaneity and freedom [*Spontaneität und Freiheit*]" and therefore are not dependent on the laws of causation – implying that there are, after all, conceivable

[3] The word *Kraft* was used by German physicists for *Energie* until about 1860. The subheading to Helmholtz's lecture, crossed out in its translated form, reads "eine physikalische Abhandlung zur Belehrung seiner theuren Olga [Helmholtz's fiancée] bearbeitet von Dr. H. Helmholtz." An English translation of the 1847 paper was provided in 1853 by Helmholtz's close friend John Tyndall (1830–1893), Professor of Natural Philosophy at the Royal Institution in London and successor in that post to Michael Faraday (1791–1867). It was published by Taylor & Francis in *Scientific Memoirs, Selected from the Transactions of Foreign Academies of Science, Natural Philosophy Series*, ed. by John Tyndall and William Francis, London, 1853. In 1889 the publishing house Wilhelm Engelmann published the 1847 Helmholtz treatise as No. 1 in its multi-volumed *Ostwalds Klassiker der exakten Naturwissenschaften*. The paper is published in volume 1 of Helmholtz's *Wissenschaftliche Abhandlungen*, Leipzig, 1882, 12–68.

[4] Helmholtz, *Wissenschaftliche. Abhandlungen, 1*, 12–13, Tyndall transl. 114–115.

limits to the range of applicability of the laws of causality. Helmholtz's position is succinct and crystal-clear:

> Whether all the processes of nature be actually referable to such – whether nature is capable of being completely comprehended or whether changes occur which are not subject to the laws of necessary causation but spring from spontaneity or freedom, this is not the place to decide; it is at all events clear that the science whose object it is to comprehend nature must proceed from the assumption that it is comprehensible, and in accordance with this assumption investigate and conclude until perhaps, she is at length admonished by irrefragable facts [*unwiderlegliche Facta*] that there are limits [*Schranken*] beyond which she cannot proceed.
>
> Theoretical natural science therefore, if she does not rest contented with half views of things, must bring her notions into harmony with the expressed requirements as to the nature of simple forces and with the consequences which flow from them. Her vocation [*Geschäft*] will be ended as soon as the reduction [*Zurückleitung*] of natural phaenomena to simple forces is complete, and the proof given that this is the only reduction of which the phaenomena are capable.[5]

The introductory section to Helmholtz's memoir on the conservation of energy has been given prominence in this paper because it touches meaningfully on Helmholtz's views concerned with the limits of scientific explanations for processes that occur in nature. In the 1847 memoir Helmholtz reminds his readers, as he does elsewhere in his writings, that although the scientist may pull out all the stops in an attempt to provide an explanation for a phenomenon or a complex of phenomena, it may not be possible to do so in a significant way for certain phenomena. A seminal instance in which Helmholtz expresses his position on the limits of scientific analysis is given in *Tonempfindungen* – a work on the *physiological* basis for the theory of music. At the end of the treatise, in chapter XIX on "Esthetical Relations," after having examined musical phenomena in relation to tones, chords, harmony, tonality, etc., he undertakes, as a scientist, to evaluate his own accomplishments in relation to music. Here is what he says:

> Here I close my work. It appears to me that I have carried it as far as the physiological properties of the sensation of hearing exercise a direct influence on the construction of a musical system, that is, as far as the work especially belongs to natural philosophy. For even if I could not avoid mixing up esthetic problems with physical, the former were comparatively simple, and the latter much more complicated. This relation would necessarily become inverted if I attempted to proceed further into the esthetics of music, and to enter on the theory of rhythm, forms of composition, and means of musical expression. In all these fields the properties of sensual perception would of course have an influence at times, but only in a very subordinate degree. The real difficulty would lie in the development of the psychical motives which here assert themselves. Certainly this is the point where the more interesting part of musical esthetics begins, the aim being to explain the wonders of great works of art, and to learn the utterances and actions of the various affections of the mind [*Bewegungen der verschiedenen Seelenstimmungen*]. But, however alluring such an aim may be, I prefer leaving others to carry out such investigations, in which I should feel myself too much of an amateur [*Dilettant*], while I myself remain on the safe ground of natural philosophy [*Naturforschung*], in which I am at home.[6]

---

[5] Helmholtz, *Wissenschaftliche Abhandlungen 1*, 13 and 16–17. Tyndall translation 115 and 117–118.

[6] Hermann Helmholtz, *On the Sensations of Tone as a Physiological Basis for the Theory of Music*, 2nd Engl. ed., Dover ed. New York, 1954, 371. Ger. ed. of 1968, 599.

References

Helmholtz, Hermann. 1954. *On the sensations of tone as a physiological basis for the theory of music*, 2nd English ed., 371. New York: Dover Publications.
Kirsten, C., et al. (eds.). 1986. *Dokumente einer Freundschaft. Briefwechsel zwischen Hermann von Helmholtz und Emil du Bois-Reymond, 1846–1896*, 78–79. Berlin: Akademie-Verlag.
von Helmholtz, Hermann. 1896. Antwortrede gehalten beim Empfang der Graefe-medaille zu Heidelberg 1886. *Vorträge und Reden*, vol. 2, 3rd ed, 314. Braunschweig: Vieweg.

Chapter 4
Early Experience in Music-Making

As a preliminary conspectus to Helmholtz's contributions to musical acoustics it will be of value to inquire what there may have been in his early cultural background and in his exposure to music and his concurrently energizing scientific career to account for the imaginative and professional approach to music that he eventually adopted as a full-fledged physiologist and physicist. The rationale for examining the issue of his early training at the level of specifics is based on an assumption. Helmholtz's emphasis on the scientific study of music from the standpoint of the listening and discriminating ear is linked, in a meaningful way, to his early music-listening and music-making experiences. It is assumed here that these musical experiences are sufficiently deeprooted and lasting to become an important component of his mature scientific endeavours in musical acoustics.

In parallel with a lifetime of scientific investigations, Helmholtz cultivated an interest in music that stretches from his early years in the family home in Potsdam where he was born and where he attended the Volksschule and the Gymnasium. These early interests in music continued to flourish in a creative way in his later years in Bonn, Heidelberg, and Berlin. Fortunately there exists in the vast corpus of Helmholtziana a large body of information about Helmholtz's involvement in music. As might be expected, this information is scattered far and wide and surfaces here and there in biographical accounts, in Helmholtz's own autobiographical reflections, in the writings of family, friends, and colleagues, and most significantly, in correspondence with family and colleagues.

The most imposing account of Helmholtz as public figure is the three-volume biography of Leo Koenigsberger (1837–1921) who for a number of years, as professor of mathematics, was Helmholtz's colleague at the University of Heidelberg.[1] The Koenigsberger biography was written in response to repeated requests on the part of Helmholtz's widow Anna (née von Mohl) and was based on the Helmholtz *Nachlass* of unpublished papers and correspondence that was made available for publication by Anna's children in 1902. A comprehensive and valuable work,

[1] Leo Koenigsberger, *Hermann von Helmholtz*, 3 vols., Braunschweig, 1902–1903.

E. Hiebert, *The Helmholtz Legacy in Physiological Acoustics*, Archimedes 39,
DOI 10.1007/978-3-319-06602-8_4, © Springer International Publishing Switzerland 2014

compiled by a colleague who was knowledgeable about Helmholtz's scientific work, the 3-volume Koenigsberger biography contains descriptions of Helmholtz's books, papers and speeches, in addition to lengthy extracts from correspondence. Koenigsberger's overall message unfortunately is hagiographic in form and not always reliable in regard to what is included and what is omitted. Three additional and complementary collections of correspondence have been published more recently and provide more immediate and more reliable access to Helmholtz's persona. In particular they reveal how Helmholtz's ideas on music and his own engagement in music-making enriched his scholarship in music and his research in the area of musical acoustics.[2]

In Helmholtz's writings on musical acoustics, the pianoforte and the harmonium were given pride of place as experimental tools in the study of sensations of tone as a basis for the theory of music. Helmholtz explored the mechanisms of these instruments and the vibrational characteristics of their struck strings and vibrating reeds in terms of the production, tone quality, intensity, and decay characteristics of sound. His investigations on consonance, dissonance, partials, and combination tones in relation to the purity of tone combinations, led to the construction of fixed-tone keyboards designed to approximate just intonation. By the time Helmholtz wrote about keyboards as instruments of experimental research he had been playing the piano for some 20 years. He was actively engaged in exploring the piano literature when he began his university studies at the age of 17. By the 1850s he had undertaken keyboard-related research investigations such as combination tones (1856) and problems involving the vibration and friction in organ pipes (1859–1861). Helmholtz's letters to his parents and friends reveal that by the time he began his university studies he had become knowledgeable about mainstream classical music for keyboards and other instrument ensembles, church music, and music written by contemporaries such as Johannes Brahms, Giacomo Meyerbeer, Gioachino Rossini, Felix Mendelssohn-Bartholdy, Robert Schumann, and Richard Wagner.

Helmholtz considered the human voice and polyphonic combinations of human voices to be the ideal instrument for providing the purest intonations and harmonics. On the other hand the piano became the instrument on which he was able, most rewardingly, to explore problems in the areas of music theory that deal with just intonation arising from the quality of their chords. The most direct information about Helmholtz's encounter with the piano while he was still a student derives from letters written to his parents over a period of 9 years while serving

[2] David Cahan (ed.), *Letters of Hermann von Helmholtz to his Parents, 1837–1846*, Stuttgart, 1993; Richard L. Kremer (ed.), *Letters of Hermann von Helmholtz to his Wife, 1847–1859*, Stuttgart, 1990; and Ellen von Siemens-Helmholtz, *Anna von Helmholtz. Ein Lebensbild in Briefen*, 2 vols., Leipzig, 1929. Olga Helmholtz, née von Velten (1826–1859), was Helmholtz's first wife. Anna Helmholtz, née von Mohl (1834–1899), was his second wife. Ellen von Siemens-Helmholtz (1864–1941) was Anna's daughter. She was married to Ernst Werner von Siemens (1816–1892) whose financial assistance and advice to the Prussian government led to the establishment in 1887 of the *Physikalisch-technische Reichsanstalt* (the *PTR*), a government-sponsored research institution first headed by Helmholtz.

as surgeon at the Friedrich-Wilhelm-Institute (Pepinière), at the Charité, and at the royal guard cavalry regiment in Potsdam. When Helmholtz began his studies at the Pepinière at the age of 17 we know that he saw to it that there would be a piano in his room. Correspondence with his parents and his brother reveal that he and his roommate were competitively engaged in practicing the piano.[3] Helmholtz assured his parents that in spite of sometimes having limited access to the piano he had not given up on his piano playing. He also wanted them to know that he would much rather play the piano than listen to others play – "What you fear, namely, that I would put aside my music, is prevented I believe, because the newer music which my roommate loves so much simply does not satisfy me; and in order to listen more deeply I must play myself. Besides, for me the expression and the performance of music by others seldom is sufficient; I always have more enjoyment with music when I realize it myself."[4] In the thick of medical studies Helmholtz complained that he and his fellow students had to live through endless evening hours learning about muscles; that it was somewhat easier for him than the others because he could devote his free daytime hours to music. "By myself I play Mozart and Beethoven sonatas, and with other fellow roommates newly acquired items for sight-reading."[5] In a letter to his parents, while on vacation at the home of his uncle in Königsberg, our 18-year-old student – who evidently was quite a good sight-reader – was spending his time at the piano and reading Schiller. "For the first 2 days I had to be satisfied with my uncle's music, which apart from a Mozart sonata consists of nothing but famous works by Strauss, Lanner, Czerny, Hünten, Auber, Ross, and Bellini, etc. I chased through all of them one after another but finally became sufficiently aggravated to return over and over to the Mozart sonata and Cramer's *Etudes* in order to somewhat strengthen my spiritual stomach." He located a second Mozart sonata, but then when he again ran out of music he was rescued by a friend who "loaded me down with a pile of music under whose weight I almost collapsed (poetic hyperbole) but most of them were piano arrangements and other artificial items [*Künstlichkeiten*] by Weber, Moscheles, Cramer, etc. Among them was a book of Mozart and Beethoven sonatas."[6]

On completion of the final government examinations to qualify as medical doctor and surgeon, Helmholtz was transferred to the royal regiment of the cavalry

[3] Cahan (ed.), *Letters of Hermann von Helmholtz to his Parents*, Letter 5, 31 Oct. 1838. Helmholtz's brother Ferdinand wrote to Helmholtz in regard to his roommate (fn. 2): "That the two of you play the Clavier so skillfully should give you the best opportunity to improve yourself. Don't be so easy-going and leave the playing to him because he plays better than you; in a similar situation I unlearned the little I had learned. But, in particular, don't let the ear-tickling and the glitter of the new Italian extravagance [*den Ohrenkitzel und das Geflimmer der neuen italienischen Ueberspanntheit*] rob you of your taste for the deep, old German intellectual music; the former is seductive, the latter constructive."

[4] Cahan (ed.), *Letters of Hermann von Helmholtz to his Parents,* Letter 6, 5 Nov. 1838.

[5] Cahan (ed.), *Letters of Hermann von Helmholtz to his Parents,* Letter 7, 1 Dec. 1838.

[6] Cahan (ed.), *Letters of Hermann von Helmholtz to his Parents,* Letter 13, 23 Aug. 1839? Johann Cramer (1771–1858), an English composer of German descent, established himself in London as a Bach, Mozart, and Beethoven concert pianist. He wrote many sonatas, concertos, and an influential didactic work entitled "*Studio per il pianoforte*", 1804, 1810.

guard in Potsdam. Soon thereafter he was released from all military service on the advice of Alexander von Humboldt (1769–1859) and Johannes Müller, who had convinced the Prussian ministry of culture that a person capable of writing an article as impressive as the Helmholtz 1847 paper on the conservation of energy should be encouraged to pursue an academic career. A short stint as assistant to Müller in the museum of anatomy and as lecturer in anatomy at the Academy of Arts in Berlin was followed in 1849 with an associate professorship in physiology in the medical faculty at the University of Königsberg. Three years as professor of physiology and anatomy in Bonn and a professorship in physiology at the University of Heidelberg in 1858 brought Helmholtz an international reputation in medicine, physiology, and physics.

Apart from the information about Helmholtz's involvement in music during his student years, noteworthy insights into the extent of his exposure to music during his most productive middle years when he was beginning to think about and experimentally pursue the field of musical acoustics, are recorded in the letters written to his fiancée, and then wife, Olga von Velten between 1847 and 1859. These were richly productive exploratory years when Helmholtz was moving freely into domains of thought and investigation that in some sense were peripheral to physics and traditional physiology. They include experimental investigations on the propagation of nerve impulses, color theory and color blindness, light accommodation and contrast, timbre in the vowel sounds of the human voice, and the physiological causes of musical harmony. In many ways Helmholtz stands out among his contemporaries and his erstwhile fellow students in the Müller circle as one who to a surprising extent had managed to build a career around the personal experiences and sensitivities that he had encountered and nourished in early life.

Biographical information on Olga (née von Velten) Helmholtz (1826–1859) is provided in an account written by Betty Johannes (née von Velten), Olga's sister, and touches on Olga and Hermann's keen interest in music.[7] Olga was born into a family of politically prominent civil servants and military doctors. In this family simple living and a high level of culture in music, language, and literature were given priority alongside training in everything for which there was talent and inclination, and for all that would foster the cultivation of a moral earnestness and an unbiased receptivity for the good life. In the von Velten family Helmholtz at first was looked upon as a somewhat strange, earnest, and contemplative guest; in time he came to be recognized as an animated and knowledgeable young man. "A very clever person but one first had to dig him out [*Sie müssen ihn erst ausgraben*]. It became a matter of treasure-seeking [*Schatzgraberei*]." In turn Helmholtz at first had the impression

[7] The account has its source in Koenigsberger, who had access to a Helmholtz *Nachlass*. It is reproduced in Richard L. Kremer, (ed.), *Letters of Hermann von Helmholtz to His Wife*, Appendix 193–201.

that the von Velten home was not real but something one might read about in a novel. He soon was able to adapt himself so well that they referred to him as the inconspicuous mediator whose opinions were so decisive, and whose suggestions were so thankfully accepted, that in his hand he held together most of the threads of the family's wide interests.[8]

One of the activities cultivated most avidly in the von Velten-Helmholtz partnership was music. Olga's sister wrote: "Much music was made in this house and outside of it by land and by sea. My sister sang beautifully. His [Hermann's] participation and his opinion soon became the standard for all of us. He taught us to thoroughly understand and love Shakespeare and Beethoven and we play-read together." The sister knew that Olga and Hermann "belonged together for life."[9] In the Helmholtz letters to Olga there are regrets about not having opportunities to sing and play together. By attending lectures or music, concerts, and operas, he was widening his musical horizons by discovering new directions in music. On one occasion Helmholtz waited in vain for his bride to join him at a symphony concert in the *Singakademie* in Berlin. He wrote to her: "You didn't come. My hearing lost its bearing. It was as if up to that point your soul with its deep and inner musical nature had been with me to guide the harmonies into my understanding. Then, my ears heard only musical figures and my soul heard nothing. Naturally I had this experience with the Mozart symphony – one of his most beautiful ones and one that caused everything around me to swim in ecstasy. I, there alone, forsaken by the better half of my soul, could just as well have been hearing scales played on the piano. Not until [Beethoven's] Coriolanus overture did I come to myself. That is a jewel, so short and concise, so decisive and proud in the midst of a mass of restlessness and tangled struggles, and then at last so sad with a number of melancholy tones – a masterwork than which nothing greater can be. Farewell my sweet and only possession."[10]

While on a visit with Faraday and Wheatstone in London, Helmholtz wrote his wife that the level of taste for music on the part of the general public in London was approximately at the level of what would appeal to street-porters (*Eckensteher*) in Berlin. "In any case the appreciation of artistry [*Kunstsinn*] in music, drama, and the performing arts seems to be almost totally absent here. The sense of taste is there only as long as it is the servant of utility; thereafter it is gone."[11]

[8] Kremer, (ed.), *Letters of Hermann von Helmholtz to His Wife*, 193–195.

[9] Kremer, (ed.), *Letters of Hermann von Helmholtz to His Wife*, 195.

[10] Kremer, (ed.), *Letters of Hermann von Helmholtz to His Wife*, 195–196. This letter is in part reproduced from Koenigsberger, *Hermann von Helmholtz*, vol. 1, 67.

[11] Kremer, (ed.), *Letters of Hermann von Helmholtz to His Wife*, Letter 29, 20 Aug. 1853.

References

Cahan, David (ed.). 1993. *Letters of Hermann von Helmholtz to his parents, 1837–1846*. Stuttgart: Franz Steiner Verlag.

Koenigsberger, Leo. 1902–1903. *Hermann von Helmholtz*, 3 vols. Braunschweig: F. Vieweg und sohn.

Kremer, Richard L. (ed.). 1990. *Letters of Hermann von Helmholtz to his wife, 1847–1859*. Stuttgart: Franz Steiner Verlag.

von Siemens-Helmholtz, Ellen. 1929. *Anna von Helmholtz. Ein Lebensbild in Briefen*, 2 vols. Leipzig: J. A. Barth.

Chapter 5
Physiological Acoustics and Combination Tones

Apart from a number of annual reports on the state of theoretical acoustics submitted to the Berlin Physical Society, Helmholtz's first substantial exploration into the field of musical acoustics dealt with combination tones. As Helmholtz discovered, this is a physiological phenomenon that turned out to be of central importance for studies on just intonation. According to the law of the independence of sound waves – a law that reduces the considerations of complex cases of sound production and reception to those of simple sound waves – each component sound wave of a vibrating object produces a sound wave that can be regarded as traveling independently of every other component of the composite sound of the vibrating object. Although the law holds rigorously only in cases where the vibrations of all the parts of the vibrating body are small in dimension, this is the case for most of the string and air-column systems utilized in musical instruments. It is not the case where the intonation hinges on combination tones and the fine-tuned analysis of consonance, dissonance, beats, or other "interruptions of harmony" as Helmholtz referred to them. Alert to the idea that combination tones are markedly sensitive to inaccuracies of intonation, Helmholtz was led to perceive that the vibration of small non-negligible dimensions represents the essential anchor point for research on the problem of intonation on fixed-tone keyboards.

It was long known that a vibrating string or column of air produces not only a tone corresponding to its fundamental frequency f, but also to a series of harmonics of frequency 2f, 3f, 4f....; that two tones of slightly different frequencies, sounded together, produce beats resulting from interference. Around the middle of the eighteenth century it was discovered that when two tones are sounded together a third tone phenomenon (later referred to as a difference tone) could be detected by a keen ear under normal conditions of music-making. Early descriptions of the third tone phenomenon occur in a treatise on composition (1745–1747) by Georg Sorge (1703–1778) and in a treatise on the acoustical foundations of harmony (1754) by the violinist virtuoso Giuseppe Tartini (1692–1770) who claimed to have discovered third tones (*terzo suono*) as he referred to them, while experimenting with double stops on the violin. Scientists at the time, Lagrange and Thomas Young for instance, postulated a beat theory for these difference tones and suggested that the beat frequency

E. Hiebert, *The Helmholtz Legacy in Physiological Acoustics*, Archimedes 39,

DOI 10.1007/978-3-319-06602-8_5, © Springer International Publishing Switzerland 2014

is large enough to be recognized as a tone itself. What was certain was that two tones with frequencies f_1 and f_2, when sounded together, produce a barely distinguishable third tone with frequency $f_1–f_2$ that does not belong to the harmonic series.

By the beginning of the nineteenth century the phenomenon of combination tones began to attract the attention of physicists such as Wilhelm Weber (1804–1891).[1] His paper of 1829 on Tartini tones led to a wealth of empirical data and a number of attempts to construct a descriptive theory to account for them. Weber supported the beat theory of difference tone, as Young and Lagrange had done earlier, but suggested that more than one beat tone was audible. Weber maintained that "once the physical basis of the Tartini tone is known, it becomes easy to determine the pitch of the Tartini tone in advance for each case, no matter what the ratio of the two beating tones might be. If such tones are equally possible, it is easy to determine which of them will be heard."[2]

After reading Weber's paper, the Finnish physicist Gustav Gabriel Hällström (1775–1844) suggested in 1832 as a challenge to Weber's beat theory that not only the fundamental frequencies of tones but their harmonic partials should produce difference tones such as $2f_1–f_2$, $f_1–2f_2$, $3f_2–f_2$, etc.[3] The challenge of demonstrating that the mechanism of the human ear, with its intricately constructed elastic and solid components, is able to interpret the observations as different real tones became for Hällström a matter of consuming interest. He calculated that second order tones should result from a blending of the first order combination tones (the Tartini tones) and the original generating tones; that third order tones should result from the second order tones, and so on. The extensive tables that Hällström drew up in comparing the data on observed combination tones with the predictions of his theory and other data from the literature were impressive, but the tones he predicted could not be observed at the time.[4]

In 1843 the Bavarian physicist Georg Simon Ohm (1789–1854) conducted a series of experiments with the polyphonic siren that led him to announce what came to be known as Ohm's acoustical law.[5] Like Ohm's famous electrical law that

[1] With his brother Ernst Heinrich Weber (1795–1878) Wilhelm wrote: *Wellenlehre, auf Experimente gegründet* (1825), the classic work on standing and travelling waves.

[2] Wilhelm E. Weber, Über die Tartini'schen Töne, *Wilhelm Weber's Werke*, *1*, Berlin, 1892, 360–364. Quotation 361.

[3] G.G. Hällström, Von den Combinationstönen, *Annalen der Physik und Chemie*, *24* (1832) 438–466. The paper was a revised version of a dissertation, *De Tonis Combinationis*, that Hällström had completed in Åbo, Finland, in 1819.

[4] Hällström, Von den Combinationstönen, 444–449 and 451–461.

[5] G.S. Ohm, "Ueber die Definition der Töne nebst darin geknüpfter Theorie der Sirene und ähnlicher tonbildenden Vorrichtungen," *Annalen der Physik und Chemie*, *59* (1843) 513–565. "Noch ein Paar Worte über die Definition des Tones," *Annalen der Physik und Chemie*, *62* (1844) 1–18. The siren, an acoustical instrument that both demonstrates the wave nature of sound and is used to determine the frequency of vibration of a sonorous body, was invented in 1819 by the Parisian physicist Charles Cagniard De La Tour (1777–1859) in 1815. The siren consists of a dish with equally spaced holes through which air blows as the disk rotates. It provides, by direct count, the frequency (pitch) of the sound emitted. The instrument was perfected in stages and took shape in Helmholtz's experiments in the 1850s and 1860s. It dominated the acoustic scene for a century. The polyphonic siren has several rings of holes concentrically arranged on the dish to allow for the variation of pitch in a known ratio without change in speed.

relates current, voltage, and resistance in an electrical circuit, Ohm's acoustical law (formulated in 1826) was structured on the work of the mathematical physicist Joseph Fourier (1768–1830). Fourier, the undisputed leader of the nineteenth century French analytical school, had shown in connection with studies on the propagation of heat in 1822 that any function can be expressed as the sum of a series of sines and cosines.[6] When applied to the phenomena of sound, as it was by Ohm in 1843, the Fourier theorem specifies that any motion of air that corresponds to a composite group of musical tones is capable of being analyzed as a sum of harmonic vibrations; that each single harmonic vibration corresponds to a simple tone sensible to the ear having a simple specific pitch. According to Ohm's acoustical law, the ear recognizes only sinusoidal waves as pure tones and is able, automatically, to perform an analysis of periodic sound waves into its component parts. This means that however complicated the motion producing the sound may be, the total sound is to be regarded as consisting of many non-interfering sinusoidal waves that are acting as if each wave is produced by itself.

Ohm's acoustical law received little attention until it was so to speak "rediscovered" by Helmholtz and applied to the characterization of tone color (*Klangfarbe*) or timbre. He demonstrated that it was the specific combination of sinusoidal components that gives a particular sound its auditory character. In reference to the notion of *Klangfarbe* Helmholtz stated the Fourier theorem as follows: "Any given regular periodic form of vibration can be produced by the addition of simple vibrations having pitch numbers which are once, twice, thrice, four times, etc., as great as the pitch numbers of the given motion."[7]

In 1856 Helmholtz presented the Berlin Academy of Sciences with the results of his own experimental investigations and theoretical interpretations on combination tones.[8] Although the 1856 papers provide ample evidence of Helmholtz's competence in the physics of vibrating systems, the cogency of his approach to characterizing and clarifying the phenomena of combination tones demonstrates in the main how firmly coupled his scientific thinking at this time in his career was with his wider concerns in the physiology of sensation. He had just completed the first volume of his handbook on physiological optics. With the decision to enlarge on combination tones and his perspective on the physiological studies of sensation by moving from optics to acoustics, Helmholtz became totally engrossed in problems unique to music theory. These interests were to capture his attention for more than a decade.

In the combination tone paper – his first research undertaking in acoustics – he was able to demonstrate the existence of the higher order difference tones that Hällström had predicted in 1832. The demonstration that Hällström's difference

[6] Jean Baptiste Joseph Fourier, *Théories analytique de la chaleur*, Paris, 1822.

[7] Helmholtz, *Sensations of Tone*, Dover ed. of 1954, 34.

[8] Helmholtz, "Ueber Combinationstöne," is published in two parts in the Academy's *Monatsberichte* and *Annalen der Physik* (1856), and is reprinted in *Wissenschaftliche Abhandlungen*, Vol. 1, 256–262 and 263–302.

tones are produced by the harmonics of the fundamental was crucial in the history of music theory because it signified that Ohm's acoustical law remained intact. At the same time the experiments that Helmholtz had designed to detect difference tones revealed the existence of another series of combination tones that he referred to as "summation tones."

The investigations of Weber, Ohm, and Hällström that have been mentioned thus far provide the point of departure for pursuing the nature of the experimental investigations and theoretical deliberations of Helmholtz's own work on combination tones. He appraised the situation as follows:

> Given m and n, whole numbers having no common divisor, it long has been known that two tones with frequencies $m\lambda$ and $(m+1)\lambda$ produce the combination tone λ. On the other hand, for two tones with frequencies $m\lambda$ and $n\lambda$, W. Weber and M. Ohm put forward the idea in a general way that the combination tone, likewise, has a frequency λ, whereas Hällström designated $(m-n)\lambda$ as the first combination. At the same time he postulated a number of higher-order combination tones with frequencies $(2n-m)\lambda$, $(3m-2n)\lambda$, etc. The higher-order combination tones were said to form lower-order combination tones with the original tones.... Poggendorff (then) correctly raised the question whether the so-called higher-order combination tones might not be the higher overtones [*Nebentöne*] that are present in practically all musical instruments.[9]

At the level of theory, several authors had offered explanations of the complexity of combination tones phenomena by assuming that the tones received by the ear are made up of a series of discontinuous impulses, the interpretation of which requires the ear to be endowed with special non-physical properties. For Helmholtz the physiologist, who in all of his deliberations had the basic principles of the physical sciences close to hand, the invention of subjective properties to explain a natural phenomenon such as the production of combination tones in the human ear was unacceptable. The suggestion in fact occasioned a blunt response: "I therefore will allow myself to present the Academy with a new explanation that is based entirely on known *mechanical* laws [emphasis added] that make it unnecessary to endow the human ear with any special properties."[10] Based on the notion that the principle of undisturbed superposition of oscillating motions is valid only so long as the motions are small, Helmholtz was able to demonstrate that combination tones should occur and be detectable when the motions become large enough for the square of the displacements to influence the motion. He calculated that among the oscillating parts of the ear only the tympanum (the eardrum) is singularly asymmetric and, with its hammer drawn inward, is able by displacement to influence the motion. "I therefore believe," he emphasized, "that I am able to put forward the conjecture that it is the characteristic form of the tympanum that determines the formation of combination tones."[11]

---

[9] Helmholtz, "Ueber Combinationstöne," 256. Helmholtz erroneously refers to G.S. Ohm as M. Ohm.

[10] Helmholtz, "Ueber Combinationstöne," 259.

[11] Helmholtz, "Ueber Combinationstöne," 261–262.

With the aid of the siren, and by employing tuning forks and resonators to isolate and reinforce frequencies, Helmholtz was able to demonstrate the existence of the higher-order difference tones that Hällström had predicted. At the same time he discovered that under optimal experimental conditions difference tones resulting from the sum of the frequencies of two primary (generating) tones could be detected, if but faintly.

> Thereby I discovered an until now unknown class of combination tones that do not fit the existing theories. I also discovered that there are objective combination tones that arise independently of the human ear. Finally, then, one is able to offer a theory of combination tones, very different from the hitherto existing theory, in which no special properties of the hearing nerves [*Hörnerven*] need to be postulated and which more adequately embrace all of the now known facts.
>
> I will designate these new tones with the name summation tone [*Summationstöne*] in contrast to the already mentioned and long known tones that we can call difference tones [*Differenztöne*] because their frequencies are equal to the difference in frequency of their primary or third lower-order combination tones. I first became aware that such tones might exist and tried to hear them with resonators activated by tuning forks. I succeeded in doing so but with great difficulty because the tones of the tuning forks have only moderate strength and the combination tones became distinct only with stronger primary tones. The summation tones are weaker than the first-order difference tones and it therefore takes much practice and attentiveness to hear them at the weaker strengths of the primary tones.[12]

Helmholtz's search for a theory to account for the various types of combination tones – one of which, the difference tones, had been recognized in its simplest form in the middle of the nineteenth century, and the other of which, the summation tones, he had discovered – eventually led him to conclude that it was not necessary to invoke non-mechanical principles to explain their existence. The anchor points that he had invoked to reach that conclusion included Ohm's acoustical law, the Fourier theorem, and the recognition that it was necessary to reckon with the fact that the motions of the various components of the inner ear lie outside the limits of validity of the principle of independence of motion for sound waves. He concluded:

> From what has been said it follows that one need not necessarily look for the cause of combination tones in the modes of sensation of the hearing nerves, but that for two simultaneously sounded tones of audible strength, the combination tones correspond, in the usual (that is, mechanical) way, to real vibrations of the tympanum [*Trommelfelles*] and the ossicles [*Gehörknöchelchen*]. Accordingly, the combination tones do not possess a mere subjective existence but would be able to exist objectively if only, to begin with, in the vibrating parts of the ear.[13]

In recent times it has been shown that the observation and explanation of combination tones are more significant in the development of musical acoustics and also are a more complicated physiological problem to unravel than anyone in the nineteenth century was in a position to realize in 1856 when Helmholtz tackled the problem. Combination tones are associated with non-linear systems,

---

[12] Helmholtz, "Ueber Combinationstöne," 264 and 282–283.

[13] Helmholtz, "Ueber Combinationstöne," 300.

i.e., systems that introduce distortion during the transmission of sound. This occurs when the cochlea of the inner ear is presented with two tones having sufficient and similar intensities, as well as a frequency difference to be audible. "Non-linear" signifies in this context that the combination of two sounds with intensities a and b is not a simple linear sum, a + b, but is given by a formula involving powers of a and b. The linear formula is valid only for small intensities of sound so that combination tones normally are heard only when the original tones are sufficiently loud. The frequency of a combination tone is either the difference (difference tones) or the sum (summation tones) of the frequencies of the two primary notes or their multiples. Difference tones are readily realized in normal conditions of music-making. Summation tones, by contrast, are heard only under special conditions.

In conclusion to this section on Helmholtz's experimental investigations on combination tones, it is significant finally to evaluate his reasons for highlighting this singularly complex phenomenon as a subject worth pursuing in some detail in connection with music. Combination tones, as Helmholtz discovered, are uniquely sensitive to inaccuracies of pitch and are therefore directly connected with intonation. The critical point at stake here is that the false combination tones of tempered intonation that show up as dissonances are more readily identifiable than they are for just intonation. The differences in terms of "pleasantness," as Helmholtz recognized, are greatest in the higher octave of the scale because there the false combinational tones of the tempered intonation are more readily observable, and the number of beats for equal differences of pitch, as Helmholtz remarked, become larger and more pronounced in their roughness (*Rauhigkeit*).

References

Fourier, Jean Baptiste Joseph. 1822. *Théories analytique de la chaleur*. Paris: Didot.

Hällström, G.G. 1832. Von den Combinationstönen. *Annalen der Physik und Chemie* 24: 438–466.

Ohm, G.S. 1843. Ueber die Definition der Töne nebst darin geknüpfter Theorie der Sirene und ähnlicher tonbildenden Vorrichtungen. *Annalen der Physik und Chemie* 59: 513–565.

Ohm, G.S. 1844. Noch ein Paar Worte über die Definition des Tones. *Annalen der Physik und Chemie* 62: 1–18.

Weber, Wilhelm E. 1892. Über die Tartini'schen Töne. In *Wilhelm Weber's Werke*, vol. 1, 360–364. Berlin.

Chapter 6
Physiological Causes of Musical Harmony

Helmholtz's effort to grapple with the study of physiological acoustics from an all-embracing point of view was undertaken the year after he had completed his comprehensive analysis on combination tones. During the winter of 1857 he presented a public lecture on the physiological causes of musical harmony "in the native town [Bonn] of Beethoven, the mightiest among the heroes of harmony [*in der Vaterstadt Beethovens, des gewaltigsten unter den Heroen der Harmonie*]." Convinced that music, its tones and sensations, and its incomprehensible, wonderful, and powerful effects merit scientific consideration as one of the arts, Helmholtz remarked:

> It always struck me as a wonderful and peculiarly interesting mystery, that in the theory of musical sounds, in the physical and technical foundations of music, which above all other arts seems in its action on the mind as the most immaterial evanescent, and tender creator of incalculable and indescribable states of consciousness [*die stoffloseste, flüchtigste und zarteste Urheberinn unberechenbarer und unbeschreiblicher Stimmungen*], that here, especially in the science of purest and strictest thought, mathematics should prove preeminently fertile.... Mathematics and music! the most glaring possible opposites of human thought! and yet connected, mutually sustained! It is as if they would demonstrate the hidden consensus of all the actions of our mind.... It is an acknowledged fact that the numbers of the vibrations of concordant tones bear to each other ratios expressible by small whole numbers. But why? What have the ratios of small whole numbers to do with concord? This is an old riddle [*eine alte Räthselfrage*], propounded by Pythagoras, and hitherto unsolved. Let us see whether the means at the command of modern science will furnish the answer.[1]

The lecture on the physiological causes of harmony was delivered while Helmholtz was in the process of vacating his professorial post in anatomy at the University of Bonn in order to take on the newly created physiology professorship and institute at the University of Heidelberg. The university he opted to go to at the time was at the pinnacle of its reputation as a center for scientific research, and the 13 years he spent there turned out to be the most productive of his entire career.

[1] Hermann Helmholtz, "On the Physiological Causes of Harmony," *Popular Scientific Lectures*, transl. of Alexander J. Ellis, New York, 1862, 22–58; "Ueber die physiologischen Ursachen der musikalischen Harmonie," Vorlesung, Bonn, Winter, 1857, in *Vorträge und Reden, 1, 1903*, 79–115.

E. Hiebert, *The Helmholtz Legacy in Physiological Acoustics*, Archimedes 39,
DOI 10.1007/978-3-319-06602-8_6, © Springer International Publishing Switzerland 2014

During these years Helmholtz published his volume on physiological music, the *Tonempfindungen* – many of whose ideas had been explored in a preliminary way in the 1857 lecture. He also published his three volumes on physiological optics and initiated investigations in hydrodynamics, electromagnetic theory, and mathematics that were completed during his following years in Berlin.

Confident that it was possible to clarify if not to resolve the apparent puzzle and age-old mystery that connects music and mathematics and that delineates what it means to speak of beauty in music from a scientific point of view, Helmholtz approached the subject by exploring how the mind and its transient states of consciousness relate not so much to mathematics directly as to the physiological and physical sciences. To this end he set out to demonstrate that music – its physical production, transmission, physiological reception by the ear, and mental interpretation – stand in need of being explored with the aid of the intellectual resources and material tools of modern science.

Helmholtz conceived that most of the requisite conceptual and material resources and tools to disentangle musical phenomena – consonance, dissonance, beats, combination tones, the tone quality or timbre of the musical instruments of the orchestra, the vowel sounds of the human voice – were close to hand. These resources included the mathematics of aural Fourier analysis and the principle of wave superposition or interdigitation (*Durchkreuzung*) according to which the total displacement of a linearly vibrating system is given by the algebraic sum of its various harmonic modes of oscillation. Other tools were invented or redesigned to meet specific needs – the resonance box with its reinforced tuning forks for pitch analysis and adjustment, ear-fitted glass and metal globes (resonators) designed for identification and reinforcement of partials, and the polyphonic siren for experimenting with complex acoustic phenomena. Not unexpectedly, the most important and most adaptable tools for experimentation in Helmholtz's acoustical toolbox were the string and air column systems embodied in the instruments of the orchestra and the keyboards. The rationale for so strong a claim is that only keyboards such as the clavichord, harpsichord, piano, harmonium, and organ, selectively and with flexible dynamic control, lend themselves to approximating and mimicking the tonal effect of an orchestra or an ensemble.; and, in effect, as in a piano-reduction, serve as virtual orchestra or ensemble.

Helmholtz's comprehensive analysis of the mystery and puzzle associated with the intersection of music and mathematics, and the identification of the intellectual and material tools requisite for the scientific illumination of that problem, brings into focus two contrasting perspectives on music. They stretch from antiquity to the present. On the one hand music deals with audible sensations that are amenable to abstract mathematical analysis and formulation and physical description within the branch of mechanics that treats of vibrating and oscillating systems composed of strings, columns of air, and solid objects. This objective outlook, as it sometimes is referred to, is important for the development of music theory, but it constitutes only a small component of the native and living phenomena that are experienced in music-making, music-listening, and the composition of music. The other perspective on music is one that explores the study of how the reception of sound by the ear is transformed into the perception of music by the human brain and its correlative neuro-physiological and neuro-psychological systems. The so-called objective aspects of the perception of music have been studied rewardingly using scientific concepts and experimentally-acquired information that draws on mechanics, anatomy, physiology, and the neuro-sciences. On the whole, however, the analysis of music is invariably undertaken against the background of subjective and culture-conditioned factors that transcend any notion of sound as objective sensation.

In the Bonn lecture Helmholtz had this to say about the dualism that straddles the reception of sound and the perception or conception of music:

> We must distinguish two different points – the audible *sensation* [*Empfindung*] as it is developed without any intellectual interference, and the *conception* [*Vorstellung*], which we form in consequence of that sensation. We have, as it were, to distinguish between the material ear of the body [*das leibliche Ohr des* Körpers] and the spiritual ear of the mind [*das geistige Ohr des Vorstellungsvermögens*]. The material ear does precisely what the mathematician effects by means of Fourier's theorem, and what the pianoforte accomplishes when a confused mass of tones is presented to it. It analyzes those wave- forms, which were not originally due to simple undulations such as those furnished by tuning-forks, into a sum of simple tones; and it feels the tone due to each separate simple wave separately whether the compound wave originally proceeded from a source capable of generating it, or became compounded on the way.[2]

In contrast to "the material ear of the body" as the objective component, Helmholtz refers to "the spiritual ear of the mind" as the subjective component that examines and interprets the objective sensations intellectually as perceptions.

No perceptions obtained by the senses [*Wahrnehmungen*] are merely sensations impressed on our nervous systems. A peculiar intellectual activity is required to pass from a nervous sensation to the conception of an external object, which the sensation has aroused. The sensations of our nerves of sense are mere symbols [*Zeichen*] indicating certain external objects, and it is usually only after considerable practice that we acquire the power of drawing correct conclusions from our sensations respecting the corresponding objects.[3]

In 1857 when Helmholtz delivered this lecture on music in Bonn – his first interpretive attempt – he was at the beginning of a decade of intensive experimental and theoretical investigations on musical acoustics. His motivations at the time were grounded, as he remarked, in searching out "the principle of artistic beauty in its unconscious conformity to law....and to lay bare the hidden law on which depends the agreeableness of consonant combinations." Helmholtz recognized assuredly that any such law or set of laws would not penetrate very deeply into the beautiful in music.

> These phenomena of agreeableness of tone [*Wohlklanges*], as determined solely by the senses, are of course merely the first step towards the beautiful in music [*sind freilich erst der niedrigste Grad des musikalisch Schönen*]. For the attainment of that higher beauty which appeals to the intellect, harmony and disharmony are only means, although essential

[2] Helmholtz (1857), "On the Physiological Causes of Harmony," 44; 103 in Ger. ed.

[3] Helmholtz (1857), "On the Physiological Causes of Harmony," 47; 106 in Ger. ed.

and powerful means....[because in] a musical work of art the movement follows the outflow of the artist's own emotions [*Strömungen der erregten Seele des Künstlers*].[4]

The distinction between the reception of sound by the ear and the interpretation of sound as music was first put to experimental test in a novel way by Helmholtz's experimental investigations of 1859 and 1860 on the tone color [*Klangfarbe*] or timbre of human vowel sounds.[5] These studies were undertaken to provide direct experimental evidence for Ohm's law. Helmholtz had sought to make the case that Ohm's acoustical law provides the essential rationale for explaining the existence and nature of combination tones – this in spite of the criticism that Seebeck had leveled against Ohm's law. Helmholtz suggested that there probably had been no meaningful dispute between Ohm and Seebeck; that the two men were merely quibbling about the use of terms. Helmholtz stated his own position as follows:

I previously already had suggested that the combined sensation produced by the outgoing motion of air from a single sounding body be designated by the term *Klang* [tone colour, timbre, quality of tone], and that the term *Ton* [tone] be restricted to designate the simple sensation that is produced by a single pendular motion of air. In that case the sensation of *Klang* is composed as a rule of several simple tones.... If one accepts all that Seebeck asserts about *Klang* in his dispute with Ohm, as well as what Ohm asserts about tone, then these two excellent acousticians are both in the right and both assertions can remain in place side by side.[6]

Prior to the Helmholtz studies on vowel sounds, the validity of Ohm's acoustical law for the explanation of the ear's ability to distinguish the individual components of a compound tone had been linked mainly with Seebeck's criticism based on phase differences. What was wanting, as Helmholtz maintained, was an experimental demonstration of the law itself. With the aid of eight tuning forks mounted on resonance boxes and activated by intermittent electrical pulses Helmholtz was able to reproduce, actually to mimic, the various vowel sounds of the human voice by eliminating all incidental noises. The pitches of the tuning forks ranged from the B_b of the lowest octave of the male voice to the highest pitch of the soprano voice. Small changes in pitch were achieved by fastening small masses of wax to the upper extremity of the tuning forks. "It was in this way," Helmholtz remarked, "that I gradually learned to mimic the various vowel sounds more or less completely, and indeed quite well and clearly in the case of U, O, Oe, E, somewhat less well in the case of J and Ue.... On the whole it is to be noted that vocal tones created by means of the tuning forks were closer to the sung tones than the spoken tones.... The artificially combined vocal tones most of all resembled the resonating tones produced by singing one of the vowels strongly into the piano."[7] Having originally set out to challenge the phase difference arguments that Seebeck had invoked to discredit

[4] Helmholtz (1857), "On the Physiological Causes of Harmony," 57; 114–115 in Ger. ed.

[5] Hermann Helmholtz, "Ueber die Klangfarbe der Vokale" (1859), *Wissenschaftliche Abhandlungen, 1* (1882) 397–407 and "Ueber Klangfarben" (1860) *Ibid.*, 408–409.

[6] Helmholtz, *Ueber die Klangfarbe*, 399–400.

[7] Helmholtz, *Ueber die Klangfarbe*, 402.

Ohm's law, Helmholtz concluded: "All the phase differences [that had been brought forward to justify the criticisms of Seebeck and others]...do not alter tone color [*Klangfarbe*] as long as the intensity of the tones remains constant.... The musical tone color depends only on the presence and intensity of overtones [*Nebentöne*] that are included in the *Klang*, and not on their phase differences."[8]

In order to examine tones that normally are too feeble to be detected aurally under ordinary conditions, Helmholtz constructed resonators in the form of glass globes that contained an extended funnel-shaped opening that fits into the outer ear. "With the aid of such resonators a number of acoustical phenomena that otherwise are difficult to examine – the objective combination tones, overtones and their beats – become exceedingly accessible."[9] Using this technique Helmholtz was able to study small effects associated with the musical voice and the musical instruments of the orchestra. In turn it provided him with an additional roster of acoustical phenomena waiting to be investigated.

In conclusion Helmholtz remarked that the claims he was putting forward in connection with the study on the timbre of vowel tones were preliminary and included no more than the 6th and 8th overtones. He remarked:

> The higher overtones produce dissonances with one another as well as beats, and when a group of beating tone pairs act together it probably will not be unimportant to consider whether or not all the rests [*die Pausen*] of the beats coincide. That depends on the phase differences. Besides, I also take it to be possible that the mass of higher dissonant overtones builds what the ear hears as accompanying noise, and what, from another approach, we already have eliminated from our consideration of tone color.
>
> In another place I have put forward the hypothesis that every nerve-fiber among the hearing nerves is fixed for the perception of a specific pitch, and is stimulated when the tone meets the ear with a pitch that corresponds with the elastic structure with which it is linked (organs of Corti or stiff hairs in the ampulles). Accordingly, the sensation of various tone colors would be reduced, so that simultaneous with the fibers that perceive the fundamental [*Grundton*] certain others that correspond to the overtones would be stimulated. This simple explanation could not be made if the phase differences in the deepest overtones were to be taken into account.[10]

The accomplishments that Helmholtz had achieved so singularly in his studies on vowel sounds were reinforced in a lecture on musical temperament delivered 8 months later in Heidelberg at the Society for Natural History of Medicine.[11] More than any of his other studies in musical acoustics, the investigation of problems connected with musical temperament set Helmholtz's mind to thinking about theoretical and practical problems connected with the design and playing of musical instruments able to sound in pure and just intonation. The temperament studies were completed 3 years prior to the publication of *Tonempfindungen*.

---

[8] Helmholtz, *Ueber die Klangfarbe*, 406.

[9] Helmholtz, *Ueber die Klangfarbe*, 404.

[10] Helmholtz, *Ueber die Klangfarbe*, 406–407.

[11] Helmholtz, "Ueber musikalische Temperatur" (1860), *Wissenschaftliche Abhandlungen, 1* (1882) 420–423.

In Helmholtz's study on musical temperament the focus falls not so much on the number ratios that define the musical intervals as on the pure-sounding chords [*rein gestimmte Accorde*] of the major scale and the analysis of the ear's sensitivity to musical intervals. He noted that when musicians saw the need to fix the intervals for both the third and the fifth rather than one or the other they built keyboards in which both were made false. "The Greeks, who it seems only performed their music monophonically or in octaves, observed correctly that an error in the advancing fifth is much more conspicuous than in an advancing third; they therefore kept the fifth pure by fixing the Pythagorean third as norm at 64/81."[12] Once the decision had been made to build a musical system on fifths it was discovered that after 12 such modulations, one arrives at a tone that almost, but not quite, coincides with the starting tone. The end tone is higher by a fraction 74/73 – a unit that came to be referred to as the Pythagorean comma. In order to bring the two tones in coincidence on a fixed tone keyboard, it became necessary to tune one or more fifths impure. As Helmholtz remarked:

> It turned out best to distribute the error equally by making all the fifths a little smaller. Deviation of the fifths in this now generally prevailing tuning system actually is exceedingly small, because the relation of the pure to the tempered fifth is 886/885. In that case the error in the third also is diminished; it sinks from 81/80 to 127/126 (1.0125–1.0029).
>
> The newer music now is decisively harmonic, and in this case it is not correct to assume that errors in the thirds are less damaging than errors in the fifths. The pernicious falsely tuned interval arises chiefly from beats of their combination tones and harmonic overtones. The vibrational frequency of the strongest combination tones is equal to the differences in vibrational frequencies of the primary tones.[13]

Expanding on the problem of beats that arise from the combination tones Helmholtz mentioned a case in which the vibrational frequency is 8 beats per second, "a decisively grating tone" that ruins the harmony.

Helmholtz was acutely aware of the fact that not all musical instruments are equally sensitive to dissonances. This presented a problem that acousticians were obliged to confront head-on.

> Singing voices are not bound to one temperament, and string instruments only to their open strings. Here a finely trained musician can sidestep the greatest hardships. The piano (Clavier) is less sensitive toward dissonances because its tones die away too rapidly, and the organ, because of the constant strength of its tones, is more closely adapted to roaring music and accumulating dissonances than to expressive soft harmoniousness. For these reasons the instruments best adapted for artistic music are those that are able to come to terms with the disadvantages of tempered tuning. Moreover the beats, if they are not very fast, are less perceptive in quickly moving music if the duration of most of the tones is smaller than the duration of the beats.
>
> The deficiencies of temperament become distinctly perceptible to all slowly moving and enduring tones, all the more the stronger they are. For that reason choirs of brass instruments are not practicable for accomplished artistic music. Especially conspicuous, as well, are the disadvantages of the currently widely disseminated physharmonika, all the more,

[12] Helmholtz, "Ueber musikalische Temperatur" (1860), 429.

[13] Helmholtz, "Ueber musikalische Temperatur" (1860), 421.

> since the combination tones in these instruments are stronger than others because of their special construction. Here the differences between pure and tempered chords are so great that the latter sound like dissonances compared to the former.[14]

Helmholtz concluded his preominary exploration of the musical temperamant problem by expressing his opinion that the production of harmonically pure-sounding keyboard instruments would require, at minimum, harmonium-type keyboard manuals tuned to thirds and fifths in relation to major chords activated by pedals, wind ducts, valves, and adjustable tuning devices. The theoretical principles and the functioning instruments designed to deal with different systems of intonation were taken up directly by Helmholtz in *Tonempfindungen* and subsequently refined and enhanced by his students and colleagues in Berlin in the 1890s and by the theoretical physicist Adriaan Fokker (1887–1972) and a group of composers in the Netherlands in the 1940s and beyond.

Reference

Helmholtz, Hermann. 1862. On the physiological causes of harmony. In *Popular Scientific Lectures*, 22–58. Trans. Alexander J. Ellis. New York.

[14] Helmholtz, "Ueber musikalische Temperatur" (1860) 422. The Physharmonica is an experimental early nineteenth-century keyboard instrument antecedent to the harmonium. Its sounds are produced by means of thin metal tongues set in vibration by a steady current of air provided by a pair of pedal-operated bellows.

Chapter 7
Sensations of Tone as the Physiological Basis for the Theory of Music

In 1860 while Helmholtz was planning the third volume of his physiological optics he wrote to F.C. Donders (1818–1889), a physiologist and ophthalmologist at the University of Utrecht: "I have set myself to compiling my acoustical works. It should become a small book, as popular as possible so that it is accessible to music lovers, since I also intend to be able to put down in writing the physical-physiological foundations of harmony."[1]

Die Lehre von den Tonempfindungen als physiologische Grundlage für die Theorie der Musik [*On the Sensations of Tone as a Physiological Basis for the Theory of Music*] was published in Braunschweig in 1863 when Helmholtz was at the pinnacle of his career as Professor of Physiology, at the University of Heidelberg.[2] The *Tonempfindungen* volume, and the three-volume *Handbuch der physiologischen Optik* (1856, 1860, and 1867), constitute the major physiological treatises that Helmholtz completed during the 15 year interval of his professorships in anatomy and physiology at the University of Bonn and in physiology at the University of Heidelberg. They were his physiology-richest years and followed on the heels of a decade of intensive scientific training, research, and intellectual contacts in the Müller circle in Berlin. To a large extent the 1857 essay on the physiological causes of harmony, published in English translation in 1862 but not available in print in German until 1865, served as preparation for the *Tonempfindungen* volume of 1863. The English-speaking public accordingly had

[1] Leo Koenigsberger, *Hermann von Helmholtz*, Braunschweig, 1902, vol. 1, 360. "Ich habe mich daran gemacht, meine akustischen Arbeiten zusammenzuschreiben; es soll daraus ein kleines Buch von möglichst populärer Haltung werden, um es auch den Musikliebhabern zugänglich zu erhalten, weil ich meine, auch die physikalisch-physiologische Begründung der Harmonielehre darin niederlegen zu können."

[2] The Preface is dated October 1862. The 2nd edition was published in 1865 with minor changes. The 3rd and 4th editions (1870 and 1877) were substantially revised and updated to include an account of investigations on the anatomy and mechanics of the inner ear. The 5th and 6th editions (1895 and 1912) are almost identical with the 4th edition and were published posthumously with Prefaces by W. Wachsmuth who mainly cited a number of new publications in the field of acoustics.

E. Hiebert, *The Helmholtz Legacy in Physiological Acoustics*, Archimedes 39,
DOI 10.1007/978-3-319-06602-8_7, © Springer International Publishing Switzerland 2014

access to Helmholtz's 1857 essay as preparation for *Tonempfindungen* whereas this was not the case for the German-speaking public. In the case of the latter the 1857 essay was delivered to the public as a lecture but was not available in print until 1865; that is, until 3 years after the publication of *Tonempfindungen*.

Tonempfindungen included an integrated and critical account of a great wealth of information on physical acoustics – information that in large part was available in scattered form in the earlier literature. More important, *Tonempfindungen* provided interpretive and supplemental coverage to experimental investigations in physiological acoustics that had been at the focus of Helmholtz's own research between the years 1856 and 1862. In the Preface Helmholtz mentioned that it had taken him 8 years to complete the volume and that this could not have been accomplished without the procurement of "new instruments that do not fit into the inventory of a physiological institute and exceed in cost the usual resources of a German scholar [*eines deutschen Gelehrten*]."

The French translation of *Tonempfindungen* by Georges Guéroult, with the assistance of piano maker August Wolff, and with a preface by Helmholtz, appeared in Paris in 1868.[3] In the preface to *Théorie physiologique de la musique fondée sur l'étude sensations auditives* Helmholtz wrote:

> Before being translated into French this book already had received a more favourable reception [*un accueil plus favorable*] among scholars and French musicians than I had ventured to hope for. It appears to me, and I was convinced of it during my two trips to Paris. . ..that the knowledge of music and the aptitude to follow excursions in the field of the exact sciences are found more frequently to be combined in France than one ordinarily finds in the other countries of Europe. . .. In being allowed myself to respond. . .to the fidelity of the text I have examined the proofs and in several points made modifications and added clarifications in the original text. Likewise at the end of the volume I have added several supplements relating to questions of physics or mathematics.[4]

The several supplements that Helmholtz referred to comprise investigations on just intonation and aural anatomy that were not to appear in the German text until the publication of the third edition of *Tonempfindungen* of 1870.

In 1875, 7 years after the French translation had been available in its first and second editions, *Tonempfindungen* appeared in English. It was translated from the third German edition of 1870 by the London-born mathematician, philologist, and phoneticist Alexander J. Ellis (1814–1890), who had translated Helmholtz's Bonn lecture into English. The work was published with the sanction of Helmholtz and supplied with additional notes and an appendix (mostly by Ellis) that increased the length of the volume by almost a third.[5] It was in this multi-appendixed format, with

[3] Helmholtz, *Théorie physiologique de la musique fondée sur l'étude sensations auditives*. Traduit de l'Allemand par M. G. Guéroult, avec le concours pour la partie musicale de M. Wolff, de la Maison Pleyel, Wolff et Cie, Paris, Victor Masson et Fils, 1868. A second French edition appeared in 1874.

[4] Helmholtz, *Théorie physiologique*, Preface, 1.

[5] Hermann Helmholtz, *On the Sensations of Tone as the Physiological Basis for the Theory of Music*, London, 1875. The 2nd English edition, translated from the German edition of 1877, appeared in 1885. Unless indicated otherwise, the edition cited in this paper is the unabridged reissue of the last English edition published in 1954 in New York by Dover Publications.

its ponderous explanatory apparatus and with information "especially adapted to the use of Music Students," that *Sensations of Tone* was presented, and continues to be accessible, to the English reader. It is noteworthy and conspicuously puzzling that the English edition of *Tonempfindungen* was not published until 10 years after the German volume came out in Braunschweig, 7 years after the French edition appeared in Paris, and a year after the French second edition was published in its revised form. Given, on the one hand, the distinctly cordial relations between Helmholtz and British scientists in the 1860s, and on the other hand the strained political relations between France and Prussia that led to the Franco-Prussian war of 1870, why did it take only 5 years to publish *Théorie pysiologique de la musique* and 12 years to publish *On the Sensations of Tone*? One might suggest that the rationale for this seeming idiosyncrasy relates in one way or another to the essential scientific and artistic extraordinariness of the France of Paris when compared with the Germany of Berlin at this time. In the 1860s, Paris, then about twice the size of Berlin, was considered to be the intellectual and architectural capital of Europe. France was a proud and united nationalistic country that could boast of unparalleled prestige in the natural sciences and hundreds of years of rich history in literature and the arts. A prominent British historian of European intellectual thought has written:

> In France during the early part of the century the foundation of nearly all the modern sciences was laid; many of them were brought under the rule of a strict mathematical treatment. It was there that scientific subjects were made so popular, and clothed with a garment of such elegant diction, that they have since that time greatly entered into general consciousness, and have promoted in literature and art an independent school – the naturalistic.[6]

In its German, French, and English versions it took almost two decades for *Sensations of Tone* to capture the fascination, first, of scientists, and then, of musicians. Owing to the timing contrasts in the translations, and the formidability of the intrusive accessory notes of the English rendition, the three versions have continued to convey somewhat different overall messages to their respective readers. Beginning around 1880, it gradually came to be recognized and acknowledged in histories of music and in handbooks, textbooks, and the research literature on acoustics, that the Helmholtz treatise had set in motion a trend in the scientific study of music that was to assume center stage in discussions on music theory for the rest of the century. In reality it more or less came to be taken for granted that *Tonempfindungen* represented the *terminus a quo* for any discussion on the role and the limits of the natural sciences in the analysis of music theory. Where Helmholtz was not explicitly mentioned in secondary works, his historical position in the field of music simply was taken for granted.

The esteem with which Helmholtz's contributions to musical acoustics were held by contemporary physicists is made evident by the eminent British physicist

[6] John Theodore Merz (1840–1922), *A History of European Thought in the Nineteenth Century*, London, 1896, vol. I, p. 75. Chapters 1, 2, and 3 deal with The Scientific Spirit in France, The Scientific Spirit in Germany, and The Scientific Spirit in England.

James Clerk Maxwell (1831–1879) who, 3 years after publication of *Sensations of* Tone, remarked in a lecture:

> No man has done more than Helmholtz to open up paths of communication between isolated departments of human knowledge. . .. By a series of daring strides, he has affected a passage for himself over that untrodden wild between acoustics and music – that Serbonian bog where whole armies of scientific musicians and musical men of science have sunk without filling it up.[7]

The late Carl Dahlhaus, musicologist at the Technical University in Berlin, made the following observation about Helmholtz's legacy in nineteenth-century music history a century later:

> Musical thought between 1848 and 1870 was indebted no less to the natural sciences than to the age's characteristic fondness for the grandiose and to the increased pace of evolution that became noticeable in virtually all areas of culture. The only work that actually proved "epoch-making" in nineteenth century music theory was Hermann von Helmholtz's *Lehre von den Tonempfindungen* (1863). It influenced practically everyone from the philosopher pondering the problems of aesthetics to the humble musician teaching how to write chord progressions.[8]

In a monograph of 1998 that examines the way in which the perceptions of consonance are related to timbre, a prominent electrical and computer engineer at the University of Wisconsin in Madison has written, "the book that started it all is Helmholtz's *On the Sensations of Tone*. . . [It] remains readable over 100 years after its initial publication."[9]

As already mentioned, *Tonempfindungen* was structured to a large extent on ideas and principles that Helmholtz already had called attention to in his 1857 Bonn lecture and that followed from the experimental investigations on acoustics that he had begun in 1855. By taking advantage of the analysis of tones and groups of tones using sympathetic resonance, and by employing instruments such as the siren, the resonator, and electrically-driven tuning forks, Helmholtz had been led to explore the nature of combination tones, the timbre of vowel sounds, and the vibrational and oscillational properties of strings and air columns.[10] Experimental research papers on musical acoustics continued to appear after the publication of the first edition of

[7] *The Scientific Papers of James Clerk Maxwell*, vol. 1, Cambridge, ed. by W. D. Niven, 1890, 754. The expression "Serbonian Bog" derives from Herodotus's account of a place in lower Egypt where whole armies were said to have perished; they had disappeared, and had sunk into the bog in the course of battle.

[8] Carl Dahlhaus, *Nineteenth-Century Music*, Berkeley, 1989, 192–193.

[9] William A. Sethares, *Tuning, Timbre, Spectrum, Scale,* Berlin/Heidelberg/New York, 1998, 47.

[10] Helmholtz's papers on acoustics through the year 1863 include: Ueber Combinationstöne" (46 pp., two parts, 1856), Ueber die Vokale (2 pp., 1857), Ueber die Klangfarbe der Vokale (11 pp., 1859), Theorie der Luftschwingungen in Röhren mit offenen Enden (79 pp., 1859), Ueber Klangfarben (2 pp., 1860), Ueber die Bewegung der Violinsaiten (10 pp., 1860), Ueber musikalische Temperatur (4 pp., 1860), Zur Theorie der Zungenpfeifen (7 pp., 1861), Ueber die arabische-persische Tonleiter (3 pp., 1862), and Ueber den Einfluss der Reibung in der Luft auf die Schallbewegung (5 pp., 1862). All of these papers are reproduced in the section on Schallbewegung in Helmholtz, *Wissenschaftliche Abhandlungen*, Leipzig, 1882, vol. 1, 256–426.

Tonempfindungen and were incorporated sequentially into subsequent editions of the work. The most important of these were his papers on the anatomy and mechanics of the cochlea, the ossicles, and the tympanum in the inner ear.[11]

In the introduction to *Sensations of Tone* Helmholtz remarked that the aim of his treatise was "to connect the boundaries of two sciences, which, although drawn toward each other by many natural affinities, have hitherto remained practically distinct – I mean the boundaries of *physical and physiological acoustics* on the one side, and of *musical science and aesthetics* on the other."[12]

In the first section of the treatise Helmholtz deals with the physical and physiological principles of acoustics that relate to tones, partials, tone color, sympathetic vibration, and aural perception. The second section deals with the identification and characterization of musical phenomena that were controversial or in process of being investigated in Helmholtz's time – notably combination tones, beats, consonance, and dissonance. The third section deals with the aesthetic, open-ended, and listening-related properties of tones: principles of style in historically diverse musics, tonality, and the progression of intervals, diatonic, chromatic, and enharmonic scales, the character of different keys, and tempered and just intonation.

In the final chapter on "esthetical relations" [*Beziehungen zur Ästhetik*] Helmholtz noted that the production, reception, and perception of tones, their harmonic and melodic combinations, and their aesthetic relations were in need of clarification and expansion using principles drawn not only, as already had been done by other investigators from mathematics and physics, but from physiology, psychology, and what Helmholtz referred to as "internal musical connections" – that is, by studying what takes place in the ear, the brain, and the human physiological makeup. He acknowledged that the task of providing an esthetic analysis of artistic musical works would "encounter apparently invincible obstacles at almost every point."

Helmholtz was convinced that:

> In the field of elementary musical art we have now gained so much insight into its internal musical connections that we are able to bring the results of our investigations to bear on the views which have been formed and in modern times nearly universally accepted respecting the cause and character of artistic beauty in general. It is, in fact, not difficult to discover a close connection and agreement between them; nay, there probably are fewer examples more suitable than the theory of musical scales and harmony, to illustrate the darkest and most difficult points of general esthetics. Hence I feel that I should not be justified in

---

[11] Helmholtz, "Ueber die Mechanik der Gehörknöchelchen" (1867), "Die Mechanik der Gehörknöchelchen und des Trommelfels" (1869), and "Ueber die Schallschwingungen in der Schnecke des Ohres" (1869). These papers are reproduced in the section on Physiologische Akustik, in Helmholtz, *Wissenschaftliche Abhandlungen*, Leipzig, 1883, vol. 2, 503–588. According to current views of the hearing phenomenon, when noise enters the ear canal as sound waves the waves strike the eardrum vibrating three tiny bones: the hammer, the anvil, and the stirrup. The vibrations create waves in the fluid-filled cochlea and are transformed into electrical impulses by hair cells in the cochlea and are transmitted to the brain where they are interpreted as sound.

[12] Helmholtz, *On the Sensations of Tone*, 1954 edition, 1.

passing over these considerations, more especially as they are closely connected with the theory of sensual perception [*die Lehre von den Sinneswahrnehmungen*].[13]

The publication of *Tonempfindungen* elicited both an immediate and a lasting response on the part of music historians, music critics, composers, music pedagogues, and scientific instrument builders. Affirmative on the whole, but also confrontational, the evaluations tended to stress the sentiment that *Tonempfindungen* was a valuable and unique synthesis of what already was known in musical acoustics; that what was seen to be new in its methodological approach and theoretical interpretation promised to offer potential for further exploration. Nevertheless, not all music historians and theoreticians were prepared to offer forthright positive appraisals concerning the merits of so formidable a work on music written by a physiologist and physicist who in any event was an outsider to the music profession. A number of critics rejected the intrinsic relevance of Helmholtz's findings. On the other hand it is evident that there were enterprising musicians who were alert to the potential importance of what the natural sciences had yet to offer music. They discovered that *Tonempfindungen* provided unanticipated insights into musical phenomena that hitherto had been unexplored, or simply taken for granted. One also is left with the impression that musicians were grappling with *Tonempfindungen* in hopes of being able to appeal to the sciences in order to validate their own theoretical views. Major reviews of the treatise continued to appear until about 1870. By then Helmholtz's own research priorities had shifted almost entirely to topics such as electromagnetic and hydrodynamic theory; by then he had accepted the physics professorship at the University of Berlin.

The most significant reviews and critical commentaries to *Tonempfindungen* appeared in a series of articles published in Breitkopf & Härtel's widely distributed and influential *Allgemeine Musikalische Zeitung (AMZ)*. The journal was published every Wednesday and in the 1860s was edited by the noted music critic and pedagogue Selmar Bagge (1823–1896), who taught composition at the Vienna Konservatorium.[14] The sharpest direct challenge to *Tonempfindungen* came from Moritz Hauptmann (1792–1868), a composer, theorist, teacher, and cofounder, with Otto Jahn and Robert Schumann, of the *Bach Gesellschaft* in 1850. Hauptmann's *Natur der Harmonik und Metrik* of 1853 was the leading mid-century German text in music theory. In his philosophical analysis of the theoretical systems of music Hauptmann stressed, in Hegelian-form, the dualism of the major and the minor. Kantor at the renowned Thomasschule in Leipzig, Hauptmann had been editor of the *Allgemeine Musikalische Zeitung* in the 1840s. When the Helmholtz volume appeared in print (the Preface was dated Heidelberg, October, 1862), it was evident among German musicians and music publishers that Hauptmann was the ideal music theorist to write the review. Apparently

[13] Helmholtz, *On the Sensations of Tone*, 1954 ed., 366. Ger. ed. of 1968, p. 588.

[14] The publishing and printing firm Breitkopf & Härtel was established in Leipzig in 1719. It achieved renown for its efficient production and improved music typography for compositions, collected editions, scores, and thematic catalogues for music by composers from the time of Telemann into modern times.

Hauptmann procrastinated. Editor Bagge became impatient waiting for the review and decided to publish his own "preliminary editorial" notice of *Tonempfindungen* in the July 1863 issue of the *Allgemeine Musikalische Zeitung*. Here, in part, is what Bagge wrote:

> The Helmholtz book offers so large a mass of material for discussions on the science of the arts that no matter how exhaustively approached from a point of view that is historically correct it seems neither advisable nor possible simply to dispose of it with a brief review. Although it draws on some ideas from Hauptmann's *Harmonik und Metric*, and even supports its author in many points, it nevertheless is basically directed against the work [Hauptmann's] since it seeks to give *Musikwissenschaft* an entirely different basis and therefore indirectly takes Hauptmann's work to be untenable. But since the qualified scholar to whom we long ago entrusted the review now tarries even longer than pleases our readers or us, we therefore, today, open the discussion with a notification [*Anzeige*] of the work that has come to us unsolicited. We believe that it will help carry the reader far enough into the subject so that later more comprehensive articles that treat its principle point and its particulars from different points of view will make the work all the more readily comprehensive.[15]

The "unsolicited" *Anzeige*, published in three successive weekly installments, had been submitted by Eduard Krüger (1807–1885).[16] An associate professor who lectured on music theory in the faculty of philosophy at the university in Göttingen, Krüger had the reputation of being "an iconoclast who, during an era defined by partisan debates on music, never aligned himself with any single viewpoint.... He considered himself part of a 'critical-historical' rather than an artistic-critical generation."[17] Apologetic in tone, and responding on behalf of other musicians, Krüger observed that it was physicists and not musicians who had given Helmholtz's work the attention it deserved. He suggested that it was too late for musicians adequately to praise the treatise since it already even without reviews had blazed a trail for itself; that musicians searching for insight into the objective foundations of subjective artistic experiences nevertheless would reap enormous benefits from its study. Krüger inferred that *Tonempfindungen* for the first time had shown the way to an "integration of the mathematical-physical and the physiological-anatomical with the characteristic artistic nature of the science of tone that is truly new in comparison with the longstanding severance of the natural-scientific from the artistic-philosophical fields of vision." He held that Helmholtz's new approach coincided with the "needs of the time, since there still are rationalists

---

[15] Bagge, Vorbemerkung der Redaction, *AMZ*, Neue folge, 1. Jahrgang, Nr. 27, 1 July 1863, col. 467.

[16] Eduard Krüger, "H. Helmholtz," *Lehre von den Tonempfindungen. Als physiologische Grundlage für die Theorie der Musik,* Braunschweig, 1863, *AMZ*, Nr. 27, 1. July 1863, col. 467–471; Nr. 28, 8 July 1863, col. 483–489; Nr. 29, 15 July, 1863, 496–501.

[17] Sanna Pederson, "Eduard Krüger," *New Grove, 13* (2000), 931/2. The work of Kurt Hoppenrath, Eduard Krüger (1807–1885). *Leben und Wirken eines Musikgelehrten zwischen Schumannscher Tradition und Neudeutscher Schule*, Inaugural Diss., Göttingen 1964 provides a commendable overview of the significance of the sciences in the mid-century development of music criticism in Germany.

[read Hauptmann] who want to dislodge the artistic from the natural with the coercion of isolated thinking."[18]

The major section of Krüger's lengthy review is given over to an exposition of those physical and physiological ideas in *Tonempfindungen* that pertain to theoretical and instrument-related matters such as consonance, dissonance, scales, and tonality. He touches peripherally on the differences of principle and interpretation exhibited in *Tonempfindungen* and Hauptmann's *Harmonik*, but seeks to portray Helmholtz as a scientist who by no means underestimated the importance of the subjective and aesthetically sensitive components of the human listening experience. In commenting on Helmholtz's analysis of the perception of tone color in the ear and the human nervous system, Krüger remarked: "It is wonderful how the most delicate sensory perceptions are linked in a precise manner with the material organs; but there is no fear that such investigations infringe on the artistic mind [*Kunstgeist*]! The mind [*Geist*] and art remain undisturbed in the secrets of their living action [*Lebenswirckung*] and to all eternity do not allow themselves to be captured by exact formulas."[19]

It is evident that Krüger was alert to recognizing that *Tonempfindungen* adequately had demonstrated adequately the extent to which the objective point of view that characterizes physics and physiology is essential and even crucial for the study of the subjective elements of "the art of making music (Tonkunst)," but he recognized as well that Helmholtz had underscored the idea that objectivity, in this case as elsewhere in music, has its limitations; that the ultimate "whys?" [*Warums?*] inevitably remain hidden. According to Krüger, Hauptmann, had taken the dark path of Hegelian metaphysical dialectics to adopt a trajectory that landed him in dogmatism. Quoting from Hauptmann's *Harmonik*: "In metaphysics, however, nothing can be given that the metaphysician has not demonstrated, and one must pursue the why, the ground of all grounds [*Grund des Grundes*]; and there it by no means is satisfactory to evade the proof by alluding to other evidence that belongs to natural or aesthetic modes of action." Kruger leaves one with the impression that Hauptmann is talking nonsense.[20]

The last installment of Krüger's triad of review articles on *Tonempfindungen* includes a section on Helmholtz's investigations on just intonation. "Our author," he writes, "has attempted to establish a just intonation system by means of a fixed-tone keyboard – an harmonium [*Physharmonica*] with two manuals that he invented and that was constructed by Schiedmayer in Stuttgart – but concerning which, short of seeing it, we are unable to form an opinion. According to the description it must have a fine artistic action and if it would become universal could lead to a turning point in the very nature of keyboards [*Clavierwesen*]."[21]

[18] Eduard Krüger, Helmholtz, *AMZ*, 1 July 1863, col. 467.

[19] Krüger, *AMZ*, 1 July 1863, col. 471.

[20] Krüger, *AMZ*, 8 July 1863, col. 485–486.

[21] Krüger, *AMZ*, 13 July 1863, col. 500.

Krüger's unsolicited 16-column review ends on a note conformable to what Helmholtz had to say in the last ten pages of his treatise on "esthetical relations [*Beziehungen zur Ästhetik*]."

> We have learned (from Helmholtz) that scales and harmonic relationships are not natural things but products of artistic invention. The reason we understand the pleasure of the beautiful, not as coincidental impressions but essentially as lawlike congruences with the nature of our mind [*gesetzliche Uebereinstimmungen mit der Natur unseres Geistes*], is attested to by the fact that we expect from every other human mind, the same acknowledgment that we give it ourselves. We have a feeling of a rationality [*Vernunftsmässigkeit*] of a work of art that reaches beyond our intellectual conception; for while what is conceived by creative genius and in which we have a part is presented to our reason as appropriate, there is and always remains the enthusiasm of an incomprehensible mystery.[22]

Several months after Krüger's review of 30 September 1863, the *AMZ* published "A letter of M. Hauptmann about Helmholtz's *Tonempfindungen*" that had been sent to Otto Jahn in Bonn on 5 March 1863.[23] Preceding the letter is a statement by Hauptmann acknowledging that he had been invited by Bagge to review the Helmholtz work with special reference to its theoretical contents. He had rejected the invitation to review the Helmholtz volume but now had consented to have the letter to Otto Jahn published in the AMZ. He noted that he had read only part of the Helmholtz treatise.

> According to what I have learned from the Helmholtz book it seems to me all that is physiological is very important and will be of great value even where it deals with tone sensation [*Tonempfindung*]; but the psychological in the sense of tone comprehension [*Tonverständniss*], that through which music is expressed in a musically-defined sense [*musikalisch-bestimmter Sinn*] and that makes music a language, is not explained in the book. It has to be something more specific than Helmholtz indicates.[24]

Hauptmann maintained that it would be as farfetched to make music with nothing but harmony as it would be to make a picture with only color and no sketch, shape or form; or [by analogy] to turn the most excellent ragout into a work of art. Although Hauptmann assumes that Helmholtz would deny that his own tonal system and his theory of music rest exclusively on harmoniousness he reasons: "In all that he as an outstanding physicist alleges, I cannot find anything that clarifies tone comprehension. I do not even understand how the first degree of the scale is determined. Like the queen in the presence of Hamlet's ghost I see all that there is and yet I see nothing. It all depends on the queen; the ghost is there, for the prince sees him." For Hauptmann, in the final analysis, *Tonempfindungen* adds up to nothing more than interesting insights into nature: "Remarkable crystalline forms

[22] Krüger, *AMZ*, 15 July 1863, col. 501.

[23] Otto Jahn (1813–1869), professor of philology and archaeology in Bonn, was a leading classical scholar and music editor. His 4-volume edition *W. A. Mozart* (1856–1859) in the revised 7th edition of Hermann Abert (1955–1966) has been published in the radically edited and revised edition by Cliff Eisen. An English translation of 1,600 pages by Stewart Spencer *W. A. Mozart* was published by the Yale U. Press in 2007.

[24] Hauptmann, "Ein Brief über Helmholtz," *AMZ*, 30 Sept 1863, col. 669.

unfold, but no architecture; only the elementary natural, not the rational human nature and not the conditions of its free artwork."[25]

Hauptmann looks at Helmholtz's world of music as one that is lacking in "phantasy," in "magic," in the "coming together of the preconceived and the perceived," and, notably, in any "inner spiritual sensation of tone." He states: "I allow the chords to grow. Helmholtz builds them together with the timber of intervals." Recognizing that some investigators, Helmholtz in particular, had reproached Hauptmann for having "only today's developed tonal system in his vision," Hauptmann remarked:

> If there is another classification of scales, and someone knows about it, then I would gladly be informed; but I have no interest in the undefinable and in mere legend. If the knowledge of chords was not there at an earlier time, that is no reason to assume that the progressive determinations of melody always have been harmonic; the feeling for something is there before the understanding.[26]

Hauptmann's final verdict is that while Helmholtz's investigations do not provide a well thought-out music theory, it nevertheless, because of the ground it covers, can provide support for a theory.

> Although I am unable to recognize a well-founded music theory in the Helmholtz investigations I nevertheless have a no less great respect for his investigations and the observational perspicacity [*Beobachtungsgenie*] with which they are carried out. As something musically theoretical the book also will impress many. While nothing specific concerning the foundation of the matter is present in terms of all of its well thought-out conditions this viewpoint possibly can promise to give support to an explanation, an insight into the inner essence of harmony.[27]

The overall impression one is left with on reading Hauptmann's assessment of *Tonempfindungen* is that, notwithstanding his admission that the work has some merits in the area of musical acoustics, there is an undercurrent of disappointment that it should be a natural scientist and an outsider who spells out for the authorities in music a theory of music whose basis derives from the physiological analysis of the sensations of tone.[28]

References

Dahlhaus, Carl. 1989. *Nineteenth-century music*, 192–193. Berkeley: University of California Press.

Helmholtz, Hermann. 1875. *On the sensations of tone as the physiological basis for the theory of music*. London: Longmans, Green, and Co.

---

[25] Hauptmann, *AMZ*, 30 Sept. 1863, col. 670.

[26] Hauptmann, *AMZ*, 30 Sept. 1863, col. 671–672.

[27] Hauptmann, *AMZ*, 30 Sept. 1863, col. 673.

[28] For additional information on the reception of *Tonempfindungen*: Erwin Hiebert and Elfrieda Hiebert, "Musical Thought and Practice: Links to Helmholtz's *Tonempfindungen*," 295–311 in *Universalgenie Helmholtz. Rückblick nach 100 Jahren*, ed. Lorenz Krüger, Berlin, 1994.

Helmholtz, Hermann. 1954. *On the sensations of tone*, 1954th ed, 1. New York: Dover Publications.

Koenigsberger, Leo. 1902. *Hermann von Helmholtz*, vol. 1, 360. Braunschweig: F. Vieweg.

Sethares, William A. 1998. *Tuning, timbre, spectrum, scale*, 47. Berlin/Heidelberg/New York: Springer.

Chapter 8
Just Intonation and the Harmonium

By mid-nineteenth century the climate of discussion had become optimum for undertaking the construction of keyboard instruments that are able to sound in just intonation. The technological expertise to realize this was readily available in the keyboard manufacturing sector. Helmholtz took the initiative for moving in this direction by transferring to the laboratory a problem that until then for the most part had been dealt with only in theory and in written works. The jumping-off point in our exploration of the construction of fixed tone keyboards in just intonation is Hermann Helmholtz.

Helmholtz was first and foremost an experimental physiologist who on his own merits had succeeded in becoming an accomplished theoretical physicist and mathematician in order to achieve his objectives in physiology. Helmholtz also was a competent pianist well-versed in music theory. In bridging theory and practice within his own musical experience and professional context as physiologist, and in choosing to adopt several perspectives when confronted with problems that a single discipline was unable to unlock, he was in a position to advance the field of musical acoustics on a frontier that reached far beyond the domain of standard physical acoustics. Long before mid-century, physical acoustics had seen the shaping of its basic principles and had found its niche well in place as a sub-branch of mechanics. The domain of musical acoustics that Helmholtz came to claim as his own had much in common with the physiological and neurological functioning of the ear, the hearing mechanism and process, and the branch of aesthetics that converges in the psychology of perception.

By the time Helmholtz published *Sensations of Tone* in 1863 he had been engaged in experimental investigations on the relationship of tones to each other, tonality, consonant chords, tempered intonation, and just intonation. Four years after publishing *Sensations of Tone* he once again became engaged in investigations in physiological acoustics, this time moving more deliberately in the direction of problems connected with the psychological aspects of hearing. In three classic papers – on the mechanics of the auditory ossicles, on the eardrum as timpanum, and on sound vibrations of the cochlea of the ear – Helmholtz examined the ear's anatomy and mechanisms and provided a mathematical description of the

E. Hiebert, *The Helmholtz Legacy in Physiological Acoustics*, Archimedes 39,
DOI 10.1007/978-3-319-06602-8_8, © Springer International Publishing Switzerland 2014

functioning of the human ear that in a preliminary way opened up the path for formulating a more complete theory of hearing. The three investigations on the physiological acoustics of the ear and on aural perception were published between 1867 and 1869 and were incorporated into the third edition of *Tonempfindungen* in 1870. The scientific instruments that Helmholtz had invented and retooled to enhance his hearing-related investigations are described in the 1870 volume and included mainly a set of resonators for separating musical tones, the polyphonic siren for determining the ratios of consonances, and the reed harmonium to study just intonation.

Helmholtz's first encounter with the functioning of the reeds that define the sounding vehicle for the harmonium appears in a study of 1861 on reed pipes (*Zungenpfeifen*).[1] He mentioned in that study that it was Wilhelm Weber who first had provided a comprehensive account of the mechanical functioning of reed pipes. He had examined the nature of the acoustic coupling of the vibrating reed tongue and the air cavity in reed organs as a means of maintaining "constancy of pitch" for reed organs under different intensities of blowing. Apart from realizing a method for maintaining constancy of pitch in his reeds, Helmholtz had acquired from Weber a method of providing an improved "standard of pitch." Both "constancy of pitch" and reliable "standards of pitch" were pivotal for Helmholtz in his study of intonation using the harmonium. Since Chladni and Weber had worked in close collaboration on problems in acoustics, Weber's work on reed pipes established an indirect historical link between the much earlier acoustical investigations of Chladni and the later investigations of Helmholtz.[2]

The term reed pipe (*Zungenpfeife*) is used to designate all wind instruments in which the path that the stream of air takes in a vibrating elastic body is alternately opened and closed. Weber chose, selectively, to investigate the mechanics of elastically vibrating metallic reeds. Due to their large mass and elasticity they are strongly set in motion by the air when the proper tone (*Eigenton*) of the pipe does not differ markedly from the proper tone of the freely vibrating metallic reed. Such metallic reeds, unlike the reeds that characterize other instruments such as the clarinet, oboe, bassoon, human lips, trumpet, trombone, and horn, constitute the sounding mechanism of the harmonium. It was the vibrating characteristics and tonal quality of the metallic reeds that became a matter of considerable interest to

[1] Helmholtz, "Zur Theorie der Zungenpfeifen" (1861), *Wissenschaftliche Abhandlungen 1* (1882) 388–394.

[2] Wilhelm Weber (1804–1891), professor of physics in Göttingen, is best known for his contributions to electrodynamics. His early interests in acoustics and wave phenomena were generated in large part by contacts with the physicist and acoustician E.F.F. Chladni (1756–1827) who for some years lived with Weber in the same house in Wittenberg. The *Wellenlehre* (Leipzig, 1825), a joint publication of Wilhelm and his older brother Ernst Weber (1795–1878), the Leipzig physiologist, was dedicated to "our admired friend Chladni, the founder of an acoustics based on experiments." Wilhelm Weber's doctoral dissertation of 1826 under Johann Schweigger, physicist and chemist at the University of Halle, was on the theory of reed organ pipes. A.E. Woodruff, "Wilhelm Weber," *DSB*, *14* (1976) 203–209; "Wilhelm Weber," Lexikon Geschichte der Physik A-Z, ed. Armin Hermann, Cologne, 1972, 397–400.

Helmholtz because of his selection of the harmonium as the ideal instrument for the study of intonation. He remarked: "As a rule pipes with metallic reeds are able to sound only a single tone, and that tone from among the theoretically possible tones is closest to the proper tone of the reed."[3]

The harmonium, also called a reed organ, is a fixed-tone keyboard instrument that unlike the pipe organ has no pipes or resonators but sounds by means of thin metallic plates that act as freely vibrating reeds set in motion by a pair of pedal-operated bellows. The tuning of reed instruments such as the harmonium, in which the reed provides the vibrating sound element, is accomplished by and depends on the mass, elasticity, and form of the reeds and not on the form or length of the pipe or funnel to which the pipes are attached, although the form and length of the connecting tube exert some – ordinarily negligible – influence on the pitch. In comparison with the pipe organ and stringed keyboard instruments, the harmonium is a very young instrument whose history can be traced to the early nineteenth century. The instrument commonly has been looked upon as a weak surrogate for the organ. As a keyboard instrument the harmonium is situated half way between the piano and the organ – externally more like the former, internally more like the latter.

The harmonium, as it first was referred to in a patent by A. F. Debain (1809–1877) in 1840, developed primarily during the first half of the century in France, but continued to be under revision and construction until the end of the century. In its many nineteenth-century variations, and as a substitute for the harpsichord and the piano, it came to be used in churches and cinemas, in salons, and in music circles. As a new instrument it experienced wide distribution but also went through a phase of critical evaluation based not so much on its new and different tonal qualities as on its unique public accessibility. The fact that the instrument appealed so strongly to dilettantes and self-taught persons led to assigning the harmonium a rank below that of the other keyboard instruments.

Not long after mid-century, the exploitation of the harmonium as an experimental tool for the study of just intonation altered the instrument's status in keyboard rankings and put it in the hands, mainly, of music theorists. Although the harmonium as a performing instrument almost has vanished from the contemporary musical scene, it is well to recall that in the roster of romantic and early modern music the instrument found its way into the compositions of composers such as Camille Saint-Saëns, César Franck, Richard Wagner, Richard Strauss, Anton Webern, Gustav Mahler, and Arnold Schoenberg.

The harmonium long has been considered to be singularly suited for accompanying the human voice or other instruments. Unlike the piano, on which the decay time of struck tones is relatively short, the harmonium player is in control of tonal duration and is able to follow and blend with the human voice or other instruments in the regulation of dynamic shading. On the other hand, since the ear is able more accurately to distinguish pitch the longer the duration of the sounded notes, the fine

[3] Helmholtz, *Wiss. Abh.*, Zungenpfeifen, 388.

tuning of pitch on the harmonium is given high priority. A listener is able more readily to tolerate an off-tune piano than an off-tune harmonium. Such are the issues that come into play in our examination of the problems associated with Helmholtz, the harmonium, and just intonation.[4]

In a number of the more venturesome music theory circles of mid-nineteenth century Europe the harmonium, with its pitch- and tone-adjustable reeds and sustaining power, came to be regarded both as the ideal instrument for study of the degree to which pitches are accurately produced and as the paradigm musical instrument for investigating the theory and practice of just intonation. Among keyboard craftsmen it was evident that the harmonium's sounding mechanism is a far less complicated machine to work with and adjust than other keyboards activated by hammer- or plectrum-operated strings, levers, and dampers. During the second half of the nineteenth century enterprising investigators and keyboard builders approached the problem of altering or expanding the standard 12-key per octave keyboard either to achieve just intonation or to encompass microtonal music by adding additional keys and multiple manuals. They invariably chose the harmonium as their preferred instrument of experimentation.

Two years after publication of his paper on reed pipes Helmholtz made arrangements with the J. And P. Schiedmayer piano factory in Stuttgart for the building of a five-octave two-manual harmonium that he believed would be suitable for his

[4] A number of works on the history and significance of the harmonium have been consulted. The earliest history was published anonymously in 1868 but is known to have been authored by W. Riehm, a Protestant pastor from Baden who was the owner of an harmonium with 12 registers and a percussion mechanism. *Das Harmonium in seiner Construction und Behandlung*, Basel & Ludwigsburg. In that volume the intention of the author is to explain, from his own experience with the harmonium, various aspects of the instrument that he believes will be of interest to the lay music learner: the instrument's dynamic output, timbre, stability of tuning and flexibility, blending characteristics with other instruments, registers, deficiencies, and the names and addresses of the German and French harmonium builders. The most informative history of the harmonium, one that was published half a century after the Riehm volume, depicts the instrument in its final modern form. L. Hartmann, *Das Harmonium umfassend die Geschichte, das Wesen, der Bau und die Behandlung des Druck- und Saugwindharmoniums nebst einer Abhandlung über das Harmoniumspiel,* Leipzig, 1913. Author Hartmann remarks in the Preface that since the piano has pushed most musical instruments and their music, especially the string quartet, out of home and family, he hopes that the modern harmonium will be reappraised and reinstated as a *Hausmusik* instrument. A third volume, a compilation of essays on specialized topics dealing with the harmonium in Germany, brings the subject into the present and treats the instrument's construction, economic significance, and utilization as an historic musical instrument. The general viewpoint expressed in this work is that "the great time of the instrument has long past...., that the harmonium today actually is, so to say, an historical instrument, appropriate and necessary at best for the performance of music of a bygone era." (8–9) Christian Ahrens and Gregor Klinke (eds.), *Das Harmonium in Deutschland. Bau, wirtschaftliche Bedeutung und musikalische Nutzung eines 'historischen' Musikinstruments*, Frankfurt/Main, 1996. The volume on the harmonium that most directly relates to topics treated in this paper will be referred to as "Michaelstein" in what follows. Conference report, volume 62 of the Stiftung Michaelstein Kloster. Monika Lustig (ed.), *Harmonium und Handharmonika*, 20. Musikinstrumentenbau Symposium, Michaelstein, 19–21 Nov. 1999, Michaelstein/Blankenburg, 2002.

own experimental investigations on intonation.[5] In *Tonempfindungen* Helmholtz explained his reasons for selecting the Schiedmayer harmonium.

> Among musical instruments the harmonium, on account of its uniformly sustained sound, the piercing character of its quality of tone, and its tolerably distinct combination tones, is particularly sensitive to inaccuracies of intonation. And as its vibrators also admit of a delicate and durable tuning, it appeared to me peculiarly suitable for experiments on a more perfect system of tones. I therefore selected an harmonium of the larger kind, with two manuals and a set of vibrators for each, and had it so tuned that by using the tones of the two manuals I could play all the major chords from F♭ major to F♯ major. The tones are distributed as follows:

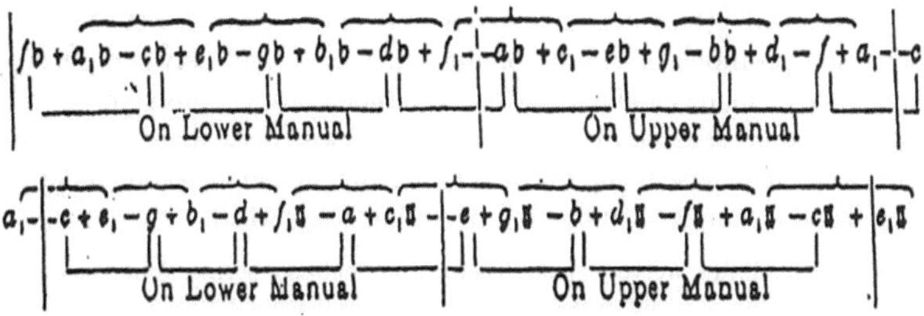

> This instrument therefore furnishes 15 major chords and as many minor chords, with perfectly pure Thirds, but with Fifths too flat by ⅛ of the interval by which an equally tempered Fifth is too flat.[6]

In comparing just intonation with equally tuned intonation Helmholtz refers to the former as possessing "a saturated harmoniousness [*ein gesättigster Wohlklang*]" that flows on "with a full stream that is calm and smooth and without tremor or beat

[5] The technical and financial arrangements for the Helmholtz- Schiedmayer transactions are spelled out in three letters (March 1861 to January 1862 with invoice for f 456), signed J. & P. Schiedmayer, Stuttgart, Piano, Harmonium, Harmonicorde. Nachlass Helmholtz, Item 426, Berlin-Brandenburgische Akademie der Wissenschaften. Akademiearchiv, Jägerstrasse 22/29, Berlin. The members of the Schiedmayer family were distinguished builders of clavichords, harpsichords, and pianos in eighteenth-century Bavaria. The modern branch of the firm was founded in 1809 in Stuttgart. In 1854 two Schiedmayer brothers, Julius (1822–1878) and Paul (1829–1890), opened the J. & P. Schiedmayer harmonium factory in Stuttgart. Julius, with training as a merchant, took charge of piano building. Paul, who had studied in Paris with A.F. Debain, the effective inventor of the harmonium, oversaw the construction of harmoniums. After 4 years in operation the harmonium factory had 40 workers and was building instruments that received worldwide recognition. At the exhibitions of 1862 in London and 1867 in Paris Schiedmayer instruments received the grand prize and the first silver prize. In the '70s and '80s the company carried on a prosperous and brisk business. The company was bombed in World War II. Exemplars of the various Schiedmayer keyboard instruments have survived mainly in German museums. Alexander Eisenmann, Schiedmayer and Soehne Hof-Pianofabrik, Vorgeschichte, Gründung und fernere Entwicklung der Firma, 1809–1909, Stuttgart, 1909. See esp. pp. 46–60.

[6] Helmholtz, *Sensations of Tone*, 1954 ed., 316. First Ger. ed. 1863, 485.

[*ohne zu zittern und zu schweben*]"; to the latter as "rough, dull, trembling, restless [*rauh, trübe, zitternd und unruhig*]" "The difference is so marked that every one, whether musically cultivated or not, observes it at once." This contrast, which Helmholtz already had commented on in his 1857 lecture on physiological acoustics, "is greatest and most unpleasant in the higher Octaves of the scale, because here the false combination tones [*falsche Combinationstöne*] of the tempered intonation are more observable, and the number of beats for equal differences of pitch becomes larger and hence the roughness greater."[7] The circumstances in which the differences between just and equal temperament become much more resolute and conspicuous are evident, as Helmholtz remarked, when one compares "the differences of effect between major and minor chords, between different inversions and positions of chords of the same kind, and between consonances and dissonances"; their "modulations become much more expressive," and "many fine distinctions are sensible, which otherwise almost disappear." As for modern musicians who are exposed only to equal temperament and therefore underestimate the inexactness of tempered intonation, Helmholtz remarked:

> The errors of the Fifths are very small. There is no doubt about that. And it is common to say that the Thirds are much less perfect consonances than the Fifths, and consequently also less sensitive to errors of intonation. The last assertion is also correct, so long as homophonic music is considered, in which the Thirds occur only as melodic intervals and not in harmonic combinations. In a consonant triad every tone is equally sensitive to false intonation, as theory and experience alike testify, and the unwanted effect of the tempered triads depends especially on the imperfect Thirds.[8]

Not wanting in a glib way to brush aside the enormous historical importance of tempered intonation, Helmholtz emphasized that the simplicity of tempered intonation is extremely advantageous for instrumental music. He recognized that "the high development of modern instrumental music would not have been possible without tempered intonation." This said, he firmly asserted that "it must not be imagined that the difference between tempered and just intonation is a mere mathematical subtilty without any practical value. That this difference is really very striking even to unmusical ears, is shown readily by actual experiments with properly tuned instruments."[9] Helmholtz pointed out that the musicians of the latter half of the seventeenth century and the early years of the 18th were accustomed to the perfect intervals of vocal music. They lived through many discussions about the introduction of different kinds of temperament. "One new method after another was invented and rejected for escaping the difficulties, and the most ingenious forms of instrument were designed for practically executing the enharmonic differences of the tones."[10]

---

[7] Helmholtz, *Sensations of Tone*, 1954 ed., 319.

[8] Helmholtz, *Sensations of Tone*, 1954 ed., 319–320.

[9] Helmholtz, *Sensations of Tone*, 1954 ed., 320.

[10] Helmholtz, *Sensations of Tone*, 1954 ed., 320.

In examining the disadvantages of tempered intonation, Helmholtz called attention to the unending difficulties that seventeenth- and early eighteenth-century music theorists and composers encountered in their attempts to achieve purer intonation. Eventually "players found themselves compelled by the use of only 12 digitals to the octave, to put up with a series of false intervals, and allow their ears to become accustomed to them."[11] Eventually they discovered that they could dispense with the few perfect thirds in the scale and make all of their thirds equally false, since it "necessarily produces more disturbance to hear very falsely tuned Thirds amidst correct intervals, than to hear intervals which are equally out of tune and are not contrasted with others in perfect intonation." As long as it is determined that the number of tones within the octave is 12, "there can be no question at all as to the superiority of the equal temperament with its 12 equal Semitones, over all others, and, as a natural consequence, this has become the sole acknowledged method of tuning."[12]

Among the circumstances favorable to the introduction, development, and perpetuation of equal tempered tuning and systems close to equal tempered tuning Helmholtz observed "that the disturbances due to beats in the tempered scale are the less observable the swifter the motion and the shorter the duration of the single notes." Short notes and their resultant few beats allow the ear no time to notice their presence.[13] In instrumental music, which unlike vocal music often is written for rapid movement, the drawbacks of tempered intonation are minimal, less conspicuous and therefore less irksome.

> We might, indeed, raise the question whether instrumental music had not been forced into rapidity of movement by this very tempered intonation which did not allow us to feel the full harmoniousness of slow chords to the same extent as is possible from well-trained singers, and that instruments had consequently been forced to renounce this branch of music.[14]

Helmholtz's rationale for so pointedly expressing and exploring the differences between tempered intonation in vocal and instrumental music is rooted in the historic fact that tempered intonation was first developed with the pianoforte and then gradually was transferred to other musical instruments. As will become evident in what follows, Helmholtz was apprised of the enormous technical

[11] Helmholtz, *Sensations of Tone*, 1954 ed., 321. In this context Helmholtz notably mentions composers such as Jean Philippe Rameau (1683–1764), Friedrich Marpurg (1718–1795), Johann Sebastian Bach (1685–1750), and the theorist and organist Andreas Werckmeister (1645–1706).

[12] Helmholtz, *Sensations of Tone*, 1954 ed., 321.

[13] Helmholtz, *Sensations of Tone*, 1954 ed., 322. The beats of the tempered fifth (9 beats in 10 seconds) are always quite audible. Beats of the combinational tones of the tempered fifth (5⅔ per second) are audible for tones that are not too weak. The beats of the major third (10½ per second) are not audible unless the tones employed have high upper partials. Beats of the minor third (18 per second) are much weaker than those of the major third and only occur for high upper partials. All these beats occur twice as fast when the chord lies an octave higher and half as fast when the chord lies an octave lower.

[14] Helmholtz, *Sensations of Tone*, 1954 ed., 323.

difficulties inherent in designing an instrument that is able to achieve just intonation; for him, nevertheless, as for many theorists and scientists before and during his time, too much was at stake to look upon the task as hopeless.

> On the pianoforte circumstances really favour the concealment of the imperfections due to the temperament. The tones of a pianoforte are very loud only at the moment of striking, and their loudness rapidly diminishes. This...causes their combinational tones to be audible at the first moment only, and hence makes them very difficult to hear. Beats from that source must therefore be left out of consideration. The beats which depend on the upper partials have been eliminated in modern pianofortes (especially in the higher Octaves where they would have done most harm), owing to the mode in which upper partials are greatly weakened and the quality of tone much softened by regulating the striking place.... Hence on a pianoforte the deficiencies of the intonation are less marked than on any instrument with sustained tones, and yet are not quite absent. When I go from my justly-intoned harmonium to a grand pianoforte, every note of the latter sounds false and disturbing, especially when I strike isolated successions of chords. In rapid melodic figures and passages, and in arpeggio chords, the effect is less disagreeable.[15]

The problem of achieving just intonation is not limited to keyboard instruments; it applies to all fixed tone woodwind instruments and valve-operated brass instruments. Their pitches can only be altered slightly by embouchure control, blowing force, and unique fingerings. String instruments, which have retained their tuning in perfect fifths, are, as Helmholtz stated, "perfectly unfettered as to intonation." All the orchestral instruments are "adapted for equal temperament, but good players have the means of indulging the ear to some extent."[16] To investigate pitch relationships on keyboard instruments, Helmholtz realized that he could call on string performers of the highest rank to assist him in the experimental investigations that he had planned. He knew that violinists and violoncellists who possess a delicate sense of harmony intuitively produce perfect thirds and sixths, especially when playing in string quartets; that whenever possible they adjust their temperaments when playing with instruments that do not so readily lend themselves to playing in just intonation.

In connection with his various publications on musical acoustics Helmholtz had developed good personal relations with a number of performers. Recognized as a fairly competent pianist among them, he was well positioned to seek the cooperation of a violinist in the circle of musicians to which he belonged. The Austro-Hungarian violinist Joseph Joachim (1831–1907) belonged to that circle.[17]

---

[15] Helmholtz, *Sensations of Tone*, 1954 ed., 323.

[16] Helmholtz, *Sensations of Tone*, 1954 ed., 324.

[17] The Austro-Hungarian violinist and composer Joseph Joachim was born in Köpcsèy, Hungary, in 1821 and died in Berlin in 1907. With studies in Vienna and Budapest, and influenced by Mendelssohn in Leipzig, Joachim's playing was in the French classical tradition. For most of his life he worked close to and in an advisory capacity with the Schumanns and Brahms. In spite of personal disputes with Brahms, he was a strong advocate of his music. From 1858 he taught in Berlin where he founded and led an influential string quartet.

In Joachim and his expertise as a violinist Helmholtz encountered exactly what he needed to make his case for just intonation:

> I was fortunate enough to have an opportunity of making similar observations by means of my harmonium on Herr Joachim. He tuned his violin exactly with the $g \pm d \pm a \pm \underline{e}$ of my instrument. I then requested him to play the scale, and immediately he had played the Third or Sixth, I gave the corresponding note on the harmonium. By means of beats it was easy to determine that this distinguished musician used b_1 and not b as the major Third to g, and e_1 and not e as the Sixth. But if the best players who are thoroughly acquainted with what they are playing are able to overcome the defects of their school and of the tempered system, it would certainly wonderfully smooth the path of performers of the second order in their attempts to attain a perfect *ensemble* if they had been accustomed from the first to play the scale by natural intervals. The greater trouble attending the first attempts would be amply repaid by the result when the ear has once become accustomed to hear perfect consonances. It is really much easier to apprehend the differences between notes of the same name in just intonation than people usually imagine when the ear has once become accustomed to the effect of just consonances. A confusion between a_1 and a in a consonant chord on my harmonium strikes me with the same readiness and certainty as a confusion between A and $A\flat$ on a pianoforte.[18]

Helmholtz maintained that the case was "precisely similar for our present singers." Here the intonation is perfectly free and "the pitch can be made most easily and perfectly to follow the wishes of a fine musical ear. Hence all music began with singing; and singing will always remain the true and natural school of all music. The only intervals which singers can strike with certainty and perfection are such as they can comprehend with certainty and perfection, and what the singer easily and naturally sings the hearer will also easily and naturally understand."[19]

By contrast, Helmholtz maintained that "the singer who practises with a tempered instrument has no principle at all for exactly and certainly determining the pitch of his voice." On the other hand Helmholtz maintained that amateurs who had not practiced with a tempered instrument but practiced much together were known to sing "in perfectly just intonation."

> Indeed, my own experience almost leads me to affirm that quartets are more frequently heard with just intonation when sung by young men who scarcely sing anything else and often and regularly practise them, than when sung by instructed solo singers who are accustomed to the accompaniment of the pianoforte or the orchestra. But correct intonation in singing is so far above all others the first condition of beauty, that a song when sung in correct intonation even by a weak and unpractised voice always sounds agreeable, whereas the richest and most practised voice offends the hearer when it sings false, or sharpens.[20]

From Helmholtz's point of view the justly tuned harmonium as an accompanying instrument was seen to be singularly appropriate for voice training. Practice, preferentially with sustained notes, he remarks, is preferred "because the singer. . .can immediately hear the beats between the instrument and his voice

[18] Helmholtz, *Sensations of Tone*, 1954 ed., 325.

[19] Helmholtz, *Sensations of Tone*, 1954 ed., 325.

[20] Helmholtz, *Sensations of Tone*, 1954 ed., 326.

when the pitch is altered slightly." Draw the singer's attention to the beats and the singer has a means of checking the voice in a decisive manner.

> This is very easy on my justly-intoned harmonium, as I know by experience. It is only after the singer has learned to hear every slight deviation from correctness announced by a striking incident, that it becomes possible to regulate the motions of the larynx and the tension of the vocal chords with sufficient delicacy to produce the tone which the ear demands. When we require a delicate use of the muscles of any part of the human body as in this case of the larynx, there must be some sure means of ascertaining whether success has been attained. Now the presence or absence of beats gives such a means of detecting success or failure when a voice is accompanied by sustained chords in just intonation. But tempered chords which produce beats of their own are necessarily quite unsuited for such a purpose.[21]

Many of the topics examined in Helmholtz's treatise on *Sensations of Tone* are interfused with historical remarks that serve to reinforce the principles here under discussion. The section on tempered intonation is no exception. Helmholtz calls attention to the influence that the introduction of tempered intonation exerted on changing styles of composition in the early eighteenth century. He observed that the first effect of tempered intonation was favorable. It allowed composers as well as players to move freely and easily into all keys, and opened up a new wealth of modulation. The alteration of intonation also compelled composers to have recourse to modulations whose consonant chords became so imperfect that the differences between their various inversions and positions were obliterated.

> It was necessary to use more powerful means, to have recourse to a frequent employment of harsh dissonances, and to endeavour by less usual modulations to replace the characteristic expression which the harmonies proper to the key itself had ceased to possess. Hence in many modern compositions dissonant chords of the dominant Seventh form the majority, and consonant chords the minority, yet no one can doubt that this is the reverse of what ought to be the case; and continual bold modulational leaps threaten entirely to destroy the feeling for tonality. These are unpleasant symptoms for the further development of art. The mechanism of instruments and attention to their convenience threaten to lord it over the natural requirements of the ear, and to destroy once more the principle upon which modern musical art is founded, the steady predominance of the tonic tone and tonic chord.[22]

The introduction of equal temperament into musical practice came hesitatingly. For Helmholtz it is not entirely clear whether it was in Bach's famous *Well-Tempered Clavichord* (24 Preludes and Fugues in all major and minor keys) or in the writings of predecessors that reference was first or only made to equal temperament or its close approximation. In any case equal temperament was not universally adopted in Germany until about 1800 or in France and England until about 1850.[23] Helmholtz's reaction to the effect of introducing equal temperament into

---

[21] Helmholtz, *Sensations of Tone*, 1954 ed., 327.

[22] Helmholtz, *Sensations of Tone*, 1954 ed., 327.

[23] In a recent article Bradley Lehman claims to have solved a 250-year-old puzzle by unlocking the significance of what appeared at the top of one of J.S. Bach's compositions as an arbitrarily scribbled design. The puzzle allegedly involves Bach's views on a method of tuning the harpsichord. Bradley Lehman, Bach's extraordinary temperament: Our Rosetta Stone – 1. *Early Music*, *33* (2005) 3–23. What Lehman's puzzle-solving exercise shows about Bach is how he thought the harpsichord should be tuned in order to play the *Well-Tempered Clavichord* the way he wanted it to be played.

modern music is expressed in a remarkably imaginative historical comparison. In reference to vocal and instrumental music he offered an evaluation of two of his most favored composers: Mozart who died in 1791 and Beethoven who outlived Mozart by 36 years.

> Among our great composers, Mozart and Beethoven were yet at the commencement of the reign of equal temperament. Mozart had still an opportunity of making extensive studies in the composition of song. He is the master of the sweetest possible harmoniousness where he desires it, but he is almost the last of such masters. Beethoven eagerly and boldly seized the wealth offered by instrumental music, and in his powerful hands it became the appropriate and ready tool for producing effects which none had hitherto attempted. But he used the human voice as a mere handmaid [*als dienende Magd*], and consequently she has also not lavished on him the highest magic of her beauty.
>
> And after all, I do not know that it was so necessary to sacrifice correctness of intonation to the conveniences of musical instruments. As soon as violinists have resolved to play every scale in just intonation, which can scarcely occasion any difficulty, the other orchestral instruments will have to suit themselves to the correcter intonation of the violins. Horns and trumpets have already naturally just intonation.[24]

To circumvent or at least to minimize the enharmonicity inherent in music performed on equal-tempered instruments Helmholtz set out to explore what it would take by way of music theory, instrumentation, compositional skills, performance practice, and sensitivity on the part of the listener, to achieve acoustical purity. In formulating such an ambitious programme Helmholtz by no means was functioning as an intransigent propagandist for just intonation. He advocated just intonation on the basis of its widespread interest and desirability among music theorists; but he recognized that whatever insights theorists might reap in regard to teaching music theory and refinement in musical performance, they nevertheless would have to strike a bargain with the musical world of tempered intonation. What this signified at the performing level was that depending on the circumstances – vocal music, instrumental music, fixed-tone instruments – Helmholtz was prepared to strike the bargain on the side of either equal temperament or just temperament.

In his just intonation ventures Helmholtz was being propelled into a situation that scientists traditionally have learned to anticipate and take advantage of. On occasion they discover that the efforts to solve or come to terms with a well formulated problem may lead to insights that shift the investigator's attention in the direction of a quite-other, and sometimes more-significant problem; the incidental by-product becomes more exploreworthy than the original challenge. Hard problems, more than easy problems, can stimulate the emergence and the invention of unanticipated outcomes and results. The challenge of a problem worthy of attack, more often than not, can demonstrate its merits by striking back and revealing a different problem that perchance turns out to be of greater significance than the original problem had posed.[25] In the case of Helmholtz, and his just intonation

---

[24] Helmholtz, *Sensations of Tone*, 1954 ed., 327.

[25] Apologia ad Piet Hein, who was a sharp-witted, limerick-handy physicist at Niels Bohr's Institute of Theoretical Physics in Copenhagen. He wrote in *Grooks*, Copenhagen, 1966, "Problems worthy of attack show their worth by striking back."

enthusiasts and compatriots, the open-ended approach to problem solving led to a wealth of musical ideas and newly constructed musical keyboards that persuaded twentieth-century instrument builders, performers, and composers to explore the domain of microtonal instruments and music encountered in Western and non-Western cultures.

In order to achieve a scale that approximates just intonation, Helmholtz indicated, as many others had done before his time, that this would require a division of the octave into many more parts than the 12-tone scale that characterizes Western music. The theoretically designed multi-toned scale therefore became the indispensable preliminary frontier on which to explore the design and the technologically demanding task of constructing fixed-tone instruments capable of delivering just intonation. It is in connection with this problem that Helmholtz first referred, specifically, to the work of R.H.M. Bosanquet, who in 1876 published *An Elementary Treatise on Musical Intervals and Temperament*. Bosanquet's *Treatise* was not the first of his publications on the subject of tuning and temperament; it was the final version of a volume that grew gradually from oral presentations and written papers that began in 1872 and most likely had been put together during the second half of 1875 and the first half of 1876.[26] The topics in musical acoustics that Bosanquet was engaged in pursuing in the 1870s overlap to a large extent with those of his British countrymen. During the second half of the nineteenth century there was a fairly strong interest in both physical and musical acoustics in England. This interest is exhibited in the publication of a series of significant textbooks on sound written by Bosanquet's colleagues in the Royal Society and the Royal Institution.[27] Although there is scant or no reference to Bosanquet's writings or ideas on acoustics in any of the musical biographies and reference books, his numerous papers on beats, temperament, just intonation, and the problem of the division of the octave, are listed in German and English catalogs

[26] The Bosanquet *Treatise* of 1876 represents a revised and enlarged version of a number of papers published in the *Proceedings of the Royal Musical Association* (1874/5) entitled: On Temperament, of Division of the Octave. 1. Points of Historical Interest; 2. Formation of Scales and Properties of Systems; 3. Instrumental Means of Control. An edited version of the 1876 London *Treatise* was published in Utrecht in 1987 by Rudolf Rasch (b. 1945) who teaches music theory, music history, musical acoustics, and musical instrumentology at the Institute for Art History at the University of Utrecht. Rasch provides extensive biographical and bibliographical information about Bosanquet and offers a perceptive commentary on the historical importance of Bosanquet's contributions to tuning and temperament. Rudolf Rasch, *MGG*, Personenteil *13* (2005), col. 1283–1284.

[27] John Tyndall, *On Sound*. Eight Lectures delivered at the Royal Institution, London 1867, 4th ed. 1883. Ger. transl., ed. by H. Helmholtz and G. Wiedemann, Braunschweig 1869; George Biddell Airy, *On Sound and Atmospheric Vibrations*, London, 1868, 1871; Sedley Taylor, *Sound and Music*. An Elementary Treatise on the Physical Constitution of Musical Sounds and Harmony, London, 1873, 1883, 1896; W.H. Stone, *The Scientific Basis of Music*, London, 1878, 1879; Lord Rayleigh (John William Strutt), *The Theory of Sound*, 2 volumes, Cambridge, 1877–1878, 1894.

and handbooks on science.[28] An examination of the titles of these papers shows that Bosanquet's main scientific interests were directed mainly toward elucidating a problem that many other investigators before his time (mainly physicists and mathematicians) had explored – namely, the design and construction of musical instruments for systems of tuning other than equal temperament or some modification of the 12-tone octave.

Robert H.M. Bosanquet (1841–1912) was born into a large French Huguenot family that settled in England after the revocation of the Edict of Nantes in 1598. His brother Bernard was the well-known British philosopher and defender of absolute idealism. Robert studied natural sciences at Balliol College Oxford, took a bachelor's degree in 1862, and gained first class honors in mathematics. As a member of the Royal Society of London, and as an appointed Fellow of St. John's College (1870), he spent the most productive period of his life teaching, conducting research in physical acoustics, and writing papers on tuning, temperament, and the beats of mistuned consonants. Bosanquet was a physicist, amateur musician, organist, singer in the Oxford University Glee Club, and member of the *Musical Association.* Scientific membership in the *Association* gave him ample opportunity to express his views and enter into discussions on questions connected with theoretical musicology and problems in tuning, temperament, organ building, and organ tuning.[29] During the 1870s the London organ and harmonium builder T.A. Jennings built three instruments for Bosanquet: an enharmonic harmonium, an enharmonic organ, and a regular organ. Appointed professor of acoustics at the Academy of Music and Fellow of the Royal Society of London in 1881, Bosanquet's last two decades were devoted largely to societal matters. In 1892 he settled in Tenerife in the Canary Islands. In his will he directed that "all his manuscripts of a scientific character should be destroyed without examination."[30]

As Helmholtz indicated in *Tonempfindungen*, the Bosanquet *Treatise* features a harmonium with a fingerboard on which the octave is divided into 53 equal intervals of which 31 supply almost perfect fifths with an error of 1/28 of the error of the usual equal temperament, and 17 of the intervals supply a major third with an error of only 5/7 of the error of the fifth of equal temperament. The fingerboard, which is arranged "in a very comprehensible and symmetrical way to make the fingering of all scales and all chords the same in all keys" is depicted in diagrammatic form in an appendix to Helmholtz's *Sensations of Tone*.[31]

---

[28] For example, Royal Society of London *Catalogue of Scientific Papers*; Poggendorff's *Biographisch-literarisches Handwörterbuch*. Passim.

[29] The Musical Association included prominent scientists such as John Tyndall, Sedley Taylor, W.H. Stone, William Pole, Lord Rayleigh, Bosanquet, and A.J. Ellis.

[30] Rudolf Rasch, Biography in Bosanquet, *An Elementary Treatise*, Utrecht, 1987, 11–23.

[31] Helmholtz, *Sensations of Tone*, 1954 ed., 328–330 and Appendix XIX, 429–430. The sections on Bosanquet here described were first included in the 4th 1877 edition of Helmholtz's *Sensations of Tone*, and published shortly after Bosanquet's 1876 *Treatise* appeared. According to a footnote on p. 328 of the Dover ed. of *Sensations of* Tone, the Bosanquet harmonium was on display in 1876 in the Scientific Loan Exhibition at the Science Museum in South Kensington.

In what follows an effort is made to show why Helmholtz chose to focus on Bosanquet and his enharmonic harmonium. It is important at the outset to recognize that all the Bosanquet papers on temperament and the division of the octave were published several years after publication of the *Tonempfindungen* volume of 1863 and either concurrently with or shortly after the appearance of the English translation of Helmholtz's *Tonempfindungen.* It is apparent, on the one hand, that Helmholtz's 1875 *Sensations of Tone* had served as a springboard for Bosanquet's decision in 1876 to publish, in one composite volume, his *Elementary Treatise on Musical Intervals and Temperament.* On the other hand it also is evident that Bosanquet pursued a decidedly physical approach in order to distance himself from Helmholtz's physiological approach. Comments to the Helmholtz volume in the *Elementary Treatise* are scarce and are exclusively limited to criticisms of Helmholtz's discussions on the enharmonic harmonium. Bosanquet first acknowledged Helmholtz's *Tonempfindungen* in 1874/5 in a one-page abstract to lectures he had given at the Royal Society in which he stated: "Throughout, the work of former labourers in the same field is reviewed: the obligations of the writer are due to Helmholtz, the late General T. Perronet Thompson, F.R.S., and others."[32] In the *Treatise* of 1876 Bosanquet wrote: "I shall not attempt to enter into the question of the physiological basis for harmony, or any of the questions discussed by Helmholtz and others in connection with this part of the subject. I avoid the controversy not because I fail to have definite opinions on these points, but because they are quite distinct from what I am dealing with here."[33]

Robert Rasch, in his extensive and insightful commentary on Bosanquet's 1876 *Treatise* and its relevance for examining English tuning and temperament practices, remarks that "the most important work on tuning and temperament of the third quarter of the nineteenth century is. . .that of A.J. Ellis" (translator of Helmholtz's *Sensations of Tone*). His works "must have had considerable influence on the development of Bosanquet's ideas about the subject" but there "is only little reference" to the Ellis translation and therefore to *Tonempfindungen* in Bosanquet's writings.[34] What Bosanquet, Helmholtz, and Ellis had in common was the problem of tuning and temperament applied to the enharmonic harmonium – a topic that in all three investigators led to important contributions to the science and technology of keyboards with more than 12-keys per octave.[35] As Rasch indicated in his commentary, the special importance of the *Treatise* was that Bosanquet was one of the few nineteenth-century authors on this subject who questioned the identification of J. S. Bach with equal temperament. Bosanquet had mentioned "that the

---

[32] R.H.M. Bosanquet, "On just intonation in music." An Abstract. *Proceedings of the Royal Society of London, 21* (1874–1875) 131–132.

[33] Bosanquet, *An Elementary Treatise*, London 1876, viii–ix.

[34] Rasch, Biography in Bosanquet, *An Elementary Treatise*, Utrecht, 1987, 24–25.

[35] The pre-Helmholtz history of this topic – the design and construction of enharmonic keyboards with more than 12-keys per octave – is extensive and merits an analysis that cannot be included here. An examination of the contributions of Ellis also would also take the reader beyond the scope of this analysis.

clavichord was Bach's favorite instrument, which, however, showed such a bad control of pitch that there was hardly any need to replace meantone tuning by equal temperament" (an argument later shown to be somewhat defective) and "that Bach's organ works use very few 'enharmonic' notes, thus not necessitating equal temperament." Bosanquet approximated the harmonic seventh (frequency ratio 4/7) from a series of fifths as either 14 that are descending or 10 that are ascending. He did this because he believed that musical intonation could be improved when "projected onto the system with 53 tones per octave with 17 units for the major third, 31 for the fifth, and 43 units for the harmonic seventh."[36]

Following a tradition that began with the work of Stevin and Huygens at the beginning of the seventeenth century, Bosanquet cast his ideas on tuning and temperament and his general theory of the division of the octave in mathematical form. The practical analogue of the theory, Bosanquet's "generalized keyboard," was designed to sound the tones and pitches of both regular or cyclical systems so that all tuning systems with equal fifths could be realized on the keyboard. This was a departure from the earlier multi-tone keyboards that had taken the 12-tone keyboard as extensions to be used for the specific tuning systems for which they had been designed. Bosanquet's generalized keyboard avoided the 12-tone bias and provided a single tuning system.[37]

Two scientifically and musically significant systems of intonation are being placed side-by-side here. One of them, based on the principle of just intonation, is represented by Helmholtz, his translator A.J. Ellis (in part), and the music theorist and harmonium builder Carl Eitz. The other is based on a system requiring multiple division of the octave and is represented by R.H.M. Bosanquet. The German authors were familiar with the work of the English authors and vice versa. The Ellis translation of Helmholtz played an important role in the process of exchange of both texts and ideas. Actually, all of these developments in intonation and their enharmonic keyboards were so complicated, and so far removed from the actual practice of music, that they were looked upon, as Rasch remarked, chiefly as "nothing more than example or experiment, however much they were admired and praised for their theoretical and scientific merits."[38] Bosanquet's generalized keyboard made room for new tonal materials that only materialized much later. His generalized keyboard became known in the German musicological and acoustical circles of the late nineteenth century largely because Helmholtz had given Bosanquet's *Treatise* notice in the German-speaking world. The Ellis translation of *Tonempfindungen* had made the idea of a generalized keyboard available to British enharmonists.

[36] Rasch, Biography in Bosanquet (1987), 30–31.

[37] Rasch, Biography in Bosanquet (1987), 51–52. In his various writings Bosanquet took note of the ideas of a number of other contemporary multi-tone keyboard proponents, and Rasch has discussed these matters.

[38] Rasch, Biography in Bosanquet (1987), 64–65.

Our examination of Helmholtz's work on the harmonium and just intonation up to this point in the early 1870s coincides in time with his move to Berlin and a new thrust in the direction of more traditional areas of physics than he hitherto had been engaged in. During the two decades of his professorship in Heidelberg his investigations had gradually shifted from physiological optics and physiological acoustics toward physics, mathematics, and epistemology. In 1870 he was appointed to the professorship in technology and physics at the University of Berlin left vacant by the death of Gustav Magnus (1802–1870). As an experimental chemist Magnus had been in that position for more than three decades. Beginning in 1870 Helmholtz's research output, teaching, and publications covered a broad range of special topics that included theoretical electrodynamics, hydrodynamics, chemical thermodynamics, monocyclic systems, the thermal theory of cyclones, the principle of least action, and the epistemology of perception. By the 1880s the University of Berlin's well-established and well-financed physics institute under Kirchhoff and Helmholtz had become a formidable research challenge for research institutions such as the Cavendish Laboratory in Cambridge to which Lord Rayleigh and J.J. Thomson had brought worldwide commendation. Having acquired universal acclaim as a physiologist who had inherited the leading post in theoretical physics in Germany Helmholtz came to be looked on by his countrymen as a universal scholar in the tradition of Leibniz.

For almost four decades, in teaching and research appointments in Potsdam, Berlin, Königsberg, Bonn, and Heidelberg, Helmholtz had been on the search of unifying principles of nature – simultaneously exploring the physiological and physical sciences and the arts. For higher-up personnel in the newly created German Empire this signified that Helmholtz was the ideal person to be singled out by university colleagues, compatriots in education, and representatives of industry and government, as a leading figure in scientific affairs of state. He came to be lionized, no less, as the German vehicle of high intellectual culture – a genuine *Kulturträger*. In the minds of representatives from government and industry there developed the thinking that in Bismarck's state the physical sciences when linked with technology had central roles to play in the future of the new German Reich. Firm financial backing came from the powerful and influential industrialist Werner von Siemens (1816–1892). To achieve the goals that had been set it was suggested by various onlookers from academia, government, and industry, that an institution devoted to state-of-the-art research, rather than only teaching, should be established as norm in Berlin.[39] This public sentiment materialized in 1887 when Helmholtz accepted the presidency of the newly founded *Physikalish-technische Reichsanstalt* (the PTR) for research in the exact sciences and technology.[40] During the same year, on the death of Robert Kirchhoff (1824–1887), who alongside Helmholtz had occupied the Berlin professorship in mathematical physics,

[39] Iwan Rhys Morus, *When Physics Became King*, Chicago 2005, Section on "Berlin's Imperial Institute."

[40] David Cahan, *An Institute for an Empire. The Physikalisch-Technische Reichsanstalt 1871–1918*, Cambridge, 1989.

Max Planck (1858–1947) became Kirchhoff's successor but acquired the new title of assistant professor of physics and Director of the newly created Institute for Theoretical Physics. The 28-year-old Planck, who had been Extraordinary Professor of Theoretical Physics at Kiel, was returning to the Berlin in which he had studied physics and mathematics a decade earlier with Helmholtz and Kirchhoff.

The problem that frames this paper – just intonation and fixed-tone keyboards – and that in large part converges at this point in the intellectual interchanges and experimental investigations of Helmholtz and Planck in Berlin, overlaps in time with the years during which the quantum theory and relativity were in process of conceptualization. Whatever interests in the new physics the young Planck and the almost 40-year-older Helmholtz may have had in common during the last decade of Helmholtz's life, our focus in this account is directed solely to their respective and collaborative activities in the field of music and musical acoustics.

In the 1860s and 1870s, as we have shown, Helmholtz, on his own, and then in consort with the violinist Joseph Joachim, explored the problem of just intonation in experimental investigations on an harmonium built for this purpose by the Schiedmayer piano factory in Stuttgart. Thirty years later in Berlin, Helmholtz returned to an interest in reexamining the problem of intonation. This time it was largely because his students and one of his own colleagues at the university, Max Planck in particular, showed a discerning interest in exploring just intonation with an Eitz harmonium that had been placed in the physics laboratory at the University. Planck had cultivated music and song and had played the piano since his youth. With Helmholtz as his closest colleague at the university he developed a keen interest in musical acoustics and became involved in the problem of just intonation with an enthusiasm that seemingly knew no bounds.[41]

Recall that in the 1880s, Helmholtz, in collaboration with students and colleagues in physics at the university, launched a programme for exploring the problem of just intonation with the aid of an harmonium that had been designed for that purpose. During the summer semester of 1892 Helmholtz, who by then had taken on the position at the PTR, delivered his last set of lectures on the mathematical principles of acoustics. Transcribed notes of the hand-written stenographic text of the lectures were made available after his death under joint editorship of Arthur König (1856–1901) and Carl Runge (1856–1927). König was one of Helmholtz's most prominent students and head of the physics division of the University of Berlin's physiological institute; Carl Runge., mathematician and physicist, was lifelong friend to Planck.[42] The lectures on physical acoustics had been organized around three main headings: the theory of small oscillations of discrete points in the state of stable equilibrium, the vibration of plucked and struck strings, elastic rods, and columns of air, and the motion of sound in open space and

[41] These matters are examined in a forthcoming monograph entitled: Max Planck: Theoretical Physics with Caution, Music with Passion, Objective Laws of Nature as Stepping Stone to the Deity of Creation.

[42] Helmholtz, *Vorlesungen über die mathematischen Principien der Akustik* is vol. III in the X-volume *Vorlesungen über die theoretische Physik*, Leipzig, 1898.

in confined areas such as open-ended pipes. In an introductory statement entitled "acoustics and its relation to the ear," Helmholtz sought to show how physical and physiological acoustics can be differentiated meaningfully without sacrificing their common theoretical identity. He referred to the branch of acoustics that is "purely physical, i.e. without regard to physiological points of view," as the science of oscillations of vanishingly small displacements that arise in bodies or systems of bodies that are close to a state of equilibrium and move under the influence of restorative forces. The nature of these forces, as Helmholtz remarked, can vary greatly and occur in all states of aggregation. To this definition he added that the relationship of these physical processes to the ear is given by recognizing that "the ear reacts to a large number of such vibrations" and "is able readily to distinguish many finer (more subtle) differences of vibration than any of the other sense organs." In any case, Helmholtz reminded the reader that the laws enunciated in physical acoustics are valid not only for audible vibrations but for the much slower and even the visible vibrations of large masses.[43]

During the early 1890s the programme that claimed the full approval and institutional support of the 70-year-old Helmholtz was carried out almost entirely at the behest of physicists and mathematicians in the laboratories of the University of Berlin and its neighboring institution the PTR. These activities created a new interest in the theory of just intonation and the construction of enharmonic keyboards that drew the attention of music teachers, educators, composers, and the public. On the one hand his supporters did not know what to make of an unwieldy physics- and technology-focused music programme originating in a university physics department, but on the other hand they welcomed the enharmonists' emphasis on vocal music in an environment in which the pure voices of a cappella chorus groups took pride of place at the expense of impure tempered keyboards and orchestral instruments.

During the years that Helmholtz was delivering his lecture course on the mathematical principles of acoustics he had, by order of the Prussian government, commissioned the construction of an harmonium designed for experimentation on just intonation. The builder was Carl Eitz (1848–1924), acoustical theorist, writer, mathematician, and singing teacher associated with the Schiedmayer piano company in Stuttgart. The Eitz harmonium had 4½ octaves. Each octave had 104 different tones activated by 52 green, blue, and red keys arranged in 8 horizontal rows as here depicted:

I	gis^{-4}	dis^{-4}	ais^{-4}	eis^{-4}	his^{-4}	fisis^{-4}	cisis^{-4}	gisis^{-4}	disis^{-4}	aisis^{-4}	eisis^{-4}	hisis^{-4}	fisisis^{-4}
II	h^{-3}	fis^{-3}	cis^{-3}	gis^{-3}	dis^{-3}	ais^{-3}	eis^{-3}	his^{-3}	fisis^{-3}	cisis^{-3}	gisis^{-3}	disis^{-3}	aisis^{-3}
III	d^{-2}	a^{-2}	e^{-2}	h^{-2}	fis^{-2}	cis^{-2}	gis^{-2}	dis^{-2}	ais^{-2}	eis^{-2}	his^{-2}	fisis^{-2}	cisis^{-2}
IV	f^{-1}	c^{-1}	g^{-1}	d^{-1}	a^{-1}	e^{-1}	h^{-1}	fis^{-1}	cis^{-1}	gis^{-1}	dis^{-1}	ais^{-1}	eis^{-1}
V	as^{0}	es^{0}	b^{0}	f^{0}	c^{0}	g^{0}	d^{0}	a^{0}	e^{0}	h^{0}	fis^{0}	cis^{0}	gis^{0}
VI	ces^{+1}	ges^{+1}	des^{+1}	as^{+1}	es^{+1}	b^{+1}	f^{+1}	c^{+1}	g^{+1}	d^{+1}	a^{+1}	e^{+1}	h^{+1}
VII	eses^{+2}	bb^{+2}	fes^{+2}	ces^{+2}	ges^{+2}	des^{+2}	as^{+2}	es^{+2}	b^{+2}	f^{+2}	c^{+2}	g^{+2}	d^{+2}
VIII	geses^{+3}	deses^{+3}	ases^{+3}	eses^{+3}	bb^{+3}	fes^{+3}	ces^{+3}	ges^{+3}	des^{+3}	as^{+3}	es^{+3}	b^{+3}	f^{+3}

[43] Helmholtz, *Vorlesungen*, 1.

As indicated in the above, linearly, each row provided fifths and fourths. Ascending rows provided major thirds and minor sixths, while descending rows provided minor thirds and major sixths. By adjusting the registers it was possible to combine any four series of perfect fifths using the variously colored keys. The exponents 0, +1, −1 refer to tonal differences in multiples of syntonic commas (ca. 81/80). The pythagorean comma (ca. 74/73) and the schisma (887/886) also were specified on the instrument.[44] It is known from the correspondence that Helmholtz had with Eitz, the builder of the harmonium, that the particular instrument designed for Helmholtz's Berlin laboratory had acquired a number of special features not present in earlier harmoniums. Eitz wrote to Helmholtz in 1889: "I have conceived of the instrument not merely as a musical instrument (to be played), but also, in a substantial way, to be used for demonstrating harmony as well as the historical development of scales. ... I have the confidence that it will be playable with a facility hitherto not achieved, and that it therefore will gain the approval of practicing musicians."[45] The functioning and musical significance of the Eitz harmonium is discussed in my paper on Planck and his contributions to music theory.

References

Bosanquet, R.H.M. 1876. *An elementary treatise*, viii–ix. London.

Cahan, David. 1989. *An Institute for an Empire. The Physikalisch-Technische Reichsanstalt 1871–1918*. Cambridge: Cambridge University Press.

Helmholtz, H. 1954. *Sensations of tone*. New York: Dover Publications.

Morus, Iwan Rhys. 2005. *When physics became king*. Chicago: University of Chicago Press.

Rasch, Rudolf. 1987. Biography in Bosanquet. In *An elementary treatise*. Utrecht: Diapason Press.

Woodruff, A.E. 1976. Wilhelm Weber. *DSB* 14: 203–209.

[44] The Eitz harmonium is described in Max Planck, "Ein neues Harmonium in natürlicher Stimmung nach dem System von C. Eitz," *Verhandlungen der Physikalischen Gesellschaft, 12* (1893) 8–9, reproduced in *Physikalische Abhandlungen und Vorträge 1* (1958) 435–436.

[45] Eitz to Helmholtz, Eisleben, 2 Sept. 1889. *Nachlass Helmholtz*, Akademiearchiv; Berlin, Item 30.

Part II
Shohé Tanaka, Just Intonation and the Enharmonium

Chapter 9
Introduction

In 1884 a 28-year-old Japanese physicist and student of music theory, Shohé/Shōhei Tanaka, received a grant from the Japanese Ministry of Education to study acoustics and electromagnetism with Helmholtz at the University of Berlin. Two years earlier he had graduated from the Imperial University in Tokyo with a major in physics with additional studies in music theory and foreign languages. The avid cultivation of music since the time of his youth, the acquisition of expertise in the field of acoustics, and the study of music theory and foreign languages at the University gave Tanaka's life the intellectual shape and form and the compelling drive to pursue the science of music as a profession. At the graduation ceremonies in Tokyo in1882 the Japanese Emperor awarded Tanaka a silver medal for academic distinction. Before going to Berlin Tanaka completed a 1-year assistant professorship at Tokyo University's Preparatory School. Taken together, Tanaka's various studies in physics, music theory, and foreign languages, in addition to the teaching experiences, embodied a well-selected and appropriate preparation for what was to come in Berlin.

On arrival in Berlin Tanaka was given an enthusiastic welcome by Helmholtz and the circle of students and colleagues in theoretical physics at the University. A number of physicists from among this group were engaged at the time in exploring problems in musical acoustics that dealt with just intonation and fixed-tone keyboards along lines of investigation that Helmholtz had pursued in Heidelberg in the 1850s and 1860s. Fortunately for Tanaka, the intonation studies had been taken up anew by Helmholtz when he accepted the professorship in theoretical physics at the University of Berlin in 1870. For Tanaka the contacts with Helmholtz and the scientists in the Helmholtz circle were to become of inestimable value – this above all from the point of view of the intellectual interchange with persons in a position to identify with his own distinctive acoustical interests. An inroad to the university also was gainful to Tanaka from the point of view of access to the material resources and experimental facilities at the University that could be put to use, as he discovered, in the construction of musical instruments.

In addition to the personal contacts, material resources, and experimental facilities that were within reach, at the university of Berlin our young Japanese scholar

E. Hiebert, *The Helmholtz Legacy in Physiological Acoustics*, Archimedes 39, DOI 10.1007/978-3-319-06602-8_9, © Springer International Publishing Switzerland 2014

was able to enhance his contacts with musicians and the cultural musical world owing to an affiliation, *as student*, with Berlin's oldest and most renowned music school, *Das Stern'sche Konservatorium für Musik*, ordinarily referred to simply as the *Konservatorium für Musik*. With the establishment of this science-music (University-Konservatorium) dualism on the horizon, Tanaka was able to use his scientific training and his intuitive manual craftsmanship to forge ahead with the problems in just intonation that he had envisioned for himself on arrival in Berlin.

During the years of the third and last German Emperor, William II (1859–1941), Berlin was experiencing a liveliness in its musical culture, and a hurriedness in the growth of its music schools, that attracted students from all of Europe and beyond. This was the case for the Japanese student Shohé Tanaka (1862–1945).

The leading historian of music schools and music education in Berlin, Dietmar Schenk, has given an account of the *Konservatorium für Musik* that captures skillfully, and with good balance, the climate of thought of this prominent music school in Berlin.[1] According to Schenk, the *Stern Konservatorium für Musik* was founded in 1850 as a privately organized and financed institution in the center of Berlin close to the huge complex of the Philharmonic and the famed Beethoven hall. It was Berlin's oldest and largest music school and by the end of the nineteenth century had become the hub of Berlin's concert life. "In the midst of the rapid development of the Prussian-German capital into a modern industrial and cultural metropolis the *Stern Konservatorium* (as it was frequently referred to) was able to stand its financial ground and fasten its grip on the life of music in Berlin." Its associates included the names of prominent students such as Bruno Walter and Otto Klemperer.

Hans von Bülow, Hans Pfitzner, and the Austro-Hungarian Arnold Schoenberg (1874–1951) were engaged as teachers at the *Konservatorium* at various times during their careers. Schönberg delivered lectures on the rudiments of music (*Anfangsgründe*) and harmony (school year 1902–1903) as well as on aesthetics and composition (school year 1911–1912). It was said at the time that Schönberg, who by 1908 had adopted atonality, cultivated no "symbiotic relationship with this very berlinische institution"; that the school represented for him "an unwieldy structure to which he was not able to adapt himself."

During the time of Kaiser William II the institutions of Berlin's official culture, namely the Royal Academy of Arts (*Königliche Akademie der Künste*) and the *Konservatorium für Musik* were accused of being saddled with an "inveterate conservatism" – a conservatism that, at least from Schönberg's point of view, was beyond the bounds, as he inferred, of any (for him) conceivable connections (*denkbare Konnexionen*). In spite of such cross-currents and idiosyncratic opinions,

[1] Dietmar Schenk, "Das Stern'sche Konservatorium der Musik: Privatkonservatorium in Berlin 1850–1915," in *Musical Education in Europe (1770–1914): Compositional, Institutional, and Political Challenges*, ed. Michael Fend and Michel Noiray, *1* (2005) 275–297. See also Dietmar Schenk, Das "Stern'sche Konservatorium der Musik. Ein deutsch-jüdisches Privatkonservatorium der Bürgerkultur Berlins 1850–1936" in *Berlin in Geschichte und Gegenwart. Jahrbuch des Landesarchivs Berlin*, Berlin (2000) 57–79.

Berlin on the whole, especially at the level of its artistic and scientific cultures, was looked upon by knowledgeable outsiders as relatively open-minded to advanced musical trends (*es zeigte sich gegenüber den avancienten musikalischen Strömungen relativ aufgeschlossen*).

It was on the invitation of the theoretical physicist Helmholtz, at the University of Berlin, that Tanaka found a niche for himself at the *Konservatorium* in which he was able to work on the problem of just intonation. His own choice of association with the *Konservatorium* rather than, for example, with Berlin's Royal Academy of Arts was deliberately planned and well-reasoned from the standpoint of a music school that was *au courant* with the novel trends in music being pursued in Vienna and elsewhere. In contrast to the conservatism of Berlin's Royal Academy, the *Konservatorium* prided itself in searching and cultivating all that was on the musical horizon even to the extent of a bold diplomatic tolerance to things in music (*es pflegte eine diplomatische Tolerance in Sachen Musik*) and to modern composers such as Schönberg and Strauss. In fact it was at the *Konservatorium* that so many ill-reputed (*verschrieene*) composers were recognized, where a discrete validation to all matters musical was judiciously and with taciturn enthusiasm kept alive. The abyss (*die Kluft*) between state and society (*zwischen Staat und Gesellschaft*) come into perspective most sharply when one compares the conventionalism of the *Königliche Akademie* with the upcoming modern trends (*aufkommende Kunst-Moderne*) at the *Konservatorium.*[2]

From among music scholars who came to the Konservatorium from faraway places the Japanese music student Shohé Tanaka is of special interest to us. According to Dietmar Schenk, Tanaka chose this particular music school, the *Privatkonservatorium*, for the high level of its academic training program. The formal reasons for the success of the institution were many. Above all, it was during the time of the Emperor (1871–1918), the so-called *Kaiserzeit*, that bourgeois music culture in Berlin reached its high point in vitality. The circumstance that the *Konservatorium* was in the hands of a Jewish family from the start and until 1933 gave rise to a cohesiveness that was both musically fertile and financially secure.[3]

During the years that Tanaka was in Berlin at the K*onservatorium* – we recall that he was there as a student – he had the good fortune to be engaged in the exchange of ideas on musical acoustics not only with other students but with a roster of outstanding teachers and composers. In addition to the *Konservatorium*, where Tanaka was in residence, and the Royal Academy of Arts, there was the *Singakademie*. They both allowed for considerable intellectual interchange with the *Konservatorium* – this in spite of the fact that there were no financial transactions or organizational arrangements between them. As we shall discover in what follows, scholars at all of the Berlin music schools felt free, on occasion, to offer opinions

[2] Dietmar Schenk, "Arnold Schönberg und das Stern'sche Konservatorium der Musik," *Journal of the Arnold Schönberg Center, 3* (2001) 71–84. See in particular 71–75.

[3] Schenk, *Das Stern'sche Konservatorium* (2005) 275–278.

(*Gutachten*) about the merits of Tanaka's theoretical views on intonation and comment on the musical, technical, and pedagogical function and advantages of the enharmonium in comparison with conventional fixed-tone instruments with twelve tone per octave keyboards.

Shohé Tanaka's contributions to the theory of just intonation and his efforts to design a fixed-tone keyboard for teaching music theory are given prominence in this study for a number of reasons. Unlike many of the just intonation keyboard ventures that had been explored by the end of the nineteenth century, Tanaka offered a remarkably convincing and theoretically cogent rationale for designing and constructing a 20-tone per octave keyboard that earned the approval of music theorists at the above-mentioned music schools in Berlin. At the same time Tanaka presented a comprehensive player-oriented account of the construction and mechanical workings of the keyboard and a description of the difficulties encountered in its manual operation. The Tanaka *enharmonium*, as it came to be called, elicited the enthusiastic endorsement of a wide spectrum of music pedagogues, music teachers, conservatory personnel, and composers. For all these reasons, Tanaka's life story, bound up as it is with the career of a young non-European who travelled to Europe in order to discover what he might bring home to his Japanese music students and colleagues, deserves full coverage. Beyond the recognition that here is an enterprising music scholar who produced a meticulously reasoned and historically documented analysis of just intonation theory and instrumentation, what merits special mention is that a young Japanese student of physics and of musicology had been able, in just a few years, to master a large body of German, French, and English literature on just intonation. He had, in fact, managed to worm his way into, and for a time to dominate discussions in Berlin on a topic in theoretical acoustics and the practice of keyboard craftsmanship that turned out to be of great interest not only to musicologists, music teachers, pedagogues, and choral directors, but to music-minded mathematicians and physicists who knew right off the bat that this was territory for investigation in which they would be able to demonstrate their unique skills in physical and mathematical reasoning.

References

Schenk, Dietmar. 2000. Das Stern'sche Konservatorium der Musik. Ein deutsch-jüdisches Privatkonservatorium der Bürgerkultur Berlins 1850–1936. In *Berlin in Geschichte und Gegenwart. Jahrbuch des Landesarchivs Berlin*, 57–79. Berlin: BMV.

Schenk, Dietmar. 2001. Arnold Schönberg und das Stern'sche Konservatorium der Musik. *Journal of the Arnold Schönberg Center* 3: 71–84.

Schenk, Dietmar. 2005a. Das Stern'sche Konservatorium der Musik: Privatkonservatorium in Berlin 1850–1915. In *Musical education in Europe (1770–1914): Compositional, institutional, and political challenges*, vol. 1, ed. Michael Fend and Michel Noiray, 275–297. Berlin: BMV.

Schenk, Dietmar. 2005b. *Das Stern'sche Konservatorium*, 275–278. Berlin: BMV.

Chapter 10
Encounter with the Helmholtz Group in Berlin

In 1890, after 5 years of intensive study with scholars at the *Konservatorium* and with Helmholtz and his associates at the University in Berlin, Tanaka published a monograph in Germany's leading quarterly for musicology entitled *Studien im Gebiete der reinen Stimmung* [Studies in the Field of Pure Intonation]. On the first page of this 90-page study, a work richly supported with historical information alongside theoretical and mathematical reasoning, there is a footnote by the eminent Austrian musicologist Guido Adler (1855–1941), who identifies Tanaka as a non-European who has succeeded in advancing the scientific study of European music.[1] "The author," he writes, "is a Japanese from Awaji who has been in Berlin for 5 years and has studied the natural sciences and music at the University. Dr. Shohé Tanaka most likely is the first of his people not only to familiarize himself thoroughly with European music (*hat sich gründlich nicht nur mit der europäischen Musik vertraut gemacht*), but has sought, with success, to advance its [i.e., European] science in his own research."[2]

The construction and the theory behind the construction of musical instruments designed to sound in just intonation can lay claim to a long and checkered legacy. Its zigzag history is riddled with clever technological inventions that are theoretically reasoned, mathematically computed, mechanically conceived, and punctuated with a liberal dose of ethnomusicological rhetoric shaped to encompass multifarious musics.[3]

---

[1] Guido Adler (1855–1941) was an Austrian musicologist who had studied law at the University in Vienna, and then music theory, which he taught first in Prague (1885) and then in Vienna (1898). Recognized for organizing various musicological enterprises (editions, congresses, institutions) he was best known for his keen thinking about the nature of music studies *per se* and for major contributions in the creation of the discipline of musicology.

[2] Shohé Tanaka, "Studien im Gebiete der reinen Stimmung," *Vierteljahrsschrift für Musikwissenschaft*, 6 (1890) 1–90. Guido Adler at this time was editor of the *Vierteljahrsschrift für Musikwissenschaft*. Tanaka's 90-page paper was made available as an offprint for the price of 3 Marks.

[3] Two monographs that chart much of this history are Wilhelm Dupont, *Geschichte der musikalischen Temperatur*. Nürnberg. Inaugural Diss. 1933, Nördlingen 1935; Patrizio Barbieri, *Enharmonic Instruments and Music 1470–1900*. Based on studies in Italian published 1983–2007, Rome, 2008.

E. Hiebert, *The Helmholtz Legacy in Physiological Acoustics*, Archimedes 39,
DOI 10.1007/978-3-319-06602-8_10, © Springer International Publishing Switzerland 2014

The Helmholtz mid-nineteenth century ventures into just intonation theory and fixed-tone keyboard construction manifestly represent the most conspicuous, most influential, and most frequently cited instance of efforts to demonstrate the principles of just intonation in connection with the musical advantage of instruments that sound in just intonation.[4] However, not many of the Helmholtz or pre-Helmholtz just intonation keyboards made their way into the world of musical performance. Those that managed to do so were welcomed by a .small number of .intonation connoisseurs anxious to demonstrate their mathematical and constructive prowess. Most of them, not surprisingly, were mathematicians and physicists. Only a small number of the keyboards that in one way or another were designed to sound in just intonation are on display in instrument museums.

Without underestimating the historical significance of earlier just intonation keyboard efforts, it is evident that what Helmholtz had accomplished in *Tonempfindungen* along the line of just intonation at mid-century was to give the subject a new supportive point of departure that not only appealed to technical instrument aficionados but to ordinary musicians. In any case, what is evident is that the Helmholtz endeavours captured the attention of a number of prominent music-devoted physicists whose contributions generated a wealth of meaningful musical-theoretic ideas and an array of somewhat heterodox instruments designed to please the discriminating ear. As will become clear, these ventures did not lead to the progress in keyboard instrumentation that had been anticipated. What nevertheless was accomplished by way of new insights and interpretations of music theory and music appreciation is well worth exploring. Toward the end of the nineteenth century, construction of keyboard instruments designed to deliver just intonation intervals, chords, and harmony became a remarkably lively subject of discussion among physicists, mathematicians, and musicians especially in Berlin and the Netherlands. The late nineteenth-century story begins with Shohé Tanaka (1862–1945).

In order to achieve temperaments that correspond to the pitch discrimination limits of the most sensitive human ear, several investigators had suggested a musical scale with as many as 53-tones per octave. Tanaka was of the opinion that this number of tones per octave was much too high for realization on keyboards that are functional. He therefore designed a two-manual selective system of 26 tonal degrees that generated 16 beatless major and minor chords. The manual for the lower c-major scale supplied major thirds. A shortened upper key inserted between e and f supplied e sharp and f sharp. The additional upper keys were divided into 2 or 3 parts so that the rear segment of the key could be shifted by means of a knee lever to the tones of either manual. Rather than having to extend the idea of intonation to its utmost theoretical limits where the ear, in any case, no longer is able to benefit from additional temperament refinements, Tanaka was content to

[4] Hermann Helmholtz, *Die Lehre von den Tonempfindungen als physiologische Grundlage für die Theorie der Musik*, Braunschweig, 1863. French ed. 1868 (2nd 1874). English ed. 1875 (2nd 1877).

achieve a degree of purity in temperament that was musically acceptable and within manual reach of the general keyboard player.

The introductory section to Tanaka's study on the harmonium in just intonation (*reingestimmtes Harmonium*) provides a résumé of the implications, limitations, and intricacies of operation of historically important just intonation systems in relation to their original development and their modification within the Pythagorean tuning system. He demonstrates, with a number of examples, to what ends, and with what degree of success, instrument builders at widely different times in history sought to model and construct harmoniums with more than 12-keys per octave. In this context he summarily discusses instruments with 31, 36, 39, 53, .and 55 keys per octave each with its own type of keyboard mechanism usually employing several manuals arranged to deliver just intonation or, at least, to bring about noticeable improvement to what is possible on a keyboard with more than 12-keys per octave. A number of the models put forward feature guidelines for transposition of the music and methods of musical notation – a bothersome assignment that entails the use of a variety of double flats, double sharps, special subscripts, superscripts, and little wiggles that tell the performer which keys to use in order to get the intended results.

The Pythagorean tuning system constitutes the formal point of departure for Tanaka's 28-tone per octave harmonium – an instrument which, following Guido Adler's suggestion, came to be referred to as the Tanaka enharmonium. As already mentioned and as will become evident in what follows, Tanaka's various enharmonic systems, generously enhanced with graphic and numerical representations, are presented in historical form. In the Pythagorean systems all fifths are pure, the major thirds are large (too flat), the minor thirds are small (too sharp), and the smallest intervals, called commas, are taken to be one-ninth of a whole step.[5] With the basic elements of these early notions about various keyboards in hand, Tanaka proceeded to demonstrate how the musical scale was interpreted and developed at different times in history to accommodate other intervals such as major and minor thirds.

In the course of examining the history of music from the standpoint of how just intonation systems had been built around intervals such as the octave, the fifth, and the fourth, Tanaka identified "something similar, conscious or unconscious, to what takes place in our hearing." He explained his idea of intervallic relationships (*intervallbeziehungen*) as follows:

> For a musician truly equipped with the right feeling for music [*für einem mit richtigen Tonartsgefühl ausgerüsteten Musiker*] the tones in and of themselves have no viable existence but are experienced and put into practice because of their intervallic relationships to certain firm tones. One almost can say that the human sense of hearing is acted upon not by tones but by intervals. To this may be added the circumstance that in just relationships

[5] The term 'comma' in music theory refers to the minute difference in pitch that occurs in the calculation of an interval when the same note is obtained by means of different combinations of octaves, perfect fifths, or pure major thirds. The syntonic comma is the difference between four perfect fifths ($C - G - d - a - e^1$) and two octaves plus a major third ($C - c - c^1 - e^1$)

> the tones gradually unite with each other in extremely small steps, the Kommata. It is here that the intervallic relationships build the only guiding rule in the endless sea of tones; and for this reason alone is it warranted (to say)….that these relationships are put into the foreground because the notes that are divested of their absolute pitch can then be regarded as mere symbols for tonal intervals.[6]

In discussing the enharmonium Tanaka refers to a modified sequence of perfect Pythagorean fifths set one above the other according to the principle of major and minor thirds, as shown in the key network diagram and its restructured keyboard.[7] In this system the symbols /, –, and \ represent, respectively, the pure major third, the pure fifth, and the pure minor third.

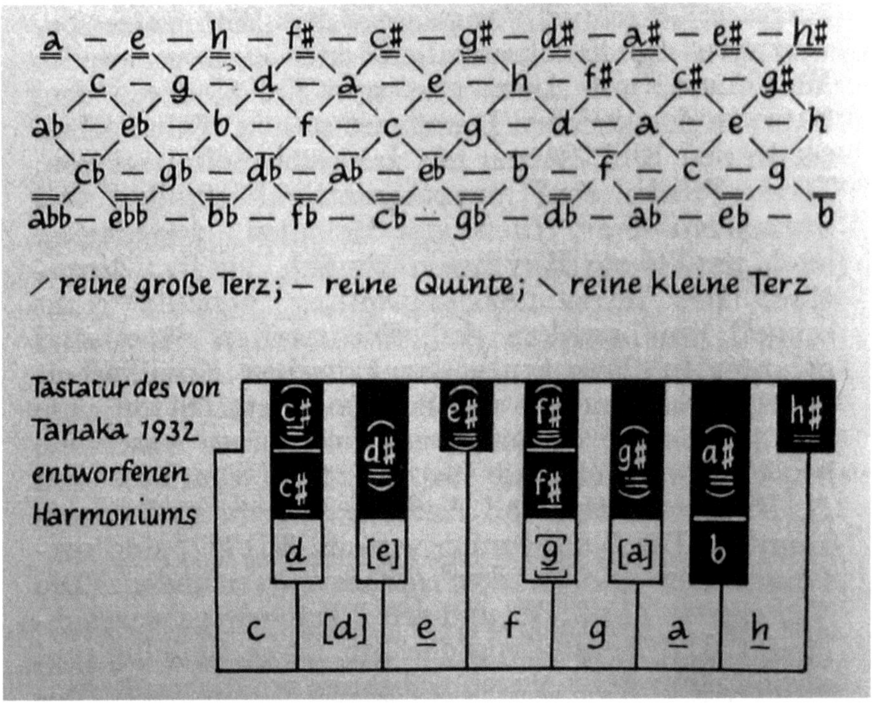

On the upper register, the tones designated ⊃ are raised two commas, e.g. c♯ – >c♯. On the lower register the tones designated [ ] are lowered by two commas and the tones designated ⊐ are lowered by one comma; e.g. [d] – >d̲, – >g̲.

---

[6] Tanaka, "*Studien*" (1890) 22.

[7] It is not feasible within the confines here to present more than a radically circumscribed arrangement of the various keyboard systems that are described in Tanaka's technically formidable 90-page analysis. As a compromise, and in line with other secondary accounts of the Tanaka enharmonium, I have followed Gido Kataoka's abridged composite version of the essential elements and keyboards that entered into Tanaka's depictions of the various just intonation harmoniums. Gido Kataoka, Tanaka, *MGG*, *13* (1966) 79.

The entire keyboard is adjustable to the left and to the right so that key c can generate 12 different tones (c, d♭, d, e♭, e, f, f♯, g, a♭, a, b, h). With this mechanical arrangement an effective instrument of 28-tones per octave is generated.

Although Tanaka's concerns with just intonation and harmoniums originally were expressed primarily within a theoretical framework, his ongoing activities and writings demonstrate that he became thoroughly engrossed in the practical aspects of keyboard construction. He was fully aware of the likelihood that the construction of keyboards more complex than the 20-tone per octave enharmonium that he was engaged in promoting might be called for to meet more specific needs than he envisioned. He shrewdly and correctly sensed that this would not come about without radical innovation on the part of the instrument-building crafts. Mindful of the idea that the 28-tone per octave enharmonium that he had invented accomplished limited goals, Tanaka left open the direction of future technological and auditory improvements and refinements. To this end he wrote: "It is true that there have been enormous advances in instrumental music in regard to temperament. And yet one should not be allowed to forget that this system, although convenient, conforms to the laws of nature only in an imperfect way and that the achievement of a more perfect approximation is not be excluded in the advancing evolution of mankind."[8]

References

Helmholtz, Hermann. 1863. *Die Lehre von den Tonempfindungen als physiologische Grundlage für die Theorie der Musik.* Braunschweig: F. Vieweg.

Tanaka, Shohé. 1890. Studien im Gebiete der reinen Stimmung. *Vierteljahrsschrift für Musikwissenschaft* 6: 1–90.

[8] Tanaka, Studien (1890) 90.

Chapter 11
The Papendick Sammlung

In 1891, a year after Tanaka's major treatise on pure intonation (*Studien*) had been published, he conceived of the idea of having G. A. Papendick, a piano instructor at the *Stern Konservatorium für Musik* in Berlin, where Tanaka was in residence, to edit and publish a small *Collection of Easy Pieces* to be performed on the Tanaka just harmonium, referred to in this study simply as *Sammlung*.[1]

Among the handful of musicians of the *Konservatorium* who had had the opportunity to become acquainted with the Tanaka enharmonium and had played it on a continuing basis for a year or more Papendick stands out prominently. A native of Dresden, Gustav Papendick had a wide-based reputation as a proficient keyboard player. He had demonstrated the Tanaka enharmonium at music societies and associations such as the *Verein der Berliner Musik-Lehrer und Lehrerinnen*.[2] It was said of Papendick, in regard to the Tanaka enharmonium, that he had familiarized himself "theoretically in its secrets and by practice had mastered its functioning and

[1] *Enharmonium. Sammlung kleiner Vortragsstücke für das reingestimmte Harmonium nach dem System des Dr. S. Tanaka herausgegeben von G. A. Papendick.* Breitkopf & Härtel, Leipzing, Brussels, London (1891). Reference is made in this essay to two Tanaka-related documents that are kept in the archives of the *Unter den Linden* branch of the *Staatsbibliothek* in Berlin (*Haus* 1). Thanks to the librarian of the music division of the *Staatsbibliothek*, a *Duplikatfilm* of these documents was made for the author while he was engaged in research at the *Max Planck Institut für Wissenschaftsgeschichte* in Berlin in 2005 and 2007. The documents are, *first*, a brochure of some 25-pages dated 1891 that includes G. A. Papendick's *Sammlung: A Collection of easy pieces for Tanaka's just Harmonium* and an accompanying four-page essay by Tanaka, *A short Explanation of the Method of Playing on Tanaka's Enharmonium*, and *second*, a brochure dated 1892 that includes an essay of five-pages, *Das Enharmonium* by Tanaka, and an addendum of ten pages of "Expert Opinions" (*Gutachten*) by music scholars from music institutions in Berlin, Leipzig, and Vienna.

[2] For information concerning Papendick's various formal opportunities to demonstrate the unique characteristics and potential capabilities of the Tanaka enharmonium: the journal *Der Klavier-Lehrer. Musik-paedagogische Zeitschrift. Organ der Deutschen Musiklehrer Vereine*, Berlin, 1880s and 1890s.

E. Hiebert, *The Helmholtz Legacy in Physiological Acoustics*, Archimedes 39, DOI 10.1007/978-3-319-06602-8_11, © Springer International Publishing Switzerland 2014

difficulties. He also had adapted and published compositions for the enharmonium. His knowledge and familiarity with the instrument were widely acclaimed. He was convinced that the sensations of the intervals were more securely installed in man by nature than any absolute tonal values."[3]

The Papendick collection of music pieces was published with an essay of four-pages, in English and German, by Tanaka entitled: *A Short Explanation of the Method of Playing on Tanaka's Enharmonium.* Taken together, the collection of piano pieces and the Tanaka essay had been designed to encourage musicians to become acquainted with the enharmonium and discover what it takes by way of finger dexterity and practice to play some of the classic piano pieces on the enharmonium.[4] It served the purpose not only of providing a brief statement of the enharmonium's structure and functioning but of introducing the instrument to musicians in search of an alternative to the standard 12-tone per octave fixed-keyboard instrument.

The Tanaka Essay of 1891 that is printed along with the *Sammlung* is prefaced with the remark that although the conventional harmonium, the reed organ, had gained a high degree of mechanical perfection by the end of the nineteenth century, it was receiving increasingly little attention from the scholarly music world. The principal reason for this neglect, as the author indicated, was that the continuous tones of its metallic reeds deliver "most unpleasantly the imperfections of the tempered scale" so that "all chords and intervals on it are invariably accompanied by beats or vibrations of mistuned consonances which interfere with the clearness and smoothness of harmony, and awake in us a perpetual nervous excitement, unbearable for delicate and musically trained ears." By contrast, it is claimed: "The Enharmonium is an instrument of the same species, on which, however, the chords and intervals are given – entirely freed from the disagreeable effects of artificial temperament – in the natural purity and harmonic consistency that is attainable only in the most exquisite performances of unaccompanied vocal and string music."[5]

Eighteen short and relatively elementary musical compositions incorporated in the Papendick *Sammlung,* all transposed into the key of C major, were presented for practice on the enharmonium and include works by Rob. Schumann, J. S. Bach, G. F. Händel, W. A. Mozart, Ludwig Beethoven, Mendelssohn-Bartholdy, Giov. Palestrina, and G. A. Papendick, among others. It is evident that all of the compositions had been chosen to demonstrate what lies within the compass of the enharmonium's fingerboard capabilities.

[3] Ph. J. Trayser & Cie, Harmonium-Fabrik in Stuttgart, July 1892, fn 16.

[4] G. A. Papendick, Editor, *Enharmonium, Sammlung kleiner Vortragsstücke für das reingestimmte Harmonium, nach dem System des Dr. S. Tanaka*, herausgegeben G. A. Papendick, Heft 1, Breitkopf & Härtel, Leipzig, Brussels, London, 1891.

[5] G. A. Papendick, *Enharmonium, Sammlung*.... Shohé Tanaka, *A Short Explanation*..., Leipzig (1893), iii.

Reference

Papendick, G.A. (ed.). 1891. *Enharmonium, Sammlung kleiner Vortragsstücke für das reingestimmte Harmonium, nach dem System des Dr. S. Tanaka*, herausgegeben G. A. Papendick, Heft 1. Leipzig/Brussels/London: Breitkopf & Härtel.

Chapter 12
The Enharmonium

In July 1892 the Philip J. Trayser & Cie Harmonium Factory in Stuttgart printed a brochure announcing that five-octave and four-octave enharmoniums were available for purchase. These harmoniums were built on the model of the pure-tempered harmoniums, called the enharmonium, invented by Shohé Tanaka in 1890.[1] The brochure that accompanied the P. J. Trayser announcement of the availability of enharmoniums on the market included an essay of five-pages on the enharmonium by Tanaka (*Das Enharmonium*) and ten-pages of "Expert Opinions" (*Gutachten*) on the merits of the enharmonium as envisioned by teachers of music theory and composition, music professor, conductors, music pedagogues, and piano teachers.[2] The five-page synopsis by Tanaka describes the characteristic features and functioning of the instrument but seeks mainly to convince the potential player and buyer that the enharmonium is technically and manually manageable for the ordinary keyboard player and is becoming an important new type of keyboard in the life of pianist and music teacher. As to the merits of the new keyboard from the standpoint of music theory, and the historical circumstance that led to its invention and construction, we turn to Tanaka's own remarks:

> Although the theory of pure harmony [*reine Harmonie*] actually has continued to be the subject of exhaustive investigation from antiquity to the present, oddly enough, until now no artistically usable instrument has been constructed that embodies these scientific and, as it were, cultural results. The reason for this remarkable situation is to be sought mainly in the circumstance that the rigorous compliance with natural purity of intervals [*Naturreinheit der Intervalle*] that the practice of music requires is a much greater number

[1] The company had been founded in Stuttgart in 1847 after its owner Philip J. Trayser had learned the business at the Alexandre company in Paris. The company was one of several that built enharmoniums on the basks of experimental investigations and concurrently designed keyboard systems of persons such as Helmholtz (foremost in its historical and outreach importance), Gustav Engel, Shohé Tanaka, Carl Eitz, Arthur von Oettingen, and others. *MGG*, Band 5 (1956).

[2] The larger instrument was playable in 12 transposition keys with two knee-operated levers for enharmonic change and crescendo and was available for the price of 1,050 Marks. The smaller instrument was playable in eight transposition keys and one knee-operated lever for crescendo at the price of 725 Marks.

E. Hiebert, *The Helmholtz Legacy in Physiological Acoustics*, Archimedes 39,
DOI 10.1007/978-3-319-06602-8_12, © Springer International Publishing Switzerland 2014

> of tones and therefore a much more refined division of the octave than we are in the habit of seeing in the twelve white and black keys of the piano and the organ. In order to avoid the great difficulties of operation of such instruments the intervals have been tempered by systematic and symmetrically falsified equalizations [*gleichmässige Fälschungen*] so that a single note fulfills the service or purpose of various neighboring tones. It is an expedient [*Notbehelf*] that seems to ease the playing at the cost of purity, and unfortunately also of the way of thinking [*Denkart*].
>
> In the now prevailing twelve-step temperament the octave is divided by means of practical approximation into twelve equal parts. The fifths and fourths are tolerably close to purity, while the thirds and sixths deviate susceptibly from it – with the consequence that the consonant intervals and chords, if not sounding false, largely forfeit their clarity. For the tender and finely built ear [*das zarte und finegebildete Ohr*] they always are accompanied by disagreeable, insulting beats and other equally troublesome side effects. On the transitory, and therefore deceptive, sounds of the piano, this deficiency in temperament almost is bearable; yet in the organ or harmonium the euphony [*Wohlklang*] is thereby so impaired that there can be no thought of enjoying the harmony.[3]

Against the background of these comments concerning the reasons why the conventional 12-key per octave keyboard cannot, on account of its construction, deliver "pure harmony," Tanaka proceeded to explain what actually can be achieved in the desired direction on his own acoustically designed enharmonium. He was convinced that in order to approximate the tonal purity that normally can be achieved only by unaccompanied singing groups, string ensembles, and to some extent the brasses – provided the performing artists are able by inner fantasy [*innere Phantasie*] to conceive the enharmonic tones – the enharmonium would be the right instrument to fill the lacunae that historically have plagued keyboard instruments when used either as an instrument for solo performance or as an accompanying instrument. Tanaka remarked:

> In order to achieve unrestricted modulatory motion in all keys, a division of the octave into 53 equal steps fulfills the practical need of music theory in relation to purity for fifths and thirds. This extreme increase in tones, that already has been realized on a number of tentative instruments such as the one by Bosanquet, understandably brings with it almost insurmountable performance difficulties. Even apart from quibbling theoretical investigations, such a number of tones is not necessary. For in order to reach the stated goal, that corresponds to representing all major and minor keys and their accompanying modulations in the most complicated harmonic combinations of the music of today, it suffices to have only 36 divisions to the octave.
>
> Nonetheless, apart from this restriction on tonal materials from a practical point of view (reducing the divisions of the octave from 53 to 36), it is no minor feat [*kein geringes Kunststück*] to perform something like a Bach fugue or a Rheinberger sonata on a keyboard, even if one provisionally, for mitigating circumstances, distances oneself from the technically difficult piano- and orchestra pieces.[4]

To realize an instrument that meets the designated requirements of acceptable intonation and provides technical maneuverability for the keyboard player, Tanaka

[3] Shohé Tanaka, *Das Enharmonium*, (1892) 3–4.

[4] Tanaka, *Das Enharmonium* (1892) 4–6.

constructed an enharmonium whose keyboard, as depicted below, combines mechanical transposition and an extension of the number of keys per octave from 12 to 20.

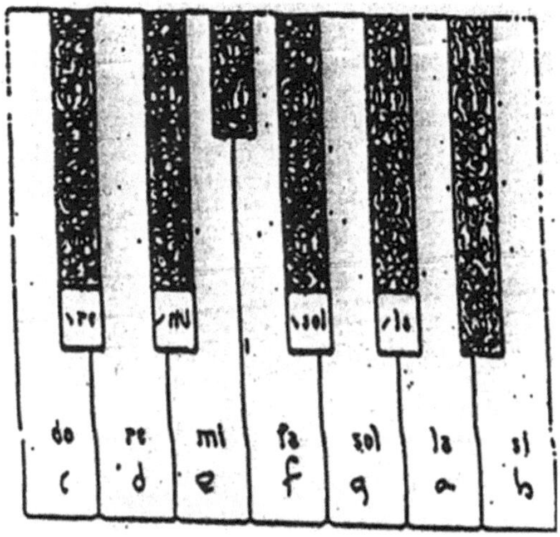

The lower white keys are able to maintain their usual form and dimension by inserting a small key between e and f. By lateral shifting of the entire keyboard and fastening devices at 12 places on a visible scale, it is possible, by virtue of the instrument's inner construction, and by means of the seven white lower keys, to obtain acoustic purity for the main diatonic major scales in advancing fifths from d sharp to f sharp.

Reference

Tanaka, Shohé. 1892. *Das Enharmonium*, 3-4

Chapter 13
Expert Opinions. Evaluating the Enharmonium

Toward the end of the nineteenth century, and several years before the Ph. J. Trayser harmonium factory in Stuttgart was in a position to place enharmoniums on the market, there were musicians and persons associated with the music profession who had had access to an enharmonium and were knowledgeable about the instrument, its construction, and its functioning. The knowledgeable group, as we shall see, included teachers of music theory and composition, conductors, conservatory directors, choir directors, and piano teachers. A number of them were enticed in one way or another to offer an expert opinion (*Gutachten*) regarding the merits of the enharmonium from the perspective of their own profession or specialty. The appraisals we will examine provide insights into the theoretical and practical importance of the subject of just intonation. They reveal the manner in which different persons had become familiar with the enharmonium, had the opportunity to play the instrument, or perhaps had participated in carrying out experimental investigations on the enharmonium at the *Konservatorium* where Tanaka was in residence. In general the *Gutachten* express the view that the Tanaka enharmonium was essential as a teaching instrument and timely for the music practice room. Despite the likelihood that a number of the so-called "experts" may have been solicited to write a *Gutachten* with the intent of getting Tanaka's enharmonium on the market and into teaching institutions, they nevertheless as a whole represent a stunning image of how a group of highly respected persons connected with the music profession and with various music institutions envisioned the extent to which the enharmonium, with its focus on intonation, was being evaluated for the teaching of music theory.

Franz Schulz, professor of music theory and composition at the *Königliche Academische Hochschule für Musik* in Berlin, observed that the "mathematical harmonium" was a major contribution to music theory and the music instrument crafts, and had been invented by a person at the *Konservatorium für Musik* whose musical tradition and artistic outlook were essentially non-Western. We recognize that the *Hochschule für Musik* discernibly was an institution that if not openly in competition with the other music schools in Berlin nevertheless would have valued having a non-Western musician of Tanaka's caliber and perspective at its institution. Schultz's generous evaluation assuredly represented a trusting rapprochement between other

E. Hiebert, *The Helmholtz Legacy in Physiological Acoustics*, Archimedes 39,
DOI 10.1007/978-3-319-06602-8_13, © Springer International Publishing Switzerland 2014

Berlin music schools. Of special note, and coming from one who was "in the know" about the technical functioning of the enharmonium, was Schulz's comment about the latest enharmonium's improved mechanical features: suitable new wind conveyors and reeds easily and accurately tuned. Recognizing at this preliminary stage that Tanaka's enharmonium could claim no more than being on the right track for the music of the future, Franz Schulz remarked: "For the purpose of teaching and instruction, the instrument in its present state already provides an excellent and indispensable aid. Its further completion will contribute to the understanding not only of the laws of harmony and intonation, but also to the clarification, strengthening, and generalization of the artistic authorization to deviate from these laws [*künstlerische Berechtigung der Abweichung von diesen Gesetzen*]." Deviation, in this context, manifestly was meant to refer to deviation from the laws established for equal temperament – thus to bolster the musical superiority of Tanaka's enharmonium and his views on the instrument as a vehicle for music theory and instruction.[1]

Schulz's colleague at the Hochschule, Joseph Joachim (1831–1907), the eminent Austrian-Hungarian violinist, string teacher, professor, and orchestra conductor endorsed Schulz's evaluation and remarked: "For institutional purposes the instrument is useful and worthy of recommendation in the highest degree."[2]

Philipp Spitta (1841–1894), professor of music at the University of Berlin and secretary of the *Academie der Künste* wrote that he personally had examined Tanaka's enharmonium on many occasions and had become convinced that it would be of great value for practical instruction. Coming from one of the most prestigious members of the Berlin community of musicians, Spitta's comments amounted to a strong endorsement of Tanaka as a creative music theorist.[3]

The director of the Berlin *Singakademie* and chairman of the music section of the *Academie der Künste* Martin Blumner (1827–1901) remarked that in comparison with all previous attempts in this area of musical acoustics, the Tanaka enharmonium represented significant progress in the theory and practice of tonal relationships.[4]

[1] Franz Schulz, *Gutachten* (1889) 8.

[2] Joseph Joachim, *Gutachten* (1889) 8.

[3] Philipp Spitta, *Gutachten* (1889), 9. Spitta was the leading figure in musicology during the discipline's late nineteenth-century foundational phase. He was professor of music history at the University of Berlin and administrative director of the *Hochschule für Musik*. Educator for an entire generation of music scholars, he had co-founded the *Vierteljahrsschrift für Musikwissenschaft*, the journal in which Tanaka's article on *Studien im Gebiete der reinen Stimmung* was published. Among performing musicians Spitta is recognized for his epoch-making studies on Bach (1873–1880).

[4] Blumner, *Gutachten* (1889) 9. Martin Blumner (1827–1901), principal conductor at the *Sing-akademie* from 1853–1876, received an honorary doctorate from Berlin University in 1891, the year he published his *Geschichte der Sing-akademie zu Berlin*. Blumner belonged to the "Berlin academics" – a group of composers committed to the ideal of Romantic-historical a cappella singing. He wrote exclusively for vocal music. [*NG, 1* (2001) 741–742.] A comprehensive history of the *Sing-akademie* has been re-edited by Gottfried Eberle and Michael Rautenberg. *Die Sing-akademie zu Berlin und ihre Direktoren*, Berlin, 1998. The work was published under the auspices of *Staatliches Institut für Musikforschung, Preussischer Kulturbesitz*.

The "expert opinions," the *Gutachten*, submitted by leading music scholars from Berlin's three music schools support the conjecture that Tanaka's enharmonium and the theoretical and practical craftsmanship embodied in its invention was an immense accomplishment. At minimum, Tanaka had fashioned an instrument that was eminently worthy of being incorporated into the training of music students. It therefore was proposed that teachers, music organizations, and music schools pinpoint the enharmonium as a teaching keyboard and take advantage of it in their training programmes.

From the time Tanaka had arrived in Berlin he had benefited from his residency at the *Konservatorium.* He had been the fortuitous recipient of financial aid from the Japanese government back home. He had taken advantage of the collegial and spiritual mentorship of Helmholtz and his physics colleagues at the University of Berlin and by 1890 had put his ideas on pure intonation in print. The following year that work, the *Studien*, brought him the doctorate in music at the University of Berlin.

Summarily, Tanaka's colleagues and compatriots had witnessed how a Japanese physicist from Tokyo had shifted his focus from non-Western to Western music. He was not yet 30 years of age at the time of his *Studien.* He had enticed Berlin's physicists to become engaged in a new branch of musical acoustics, and, as Hans von Bülow noted in a footnote on the first page of the *Studien*, he not only had been the first of his own people [*der erste seines Volkes*] to have thoroughly familiarized himself with European music, but successfully had advanced European science with his own research. Tanaka's experimental investigations and writings on music theory, correlated as they were with the construction of an enharmonium, captured the attention and the admiration of many of his music superiors – predominantly teachers, pedagogues, conductors of vocal and choral music, composers, and music publishers.

In spite of all these positive appraisals there were those who felt that the *Gutachten* had not adequately recognized the significance and extent of Tanaka's contributions to musical acoustics. Hans von Bülow, no Berliner, was one of them. As if to infer that the Berlin coterie of *Kleinmeister* who had provided the *Gutachten* had not adequately extolled Tanaka's accomplishments, Bülow, the emerging Wagnerian of his generation, took his chance to have a say about Tanaka's contributions in the *Hamburgische Musik-Zeitung*. By this time Bülow's own keyboard performances had brought him the accolade of being the "passionate intellectual."

In anticipation of the publication of Tanaka's 1890 study on *reine Stimmung* (which he apparently had seen in its pre-publication form), Bülow offered the following comments to the press:

> For men destined to greatness, representatives of the natural sciences such as the illustrious Helmholtz, it would be incumbent, in first place, and in mathematical and physical terms, to verify and to offer an opinion concerning the impressive and ingenious discovery of a transposing- and modulating-harmonium-type keyboard instrument that I herewith allow myself to give the misunderstood name *Enharmonium.* In its structural refinement and mechanical simplicity this instrument surpasses all earlier experimental efforts. My lay

> consciousness forbids me to have a say in such an Areopage. On the other hand, as an older and not totally inexperienced practicing musician I consider myself competent and entitled to offer my guildsmen [*Zunftgenossen*], and also all earnest lovers of the fine arts, a vote for the worth and practical value of this discovery.[5]

To underscore his enthusiasm for the Tanaka instrument, von Bülow mentioned that the instrument builder Johannes Kewitsch in Berlin would be supplying him with an enharmonium for his own domestic use (*Hausgebrauch*).[6]

In order to be able to "make music that is pure [*rein musiciren*]," von Bülow emphasized that "it is necessary to think pure tones on the basis of pure sensation; that is, on the basis of pure hearing." This means that "the *sensus* needs to precede the *intellectus*." Bülow suggested that not only almost all auditory corruption (*Gehörcorruption*) can be traced to ignoring the syntonic comma but that most piano players, and that includes almost all musicians, have allowed themselves to be seduced and intoxicated by an imaginary piano sound, a piano-lie. "This piano lie [*Clavierlüge*] may also be regarded as the nurturing mother of all the innumerable homo- and poly-cacophonic monstrosities that continue to be hatched out in the receptive but immature brains of a German, a French, or, worst of all, a misunderstood [*misverstandener*] Russian Wagner or Berlioz."

In his concluding appraisal of the Tanaka enharmonium and its role in the revival of just intonation as a technically achievable goal – an appraisal typically loaded with suggestive and hidden Bülowian innuendo – von Bülow exploited the theme that it took an Asian, a non-European, to rescue Germany from being swallowed up in Western-style equal temperament.

> And so I bid a lordly welcome to the ingenious Asian who is able to boast that he is allowed to walk in the footsteps of the sublime religious reformer, moral philosopher, and teacher of his people who died two thousand and several hundred years ago. We Europeans may be permitted, without a struggle or ingratitude, to accept the good deeds of a worthy son from a civilized country of the world that is called the cradle of mankind. Through him the sacred Confucius can directly become for us a helper by profane Contorsion, Convulsion and Confu-sion! Amen! [*Der heilige Confucius werde durch ihn uns mittelbar zum Mithelfer aus unheiliger Contorsion, Convulsion und Confu–sion!*][7]

These flamboyant endorsements (were they really genuine?) for the young Tanaka and his enharmonium, and their importance for the development of refined musicianship, were written by a commanding artist who was known for his

[5] Hans von Bülow, "Das Enharmonium des Herrn Doctor Shohé Tanaka," *Hamburgische Musik-Zeitung. Organ für die musikalische Welt, III*, (29 Sept. 1889–1890) 155–156. The Areopagus is the oldest and most sacred meeting place of the prime council of Athens. Von Bülow, *Gutachten* (1889) 9–10.

[6] Johannes Kewitsch was a fixed-tone keyboard builder in Berlin whose firm had been called upon numerous times to build enharmoniums for special clients. The Kewitsch firm was established in 1836 and produced pianos and harmoniums; and after 1890, enharmoniums. See the discussion above for Tanaka's demonstration of the enharmonium's capabilities on the request of their Imperial Majesties.

[7] Hans von Bülow, Das Enharmonium, *Hamburgische Musik-Zeitung* (29 Sept., 1889–1890), 156. Von Bülow, *Gutachten* (1889) 9–10.

expressive abilities at the piano (Liszt had accepted him as a pupil at Weimar in 1851) and his acute intellectual power of interpretation. He also was known among his colleagues as a music critic who both impressed and offended his readers by offering conductors and other musicians insults and too much advice. According to one of his biographers von Bülow was quarrelsome, passionate, and given over to extremes of mood. "Yet Richard Strauss had the highest regard for his intellect, his analysis of phrasing, and his grasp of the psychological content of the music of Beethoven and Wagner."[8]

Hans von Bülow (1830–1894) occupied a commanding position in the life of music culture in Germany especially in Berlin where he was head of the piano department at the *Konservatorium für Musik*. His illustrious, productive, and stormy career merits comment. In a recent and definitive biography Alan Walker wrote concerning Bülow:

> He strode across the world of nineteenth-century music like a colossus. Bülow was music's great reformer. He set out to make a difference. His career... .was epoch-making, and it unfolded in at least six directions simultaneously. He was a renowned concert pianist; a virtuoso orchestra conductor; a respected (and sometimes feared) teacher; an influential editor of works by Bach, Mendelssohn, Chopin, and above all of Beethoven, in the performance of whose music he had no rival; a scourge as a music critic whose articles resembled the spraying of antiseptic on bacteria.

A confirmed Wagnerian, Bülow premiered *Tristan und Isolde* and *Die Meistersinger von Nürnberg*. The two musicians interacted at various points in their careers, an interaction culminating in Bülow's marriage to Liszt's daughter Cosima, who later left Bülow for Wagner.

On the podium Bülow was referred to as wielding "a singing baton." His memory was infallible; he conducted and played everything from memory. "He understood that a good programme, like a good piece of music, is greater than the sum of its parts, that each piece could and should throw light on the others.... Bülow's concerts became memorable for their musical logic. The individual pieces were like members of one family, bound together by biography, history, genre, and sometimes by their key-schemes – golden threads that ran from one composition to the next."[9]

No discussion on the history of just intonation theory and practice is conceivable without a consideration of its central relevance to vocal music. It therefore comes as no surprise to discover that a nimbus of the most enthusiastic endorsements for Tanaka's theoretical and practical efforts in Berlin and elsewhere in Europe came from the musical community most closely allied with vocal music and the personnel associated with the *Sing-Akademie*. By the 1890s, choir organists, a cappella choir directors, vocal teachers, and composers of vocal music had become so convinced of the significance of the enharmonium in their professional work as to

[8] Christopher Fifield, "Hans von Bülow," *NG*, *4* (2000) 599–601.

[9] Alan Walker, *Hans von Bülow. A Life and Times*, Oxford, 2010, pp. 5 and 10.

seek support from the Prussian ministry of culture for financial aid to commission and install Tanaka enharmoniums in their own respective institutions.[10]

In an endorsement of 1890 from Karl A. Haupt (1810–1891), director of the *Institute of Church Music in Berlin* it was argued that Tanaka's accomplishments with the enharmonium were not so much that he had furthered the virtuosity of keyboard playing or even enhanced it, but that the enharmonium had given support to a cappella singing in pure intonation and in particular had accentuated the playing of classical music and deepened its enjoyment. An organist famous for his improvisations on Bach, Haupt's evaluative statement was co-sponsored by the esteemed Anton Bruckner in Vienna. Not content to restrict the realization of pure intonation to the enharmonium, Haupt maintained that it was feasible to transfer the idea of pure intonation to the organ, using, as Tanaka apparently had suggested, a special manual with 16., 8., and 4. pedal tones. Haupt added in conclusion: "May the young talented scholar find the necessary interest and support to peacefully and in an unhindered way promote the work that anticipates further development."[11]

In a footnote to the Haupt *Gutachten*, it is mentioned by publisher Breitkopf & Härtel that Haupt was a highly recognized advocate of classical organ playing and that his idea of constructing an organ on the model of Tanaka's pure intonation enharmonium had been conveyed to the *Royal Prussian Ministry of Culture* and handed over to Tanaka for construction. According to Haupt:

> [The organ] currently is being built in the time-tested *Atelier* of Walcker & Co. in Ludwigsburg [Württemberg] under the personal supervision of the inventor [*Erfinder*] The organ's mechanical completion was made possible only with the aid of an electro-magnetic traction device [*Traktur*] and when completed in Berlin will be made accessible to the interested public.[12]

A uniquely oblique twist to the status of the keyboard as an accompanying instrument for vocal music was given by Heinrich von Herzogenberg (1843–1900), composer, would-be friend of Brahms, and co-founder with Phillip Spitta of the Bach Society. Herzogenberg wrote church and chamber music in the spirit of Bach and Schütz. As professor of music at the master school for composers in Berlin he encouraged the teaching, performance, and composition of music at the level ordinarily achieved by singing voices, string instruments, and, in a limited sense, brass instruments. Herzogenberg recommended that his students and colleagues study Tanaka's theoretical treatise on intonation as a correction to the widespread dominance of equal temperament. "In my opinion," he remarked, "the overestimation of equal temperament derives from the overestimation of the

---

[10] See e.g. the statements of A. Haupt (Director, Institute of Church Music, Berlin), Anton Bruckner (Composer, Vienna), J. Hellmesberger (Director Vienna Conservatory), Oscar Paul (Prof. Harmony, Leipzig), Carl Reinecke (Conservatory and Director of Gewandhaus concerts in Leipzig). Trayser & Cie., *Gutachten, loc. cit.*, 11–13.

[11] Haupt, Bruckner, *Gutachten* (1890) 11.

[12] Trayser & Cie., 11. E. F. Walcker & So., a family firm founded by Eberhard Friedrich Walcker (1794–1872) constructed harmoniums from 1838 to 1914; after the 1890s Walcker built an enharmonic pipe organ for Tanaka.

piano (as a musical instrument). Harmonists [*Harmoniker*] who base their theory on equal temperament have neither thought beyond nor listened beyond the keys of the piano [*haben über die Tasten des Claviers weder hinausgedacht noch hinausgehört*]. An especially constructed harmonium for teaching purposes, based on the principle of the Tanaka enharmonium, has become a serious necessity for many. I may express the hope that it will not take too long to wait for."[13]

As if to suggest that the musical world was anxiously waiting for an interim resolution to the intonation problem, while Tanaka-type harmoniums still were in process of being improved, Herzogenberg suggested that the construction of double-manualled organs and harmoniums would allow one manual to be used for music in equal temperament and another as accompanying instrument for vocal and string ensembles.

> But this is a glance into the future, so let us return to the present and not neglect to render thanks for what has been achieved and technically solved so brilliantly. Japan herself must decide how much to thank us for the fruits of the tree inoculated with European culture. Here we experience, but hopefully not for the last time, the transfusion of Japanese intellectual sagacity and technical dexterity [*japanischer Velrstandesschärfe und technischen Geschikes*] into the arteries of an eminently Occidental question.[14]

G. A. Papendick, whose musical activities already have been discussed in connection with the *Sammlung* of easy compositions for the enharmonium, was a piano teacher at the Stern Conservatorium in Berlin. He devoted an entire year to practicing and performing on the Tanaka enharmonium, writing music for it, and making an effort to become thoroughly knowledgeable with its characteristic features and practical difficulties. He was convinced that the instrument represented "the good and fortunate but not entirely final result of mathematical-physical calculations." Papendick's year-long preoccupation playing the enharmonium, as well as his efforts to master the intricacies and the difficulties of the instrument's functioning, signified to knowledgeable musicians that he was the right person to be called on to demonstrate the instrument's potential on special occasions as they arose. That occasion presented itself for Papendick almost immediately after it had been proposed In April 1890 the Berlin daily newspaper, the *Norddeutsche Allgemeine Zeitung*, carried the news that the Japanese physicist Shohé Tanaka had exhibited his enharmonium to an audience of high-level government dignitaries assembled by the German Emperor and Empress. In a lecture that followed, and that was delivered in fluent German, Tanaka spoke about and demonstrated the mechanical operation and transposition facilities on his instrument and explained the basic principles of the pure tonal system that underlie the differences between the conventional keyboard instrument and the enharmonium's unique construction. According to the newspaper account:

> In the musical performance of Händel's Prayer, the Beethoven Serenade for String-Trio, and the Finale from Mendelssohn's last organ sonata, Mr. G. A. Papendick, who was in the

[13] von Herzogenberg, *Gutachten* (1891) 14.

[14] von Herzogenberg, *Gutachten* (1891) 14.

> audience, had the honor to participate in the session and exhibit the artistic advantages of the Tanaka enharmonium so effectively that their Imperial Majesties expressed complete satisfaction. The Emperor's deep interest in the new discovery was made manifest in that he enticed Johannes Kewitsch, builder of the enharmonium, to completely disassemble the instrument in order to thoroughly explain its inner workings. The Emperor expressed the wish that the same procedure might be put in practice with a larger church organ. In conclusion the Emperor congratulated Dr. Tanaka in the most benevolent way for the recognition he had encountered among the musical authorities.[15]

It's noteworthy to acknowledge that the Emperor, who became so enamored with Tanaka's enharmonium, was a person who had endeavored to maintain and extend the royal prerogative to make Germany a major naval, colonial, and commercial power. Not as liberal-minded as his parents, William was noted for his fondness of military displays – a characteristic entirely consistent with his remarkable interest in Tanaka's "new discovery" as he termed it. Moreover, the fact that the discovery was by a foreigner, who had taken on the German ethos that led him to publish his most important ideas on music theory in German, and that he spoke the language fluently, greatly impressed the Emperor.

References

Fifield, Christopher. 2000. Hans von Bülow. *NG* 4: 599–601.

von Bülow, Hans. 1889–1890. Das Enharmonium des Herrn Doctor Shohé Tanaka. *Hamburgische Musik-Zeitung. Organ für die musikalische Welt* III: 155–156.

Walker, Alan. 2010. *Hans von Bülow. A life and times*, 5–10. Oxford: Oxford University Press.

[15] "Aus Berlin," *Norddeutsche Allgemeine Zeitung*, Abend-Ausgabe, *29*, Nr. 178, (17 April 1890), 2. *Gutachten* 17. The Emperor mentioned in the above was William II, Emperor and King of Prussia from 1888–1918. He was the last of the German Emperors and was known in part as a person whose constant theme was German imperialism.

Chapter 14
With Bruckner in Vienna

Following the years in which Tanaka was engaged in investigating problems in just intonation in collaboration with music students and teachers at the *Konservatorium* and in consultation with Helmholtz and his physics colleagues at the University in Berlin, he was able to extend his study in Europe by establishing himself as a delegate at the Japanese Embassy in Vienna (1893–1899). From among the various music centers in Europe, Vienna at this time provided Tanaka with by far the best opportunity to extend the music-theoretic studies he had cultivated in Berlin. He originally had, in fact, chosen to study in Berlin not on the basis of its reputation as a paramount center for music but because he wanted to study and work with the famous Helmholtz. As it turned out he was able in Vienna to extend his intonation studies and explore the field of electromagnetic studies (not examined in this paper) for which he originally had in part received the grant from the Japanese Ministry of Education.

By the turn of the century Vienna had become the recognized metropolis for the new trends in music and culture that were to hold a grip on Europe – a grip that she was destined to maintain well into the twentieth century.[1]

[1] The main institution for the promotion of the new music in Vienna was the *Gesellschaft der Musikfreunde* founded in 1870. Its conservatory, the *Hochschule für Musik und Darstellende Kunst*, became Vienna's main music school. In addition there was the *Verein für Musikalische Privataufführungen* (1919–1921) founded by the Austro-Hungarian composer Arnold Schönberg (1874–1951) with the intention not only of narrowing the gap between composers involved in the new music trends and the public, but by engaging his own students in the presentation of new music under the most favorable conditions.

E. Hiebert, *The Helmholtz Legacy in Physiological Acoustics*, Archimedes 39,
DOI 10.1007/978-3-319-06602-8_14, © Springer International Publishing Switzerland 2014

Tanaka's most direct and most important contact with musicians in the metropolis of Vienna turned out to be Anton Bruckner (1824–1896).[2] It was Bruckner who first provided Tanaka with an opportunity to demonstrate the enharmonium and to elicit support in Vienna for his just intonation ideas. He simultaneously was able to make available the acoustic know-how and craftsmanship that he had been pursuing and cultivating with the Helmholtz group in Berlin. The original Tanaka-Bruckner encounter was followed by subsequent meetings with other musicians and instrument builders who were anxious to experience the enharmonium and its novel features as a fixed-tone keyboard instrument different from, and conceivably superior to, the 12-tone per octave harmonium.

Early in his career Bruckner had made a name for himself in London and Paris as an organ virtuoso and improviser. By the time Tanaka met him in the early 1890s he had become an internationally recognized composer. It appealed to Tanaka that as a music teacher at the Vienna Conservatory Bruckner was known for holding to the principle that whenever possible "mathematically pure intonation" should be preferred over customary tempered tuning.

During Bruckner's lifetime it was said that conservative critics recoiled from his music. He was identified as both simpleton and mystic and had been criticized for his Wagnerian leanings during the bitter Brahms-Wagner rivalries. On the other hand Eduard Hanslick (1825–1904), the Vienna University professor, music critic, aesthetician, and pioneer in music appreciation, referred to Bruckner as the "gentlest and most peaceable of men who becomes an anarchist during the act of composition." Others argued even more forcefully that the line of musical mastery stemming from Bach and Beethoven found its rightful culmination in the symphonies by Bruckner.[3]

Friedrich Eckstein (1861–1939), one of Bruckner's theory students, relates the following story: "One day, we were just in the middle of an interesting work, when a distinguished Japanese who called himself Tanaka, a member of the Japanese delegation [in Vienna], presented himself and informed the somewhat astonished master that he had studied acoustics and music theory with Helmholtz in Berlin;

[2] Son of a village schoolmaster and organist, Bruckner was sent as a chorister to the St. Florian monastery at the age of thirteen when his father died; there he was able to continue his organ studies in conjunction with music theory and lessons on the violin. In 1855 he was appointed organist at the Linz Cathedral. As his Masses and first symphony reveal, he early on had come under the influence of Wagner (*Tannhäuser* and *Tristan und Isolde*). He was criticized by his music friends for his Wagnerian leanings during the heated Brahms-Wagner rivalries. From 1875 Bruckner taught at the University of Vienna but only after the initial opposition from his music colleagues at the university had been put aside. The main influences behind the distinction and lofty objectives, and the basic model for the scale and shape of Bruckner's symphonies, mysterious in the way their introductions fade in from silence, are Beethoven and Wagner. Textually complicated (as first written down and as revised and cut), Bruckner's symphonies have been published in two versions in the *Sämtliche Werke* edited by R. Haas et al. *Concise Encyclopedia of Music*, London/New York (1988) 111–112.

[3] Benjamin Korstvedt, "Still Searching For Bruckner's True Intentions," *New York Times*, July 10., 2011, Ar 19.

that after difficult and arduous efforts he had managed in a reasonable way to improve and, by affixing lever-like drawknobs, to simplify a Bosanquet-type 53-tone per octave instrument."[4] After bows and expressions of respect, Tanaka handed Bruckner a German monograph (most probably his *Studien* of 1890) that deals with pure intonation and provides a brief description of the newly constructed enharmonium. He implored Bruckner and Eckstein to see the new instrument that had been installed at the Japanese delegation; that a wagon from the delegation was ready to take them there. After initial resistance Bruckner consented, was received with ceremonial flourish, and shown the "remarkable organ-harmonium." According to the Eckstein account, a demonstration followed in which Tanaka played the prelude to Lohengrin from a musical score whose notation was incomprehensible to Bruckner. Although the player occasionally faltered, the impression was highly "surprising and overpowering." When he played the same piece on a nearby ordinary harmonium in equal temperament Eckstein remarked "we had the impression that the ordinary instrument was intolerably out of tune." Bruckner, who became ever more interested, wrote down several enharmonic chord combinations and several diatonic modulations that can be regarded as enharmonic. As Eckstein relates, Tanaka first reworked the Bruckner-notated chord combinations in his own way using all kinds of symbols and numbers, sat down at his instrument and then, as Bruckner's student reminisced, "we had the unique pleasure of having those otherwise ideal and purely intellectual reinterpretations demonstrated for us in full actuality and in clearly audible tonal steps and digital alternatives." Bruckner became so taken up with the demonstrations that he was reluctant to leave the embassy and revelled in listening to more enharmonic chord combinations.[5] On telling his students about the Tanaka enharmonium in 1891 Bruckner remarked, "Ah hat das wunderbar geklungen [Ah, did that sound wonderful]."[6]

The episode in which Bruckner had exhibited so much enthusiasm for the ideas of a Japanese musician, who in turn had been influenced in his music-theoretic ideas by Helmholtz, is given an odd twist by Eduard Hanslick who reported what had taken place at an earlier time in Bruckner's career. Recall that according to the Eckstein account, Tanaka, as a Helmholtz student, had made his way into Bruckner's studio in Vienna; and, in effect, had convinced him at the keyboard that what he had learned from the author of *Tonempfindungen* had important consequences for music theory and music performance.

By the 1890s most music teachers in Vienna would have been familiar with Helmholtz's *Tonempfindungen* (1863), which by then had seen its fourth edition and had been popularized in Vienna by Ernst Mach's *Einleitung in die Helmholtz'sche Musiktheorie. Populäre für Musiker dargestellt.* Prag, 1866.

[4] R.H.M. Bosanquet (1841–1912), professor of music at the Academy of Music in London, devised an harmonium with a "generalized keyboard" using the 53-system of division of the octave.

[5] Friedrich Eckstein, *Erinnerungen an Anton Bruckner*, Vienna, New York, 1923, 37–40.

[6] Anton Bruckner, *Vorlesungen über Harmonielehre und Kontrapunkt an der Universität Wien*, ed. Erich Schwanzara, Vienna, 1950, 63.

Bruckner's biographer, Alfred Orel, reported that when Bruckner was a lecturer in music theory at the University of Vienna in the late 1870s he applied for a professorship in music theory. The decision of the committee that was given the task of evaluating Bruckner's credentials in music theory rejected his application. When one of the professors sought to defend Bruckner by explaining what was involved in the aesthetics of the art of music a member of the search committee countered with the remark that Bruckner probably had not even kept abreast of the newest research on acoustics provided by Helmholtz's *Tonempfindungen*, which was required reading for music students at the university.[7]

References

Bruckner, Anton. 1950. *Vorlesungen über Harmonielehre und Kontrapunkt an der Universität Wien*. ed. Erich Schwanzara, 63. Vienna: Oesterreicher Bundesverl.

Eckstein, Friedrich. 1923. *Erinnerungen an Anton Bruckner*, 37–40. Vienna/New York: Wiener Philharmonischer Verlag.

Haas, R., et al. 1988. *Concise encyclopedia of music*, 111–112. London/New York: W.W. Norton.

Korstvedt, Benjamin. 2011. Still searching for Bruckner's true intentions. *New York Times*, July 10, 2011, Ar 19

Orel, Alfred. 1925. *Anton Bruckner. Das Werk. Die Künstler. Die Zeit*, 143. Vienna/Leipzig: A. Hartleben.

[7] Alfred Orel, *Anton Bruckner. Das Werk. Die Künstler. Die Zeit*, Vienna/Leipzig 1925, 143. Orel was Privatdozent for music history at the University of Vienna in 1925.

Chapter 15
Encounters with Music Theorists in Japan

Tanaka returned to Tokyo in 1899 after 15 years of study and research in music theory and music instrumentation in Berlin and Vienna. As in the case of other Japanese scholars who had been selected to study abroad and had immersed themselves in Western culture during the Meiji period, Tanaka returned home just as his country was opening its borders after the closure that had been brought on in 1894 by the war with China. By the turn of the century the last territorial restrictions were abolished and Japan was accepted as an equal with the major powers. The favorable situation in music that Tanaka inherited on his return to Japan was intimately linked with the political modernization and liberalization of Japan and the return of governmental power with the reign of Emperor Meiji (1868–1912) and his council. During the decades in which Japan had abandoned its isolation and experienced drastic reform and modernization, there had developed a widespread interest in all that was Western. While certain characteristics of traditional Japanese culture had been displaced temporarily, the liberalizing initiatives in music that Japan had fostered during these years allowed Tanaka to meet the goals he had established for himself in Berlin and Vienna. By example, and by mandate, the Emperor had promoted the spread of European culture and coextensively had been intent on protecting Japanese culture from decay and oblivion. It was a cultural alignment that was much to Tanaka's liking.

Shortly after the Meiji takeover during the 1870s, a number of organizational factors had contributed to the Europeanization of music in Japan. Concurrently the bureau of royal court music (*Gagaku-kyoko*) was established in the court ministry. Descendants of the older families reassembled and led *Gagaku* (literally "elegant music") not only in the courts but on public occasions. *Gagaku* music ordinarily was sung with natural voice and very little instrumentation. Music pieces were classified according to length as well as tempo and meter. Performed in various genres on combinations of plucked, wind, and percussion instruments, the music was characterized by smoothness, serenity, and precise execution.[1] During the

[1] *The Harvard Dictionary of Music*, 4th ed. Cambridge, 2009, 340.

E. Hiebert, *The Helmholtz Legacy in Physiological Acoustics*, Archimedes 39,
DOI 10.1007/978-3-319-06602-8_15, © Springer International Publishing Switzerland 2014

Europeanization the court musicians traditionally learned to master a European music instrument in addition to their own classical instrument and played under the direction of foreign composers. Alongside open concerts the "Music-enthusiast Union" of the music and dance groups continued under the Meiji to function in large shrines and temples without interruption. The ministry of culture in turn established a division for European music and introduced the European tone system in the schools for singing.[2]

The Japanese historian of music Kishibe Shigeo has observed that the modernization of music in Japan did not bring into play a change in intonation as it did, for example, for the modernization of music in India and elsewhere in the Middle and Far East. Accordingly, the tasks of contending with the subtle refinements of traditional Japanese music were considerably less formidable than for other Asian countries because in Japan as in China "the twelve unequal-tempered system was used." By contrast to the Western system, "one octave is divided into 22 tones in (ancient) India, into 17 tones in Arabia, into 7 tones in Thailand, and into 5 tones in Indonesia... and Melodies are based upon these different tone systems, as well as on the different 5 or 7 tone scales or modes that sound quite different from one another."[3]

In seeking to explore the characteristic features in the musical fabric of the Japanese music that Tanaka confronted on his return to Japan, we begin with some perspectives on Japanese music sketched by the prominent German scholar of Japanese music and culture, Eta Harich-Schneider.[4] In her history of Japanese music, Professor Harich-Schneider has identified and examined three divergent trends active in musical Japan. There is first the trend toward total internationalization promoted almost unanimously by the wealthy class in contact with the West. This group treats its own traditional Japanese music either with animosity or indifference. Then there is an anti-foreign traditionalism associated with strong nationalism and political conservatism. It is distinct among the old aristocracy, the Buddhist clergy, the Shinto priests, and the lower middle classes in the cities and rural populations. "The third group, the compromisers, was strongest during the Meiji and Taishō years (1912–1926). To this group belonged Tanaka Shohé,

[2] Hans Eckardt, Japanische Musik, Section F. Neuzeit (seit 1868), *MGG*, *6* (1957) 1749–1751.

[3] Kishibe Shigeo, *The Traditional Music of Japan*, Tokyo, 1969, 14. See also Mark Levy, *Intonation in North Indian Music. A Select Comparison of Theories with Contemporary Practice*, New Delhi, 1981.

[4] Eta Harich Schneider (1897–1986) was a German historian of Japanese music and culture who launched her career in Berlin as pianist in 1924 and as harpsichordist in 1931. She was dismissed from her post as professor of music at the *Hochschule für Musik* in Berlin in 1940 for refusing to join the Nazi party. Her flight to Tokyo to become director of the music department of the U.S. Army College in Tokyo was the initiative for an ensuing professional career in Japanese culture. After extending her Japanese studies at Columbia University, she published works on the harpsichord, Japanese court music (*Gagaku* and *Bugaku*), and a pivotal *History of Japanese Music*, London/Oxford 1973. She also wrote a much quoted paper on "Renaissance Europe in Japanese eyes. Record of a triumphal journey," *Early Music*, *1* (1979) 19–25. Lionel Salter, *New Grove*, *10* (2001) 19–25.

Yamada Kosaku [featured below] and all the practical traditional musicians who attempted to combine Western and Japanese music."[5]

Over Japan's almost two-thousand-year history there have been many periods of both continuity and change, a great deal of experimentation, and a continually shifting attitude toward music. In the case of Shohé Tanaka, who is identified here as belonging to Harich-Schneider's group of practical traditional musicians who sought to combine Western and Japanese music, the penchant for experimental change looms large. This showed up foremost in Tanaka's inventions and experimentation with a unique and unconventional keyboard, the enharmonium. The invention not only parallels Tanaka's recognition that keyboard instruments occupy a position of essential primacy in the development of modern classical music, but complements the Japanese talent for acrobatic feats, descriptive dramatization, and stupendous manual skills at the keyboard. These features are all the more remarkable seeing that it was not until 1882 that the piano was introduced into Japan just 2 years before Tanaka became preoccupied with keyboard instruments and their intonation in Berlin.

The pianos that were extant in the 1880s in Japan were largely handmade and were imported to Japan mainly from the factories of Julius Blüthner and Carl Otto.[6] Among persons who were sent abroad to care for the Japanese pianos, the Japanese reed organ builder Nishikawa learned the piano-construction trade and by 1885 was involved himself with his fellow technicians in building pianos and harmoniums in Yokohama. The trade was passed on to relatives who went abroad to learn about American pianos. On return to Japan they founded the Matsumoto piano company that in 1965 still was operating a small craftsman business.[7]

According to Harich-Schneider, Shohé Tanaka was the most prominent Western-trained person to master European music while simultaneously achieving distinction in the field of musical acoustics within the European Helmholtzian tradition. At home in Japan he came to be recognized as a brilliant and talented scientist who had worked with the famous Helmholtz in Berlin and had invented a technologically ingenious keyboard instrument, the enharmonium, on which, it was

[5] Eta Harich-Schneider, *A History of Japanese Music*, Oxford, 1973, 595–596.

[6] Blüthner pianos are made by a German firm of piano builders founded in Leipzig in 1853 by Julius Blüthner (1824–1910). His grand pianos achieved renown after 1873 when aliquot scaling (the addition of a fourth sympathetic string to each trichord group in the treble) was introduced to enrich the upper register. Distinguished by a round, slightly romantic tone, the firm's instruments are still made largely by hand. The Carl Otto piano firm was founded by Gustav Carl Otto in Berlin in 1866 and at the time employed twenty workers who constructed each year about 200 small upright pianos, the so-called pianinos. By the turn of the century the firm was overseeing a mature piano business. For a comprehensive source of information on keyboard builders: Hubert Henkel, *Lexikon deutscher Klavierbauer*, Frankfurt am Main, (2000) 457.

[7] Sumi Gunji Hamamatsu, "Japan und seine Klaviere," *Dokumentation Europiano Congress Berlin 1965*, ed. By H. F. Herzog et al., Berlin 1966, 213–216. Shinkichi Matsumoto (1865–1941) started his career as a carpenter, was apprenticed to the read organ builder Nishikawa (see above), started making his own organ reeds and American reed organs in 1893, began piano making in New York, and formed Matsumoto & Company Ltd in 1907.

said, "the partials of harmonic series can be eliminated at will." Although for a number of years there were Japanese musicians who were highly critical of Tanaka's "scientific approach," he was in a good position, notably, to elicit strong approval from his countrymen for a number of other accomplishments that were universally commendable. He had introduced the Western system of notation for Japanese music, he had activated a number of traditional music guilds, and he had pioneered in the transcription of theatrical and vocal music. Above all Tanaka had initiated a trend for establishing Japanese concert music as an acceptable art form of musical expression.[8]

A provocative contrast to Tanaka's resolute science-focused approach to music is provided by the Japanese scholar Yamada Kosaku.[9] Like Tanaka, Yamada belonged to a group of Japanese musicians who sought to combine the best elements of Western and Japanese music. After 4 years of vocal and cello music and music theory at the Tokyo Music School, Yamada went to Berlin to extend his musical horizons, first in the study of composition with Max Bruch (1838–1920), composer of operas and secular choral music at the Berlin Academy, and then with the musicologist Johannes Wolf (1869–1947), pioneer in source studies, and professor at the University of Berlin. Returning to Tokyo, Yamada lived the life of conductor, composer of symphonic, orchestral, and opera music, and teacher of Western harmony and composition. He was actively engaged in founding the Tokyo Philharmonic Society Orchestra.[10]

A comparison of Tanaka's and Yamada's European training and contrasting contributions to the music of modern Japan is instructive. As a student of physics at the University of Tokyo, Tanaka had graduated with a specialty in acoustics. When the opportunity for foreign study arose, Tanaka, who had actively cultivated music since his youth, elected to study with Helmholtz in Berlin and focus on a problem in musical acoustics. As in the case of his teacher Helmholtz, the problem orientation for Tanaka was intonation – a subject that had received its motivation mainly from the side of theory but was thought to be resolvable, if at all, by designing and constructing keyboard mechanisms that are capable of delivering pure or just intonation. On the other hand Yamada undertook studies in Berlin with the intent of taking home to his countrymen as much of Western music and its harmonic structure and melodic emphasis as was possible – especially in the tradition of

[8] Harich-Schneider, *A History* (1973), 544–545. Harich-Schneider recalls, "In 1943 I had the honour to be received by the old gentleman who personally demonstrated his invention to me."

[9] Yamada Kosaku (1886–1965), one of the best-known Japanese composers, was the child of poor Christian parents who converted to Buddhism. His striking musical talents won him a scholarship to the Berlin *Hochschule für Musik*. Extraordinary linguistic abilities (German, French, Italian, and English, unusual among Japanese scholars) gave him an access to Western culture that was unique. His interests included literature, the theatre, the pictorial arts, and opera as a main theme. Yamada also had great merits as a teacher of Western harmony and composition. Harich-Schneider, *A History* (1973) 545.

[10] Masakata Kanazawayo Akioka, Yamada Kōsaku, *NG*, *27* (2001) 633–635; Harich-Schneider, *A History* (1973) 545.

German romanticism and the works of composers such as Wagner, Strauss, and Skryabin. By contrast, Tanaka never lost his Japanese identity and constantly was on the lookout for theoretically sound and technologically feasible ideas for improving the temperament of tones and intervals for musically sensitive ears. Yamada on the other hand had assimilated Western culture without giving up an inch of his Japanese culture. He never lost his identity as a Japanese composer and wanted above all to enlarge and enrich the Japanese appreciation of music. Tanaka wanted, rather, to tackle the problem of supplying the discriminating ear – Western or Japanese – with the purest form of music. In summary, Tanaka's objective was twofold: to search out opportunities that would foster the vitality of Western music and its harmonic structure while simultaneously enriching the subtleties of traditional Japanese music in such a way that it would be able to detach itself from those particular Japanese elements that hindered its development as an autonomous art form.

In the course of attempting to meet the twofold objective of simultaneously fostering both Western and Japanese music, Tanaka encountered a number of unanticipated barriers indigenous to the music of his own Japanese tradition. As ethnomusicologists were quick to pinpoint and caution almost instinctively, the customs and concepts in the domain of aesthetics in one culture are not in most cases readily transformed or reshaped without the loss of essential characteristics and nuances from the other culture. To preserve Japanese music, to comprehend the conditions of its emergence and intellectual conceptions, and to understand its psychical and social implications ultimately entails experiencing the music in its own context and listening to the music with the ears and aesthetic sensibilities that belong to Japanese culture. Writing in this frame of mind in 1948, the historian of music Eishi Kikkawa remarked:

> One needs to know with what kind of resources, with what intentions, under what circumstances, and for which public the music was composed. It means that one needs to set aside the prejudice that "the music of the West is the music of the world," and acknowledge that in Germany a German, in India an Indian, and in Mongolia a Mongolian music exists; and that all of them for the time being are confronted with various music cultures determined by the natural circumstances of their respective races and so forth. In other words an evaluation based on the vital consciousness of the Japanese, the aesthetic sensations of the Japanese, the world view [*Weltanschauung*] of the Japanese – in brief, the contact with the spirit of Japanese music – that is the alpha and the omega for understanding Japanese music. The reproach that Japanese music is simple, the lament that Japanese music has no harmony, the discontent about the meager scale of Japanese music, and the censure of the lack of pure instrumental music in Japanese music – all, I believe, have become intelligible and are not just the shortcomings of Japanese music.[11]

---

[11] Eishi Kikkawa, *Vom Charakter der japanischen Musik (Nihon ongaku no seikaku)*, transl. into German from the 1948 Japanese version. Basel, London, 1984, 8 and 200. Eishi Kikkawa (b. 1909), a graduate of Tokyo University (1933), was known for his history of Japanese music (1965) – "the best-documented discussion of the subject." His research speciality was musical instruments that accompany the voice, namely the various genres of the shamisen and the koto. *NG 13* (2001) 588–589.

According to Komiya Toyotaka, the institutional and organizational structure of the Japanese musical world that Tanaka inherited when he returned to Tokyo in 1899 was one that made his endeavours far more puzzling and probationary than what he as a brilliant young musical theorist and part-time engineer had encountered in Berlin. To cite an example of the kind of problem Tanaka faced, Toyotaka remarked:

> In the world of Japanese music the names of the composers have almost all been lost. What remains is the names of the performers, and even then it is usually the names of the singers rather than of the samisen players which remain. Most of the composers were samisen players.... It usually happened that the composers were at the same time performers, and their fame as composers amounted to the same thing as their fame as performers, but in the *nagauta*, although all the composers were samisen players, theirs was an unthanked task, for the tendency was for the singers to get all of the credit. This was particularly true in the theatre because the singer's role is the most conspicuous one and it is he who can most easily move people. He thus tends to carry off the popularity. The problem stems basically from the fact that although there is theoretically an equal division of responsibility between the singing and the samisen, the vocal part always predominates.[12]

Conscious of the situation in the theatre and of the urge to study Japanese music with the help of a teaching instrument such as a keyboard, Tanaka, on his return to Japan, founded a Research Institute (Hôgaku Kenkyûko) in Tokyo charged with the task of promoting the younger generation's nurture of traditional Japanese music. This led in 1902 to the establishment of the Nagauta Study Group (*Nagauta Kensaikai*) whose objective it was to give traditional Japanese music a form of existence independent from Kabuki; that is, free from the limitations imposed by the samisen.[13] In this context the samisen is used to accompany various vocal genres, each of which uses a different size of instrument and plectrum. Vocal music that uses the samisen has a history of being connected mainly with the acting and dancing of Kabuki actors. The samisen in essence prevented instruments, and especially instrumental music, from becoming an independent entity and thus restricted the expansion of music as an autonomous domain.[14]

By contrast, the Nagauta Study Group that Tanaka had established in 1902 encouraged the study of vocal music that could be sung Western style with piano and violin, or more often, because of the shortage of instruments, with the koto and the kokyū.[15] According to Toyotaka:

[12] *Japanese Music and Drama in the Meiji Era*. Compiled and Edited by Komiya Toyotaka. Translated from the Japanese by Edward G. Seidensticker and Donald Keene, Obunsha Tokyo, 1956, 51–52. Toyotaka (1884–1966) was the sometime director of the Tokyo Academy of Music.

[13] The samisen, or shamisen, is a three-stringed, banjo-like, long-necked lute held by the left hand and activated by the right hand.

[14] Don Michael Randel (ed.), *The Harvard Dictionary of Music*, 4th ed. Cambridge, 2003, "East Asia," Meiji period (1869–1912), 270–273.

[15] The koto is a long slender rectangular wooden zither with 13 strings tuned to various modes of the pentatonic scale. It is played with plectra worn on the thumb and takes the place of the piano. The kokyū is a three- or four-stringed bowed lute, and takes the place of the violin.

> Success of the Nagauta Study Group in achieving independence from the Kabuki marked the first step in the development of the *nagauta* as pure music. At the same time one could not say that the possibility of true samisen music had been adequately investigated as long as no step had been taken in the direction of freeing the samisen from the singing and the making of it into an independent instrumental music.[16]

Although Tanaka seems not to have had extensive direct person to person influence on the development of music in Japan, a number of his Japanese students helped him put his advanced theories on intonation and his pioneering use of the Western five-line notation system into practice. Others joined forces with him in the practice of writing Japanese music in the Western style of notation.[17] One of Tanaka's students, Yamada Schumpei (1886–1933), a samisen musician who helped put Tanaka's theories into practice, published the *Nagauta Shin-keikobon* (New Nagauta Practice-Book). On the suggestion of Tanaka, Shumpei wrote down in Western notation the complete repertoire of a number of Japanese composers whose music survived in the scores of the *Music Study Committee*.[18] A number of others who were associated with the Nagauta study group launched a Society for refining the Nagauta (the *Nagauta Kensei-Kai*), an organization, as Toyotaka remarked, that had as its objective the liberation of the *Nagauta* from its subordination to the theatre and its establishment as an independent concert art. The achievements of the Society were crucial in the spreading of an appreciation of *Nagauta* beyond the narrow circle of devotees on which it had until then depended. It served to remove from the samisen the stigma of being associated with "pleasure quarters" and make it an acceptable instrument for the proper bourgeois.[19]

Tanaka's departure from Vienna and return to his home base in Tokyo in 1899 marks the mid-career juncture for a 37-year-old Japanese scholar who was remarkably well prepared to engage his own countrymen in the ins and outs of Western music, its styles and methods of composing and performing, its notational system, and, most significantly, the idea of a just intonation system capable of demonstrating the musical advantages of fixed-tone keyboards designed to deliver just intonation. The promising political situation in relation to music in Japan, and the favorable state of music being cultivated in Japan's musical institutions at the time of Tanaka's return, provided him with a timely occasion for promoting a new music that sought according to his own well calculated equation to bring about a fusion of the best in traditional Japanese music with the classics of the Western world.

In what follows, no more than a cursory attempt is made to examine the literature, in Japanese, that deals with Tanaka's career and contributions in the field of music theory for the post-Berlin, post-Vienna years in Tokyo. The emphasis will be twofold: first, to mention in brief outline what Tanaka had to say about the

---

[16] Toyotaka, *Japanese Music*, (1956) 521.

[17] A list of the associations and persons involved in fashioning different styles of Japanese music is given in the Toyotaka volume, 37–38

[18] Komiya Toyotaka, *Japanese Music and Drama in the Meiji Era*, Tokyo (1956), pp. 368, 388, and 396.

[19] Toyotaka, *Japanese Music and Drama*, (1956), 387–388.

Japanese foundations of harmony after he returned to Tokyo; and, second, to offer a few reflections on how other Japanese musicians, both colleagues and outsiders, evaluated Tanaka's ideas on just intonation and his efforts to design and construct enharmoniums. The two most important post-1899 monographs on Tanaka and his life's work, both in Japanese, are, first, a volume by Tanaka published in Tokyo in 1940 entitled *The Foundations of Japanese Harmony*, and, second, a volume by Sadao Ito on the evaluation of Tanaka's views on music and just intonation published in Tokyo in 1940 and entitled *Tanaka Shohé and Just Intonation.*[20]

Additional information about Tanaka's own contributions to musical acoustics is given in a Japanese handbook on the history of physics published in 1978.[21] Tanaka is credited in this historical handbook with having been a major contributor to the Helmholtz theory of sound. It is mentioned that he invented an organ according to the Helmholtz theory, that he played it in the presence of Emperor William II, and that the Emperor supported further investigations that led to Tanaka's creation of several just intonation pipe organs, one of which is kept in the Deutsches Museum in Munich. This observation is of particular interest in that it shows that Tanaka's work with Helmholtz on just intonation was recognized by professional Japanese physicists in their general volume on the history of physics and namely in the section that covers Japanese contributions to acoustics during the Meiji period.[22] The reason that Tanaka had undertaken to investigate just intonation, we are told, was to develop a suitable foundation for dealing with harmony, that his contribution to the Japanese music community was considerable, and that Japanese musicians, in particular, had respected him highly – this in spite of the observation that they questioned both the ease with which just intonation was achieved theoretically and controlled manually on enharmonic keyboard instruments.[23]

Tanaka's *The Foundations of Japanese Harmony* was published towards the end of his life in 1940 shortly before Japan opened hostilities against the United States and Great Britain. In the Preface, which is reticent about his own earlier experience in Germany, Tanaka remarked that his uppermost interest in Japanese music had been to discover a unique manner of expressing traditional Japanese harmony. He had wanted Japanese music to be an appealing component of Japanese culture. He cautioned that this would not be accomplished by imitating Western harmony since the historical forms and structure of Western harmony did not suffice to capture and express the fine nuances and unique native characteristics of Japanese music.[24]

[20] Tanaka Shohé, *Nihon wasei no kiso* (*The Foundations of Japanese Harmony*), Tokyo, 1940; Sadao Ito, *Tanaka Shohé to junseicho* (*Tanaka Shohé and Just Intonation*), Tokyo, 1940.

[21] *Nihon no butsurishi* (*History of Physics in Japan*), vol. 1, edited by The Physical Society of Japan, Tokyo, 1978. This handbook is a valuable source of information about Tanaka that was deemed important for Japanese scientists as they envisioned the area of acoustics. Part III, Chapter 9, Meiji Period, 184–188.

[22] The Physical Society of Japan, *The History of Physics in Japan*, vol. 1, Tokyo, 1978, 185. For reference to a demonstration of the enharmonium to the Emperor in 1890, 26–27.

[23] The Physical Society of Japan, *The History of Physics in Japan*, vol. 1, Tokyo, 1978, 186.

[24] Tanaka Shohé, *The Foundations of Japanese Harmony*, Tokyo, 1940, 1.

As might have been expected, the Tanaka volume reveals an element of personal pride on the part of a young scholar who had impressed Western musicians with his extraordinary reserve of intellectual, technological, and organizational prowess. The work touches on nationalistic sentiments about the existence of a rich but untapped reservoir of traditional non-Western music that Tanaka deemed worthy of recognition as an important component of the world's musical heritage. Following the title page of Tanaka's *Foundations* there is a photograph of Tanaka and a statement in hand-written German Gothic from the Emperor:

"William, German Emperor and King of Prussia Berlin, 15/IV, 1890. I take the harmonium to be one of the most important inventions of our century and epoch-making for the field of tonal art." This statement "was made in reference to the instrument in pure intonation that was named the enharmonium, and invented by Shohé Tanaka." The particular Tanaka organ to which the Emperor and Empress were introduced in 1893 in Berlin was built in the workshop of the organ building firm E. F. Walcker in Ludwigsburg. A concert organized by the director of the cathedral choir indicates that on 21 September 1893, in Berlin, a number of prominent organ players, in collaboration with orchestral instruments, vocal soloists, and a cappella groups, demonstrated the "Tanaka syntonic purely tuned organ."[25]

In a volume on Tanaka and just intonation written several decades after Tanaka's death, Sadao Ito published a comprehensive chronology that includes pertinent information about Tanaka's professional career in musical acoustics and railway engineering. It covers his life in Japan from 1899 until his death in 1945.[26] In the Ito volume Tanaka's life is built around three contextually distinct time periods and cultural environments. Ito recalls that at the age of 20 Tanaka graduated from the physics department of the University of Tokyo with a strong additional commitment to music. For academic distinction he received a silver medal from the Japanese emperor. After a year of teaching as an instructor in the Preparatory School of Tokyo University he was sent to Germany by the Japanese Department of Education to study acoustics and electromagnetics. Although the original appointment was for 3 years he spent several additional years in Europe before returning to Tokyo. On the basis of his *Studien im Gebiete der reinen Stimmung* (1890) he was awarded the doctoral degree of science at the university of Berlin in 1891. On several occasions prior to his return to Japan Tanaka was received by the German Emperor in order to demonstrate his enharmonium as well as a just intonation pipe organ with 20-notes per octave and 36 pipes that had been completed in 1893.

[25] The organ building firm of E. F. Walcker (1794–1872) was established in 1780. The firm made harmoniums from 1838 to about 1914 and constructed an enharmonic pipe organ and an enharmonic harmonium for Tanaka. The name *syntonia*, or *syntonic comma*, was introduced by Ptolemy and represents the interval by which the ditone exceeds the pure major third (5:4); that is $(9.8)^2 - 5:4 =$ 81:80. Tanaka Shohé, *The Foundations of Japanese Harmony*, 53–55.

[26] Sadao Ito, *Tanaka Shohé and Pure Intonation*, Tokyo, 1968. See especially 90–99.

Continuing the Ito account, we read that since the Japanese Department of Education originally had sent Tanaka to Berlin for the study of acoustics and engineering, he had decided, after his studies in acoustics, to spend the remaining years in Germany studying railway mechanics at his own expense. On his return to Japan in 1899 Tanaka worked for the Japanese Railway Company and the Imperial Railway Service (1899–1907). In 1909 He became the Director of the Japan Railway Service and in 1911 Head of the Japan Railway-Testing Laboratory. He retired from the Japanese railway positions in 1913 just before World War I. The 4 years of war and its disastrous aftermath left Tanaka and many scholars in Japan and elsewhere intellectually and professionally completely stranded.

In 1921 after the war, the Japanese Department of Education enticed Tanaka to return to the study of traditional Japanese music and resume his earlier interests in the subject of just intonation. Over a period of two decades he was able to establish a network for further study in this direction, but he also encountered opposition from prominent Japanese musicians who claimed that a number of their best music students were not able to master the theoretical complexities of just intonation or the instruments designed to play in just intonation. The musician Tamura Naragoshi, who at the time was considered to be an authority in Japanese music, stated outright that Tanaka's instrument was unplayable and probably erroneously constructed.[27] A decade later the situation had shifted in Tanaka's favor. When the Tokyo Music Society chose Tanaka as its chairman, Tamura gave him his full support – showing that Tanaka's just intonation enterprise and the enharmonium had not been entirely without success among Japanese musicians.[28]

In the 1930s, when he was in his late 70s, Tanaka once again turned his attention to the construction of the just intonation harmonium, this time in hopes of being able to improve on some of its impractical and complicated performance techniques. This led to the construction of several instruments with 21-keys per octave (assignable to 46 notes, almost four-octaves) and the establishment of a network for their study. In recognition of these contributions to musical acoustics, Tanaka received the cultural prize of the Asahi newspaper in Tokyo in 1938. On the occasion of the receipt of the prize, Tanaka met the famous Japanese physicist Hantaro Nagaoka (1865–1950), Professor of Physics at Tokyo University, who the previous year in 1937 had been the recipient of the National Cultural Prize. It was most appropriate for the two men to meet in Tokyo on the occasion since Nagaoka's years of study in Berlin, Munich, and Vienna (1883–1896) had overlapped to some extent with those of Tanaka in Berlin and Vienna.[29] On a later occasion a student of Nagaoka relates: "I was enrolled in the Department of Physics and also went to Tokyo Music School to study music.... As graduation approached I told Professor

[27] Ito Sadao, *Tanaka* (1968), 92.

[28] Ito Sadao, *Tanaka* (1968), 138.

[29] Ito Sadao, *Tanaka* (1968), 94.

Nagaoka that I would like to study Japanese traditional music. He told me that there is a true authority in the subject and advised me to visit Dr. Tanaka."[30]

As we have seen, from the Ito perspective, the life story of Shohé Tanaka can best be told in a lop-sided double-column ledger that deals with the activities of a scholar who, during his most productive 15 years as a foreigner in Berlin and Vienna, made a name for himself in the Helmholtzian musical world but who, without making much fuss about it, was studying railway mechanics on the side. During the last 40 years of his life in Japan, Tanaka came to acknowledge that neither he nor his musical colleagues had been aware of the difficulties inherent in adapting what one can learn about the structure of Western music to traditional Japanese music. On the other hand Tanaka had discovered that his knowledge and training in Western railway mechanics had carried him to the crest of the Japanese railway engineering world, whereas progress and lasting success came but slowly and sporadically in the inchoate and undeveloped domain of the humanities. Not much of note seems to be on record regarding the final Tanaka years leading up to the ill-omened Second World War. We know that most of Tanaka's manuscripts were destroyed in 1945. Three months after the Japanese surrender Tanaka died (Oct. 1945) in the Chiba farmhouse in which he had taken refuge from the Tokyo bombings.[31]

Tanaka's *Studien im Gebiete der reinen Stimmung* of 1890 embodies the essential work on which the reputation of the 28-year-old Tanaka, Japanese student of physics and music theory, came to be based. Against the background of arguments that Tanaka brought to bear on the tonal preeminence of the enharmonium over against 12-tone per octave keyboards, we have explored a number of secondary works and supportive documents that were published in connection with Tanaka's ideas on just intonation and on the construction, functioning, and manual operation of the enharmonium. A number of the support documents fortunately were published in English and therefore were more accessible to Japanese scholars than his German *Studien*. The visionary focus and intensity with which the youthful Tanaka "as an alien" in the European musical world promoted his ides on just intonation and the enharmonium during the early years of his studies in Berlin is the central component of our Tanaka narrative. His overall message is notably put in evidence most directly and with a visceral grip on his subject by the way in which Tanaka sought to make the ideas of just intonation and their practical implementation in the enharmonium known and accessible to Japanese music students, music teachers, music pedagogues, and composers. From the superabundance of positive endorsements of the enharmonium, coming from pedagogs and teachers from Berlin, Vienna, and elsewhere in Europe, it is evident that the enharmonium was an instrument that had found its greatest use as a teaching tool for the study of music theory.

[30] Ito Sadao, *Tanaka* (1968), 180. Nagaoka was known in physics for having proposed the Saturnian model of the atom (1904). Scientists at the University of Tokyo claimed that it was Nagaoka who was primarily responsible for promoting the advancement of physics during his teaching years (1901–1925). Eri Yogi, Nagaoka, *Dictionary of Scientific Biography IX* (1974) 606–607.

[31] Toyotaka, *Japanese Music* (1956), 368.

References

Eri Yogi, Nagaoka. 1974. *Dictionary of scientific biography*, vol. IX, 606–607. New York: Charles Scribner's Sons.

Hamamatsu, Sumi Gunji. 1966. Japan und seine Klaviere. In *Dokumentation Europiano Congress Berlin 1965*, ed. H.F. Herzog et al., 213–216. Berlin: Europiano Congress.

Harich-Schneider, Eta. 1973. *A history of Japanese music*, 595–596. Oxford: Oxford University Press.

Henkel, Hubert. 2000. *Lexikon deutscher Klavierbauer*, 457. Frankfurt am Main: Bochinsky.

Levy, Mark. 1981. *Intonation in North Indian music. A select comparison of theories with contemporary practice*. New Delhi: Biblia Impex.

Randel, Don Michael (ed.). 2003. *The Harvard dictionary of music*, 4th ed. Cambridge, MA: Belknap Press of Harvard University Press.

Schneider, Eta Harich. 1979. Renaissance Europe in Japanese eyes. Record of a triumphal journey. *Early Music* 1: 19–25.

Shigeo, Kishibe. 1969. *The traditional music of Japan*, 14. Tokyo: Kokusai Bunka Shinkokai.

The Physical Society of Japan. 1978. *The history of physics in Japan*, vol. 1. Tokyo: The Physical Society of Japan.

Toyotaka, Komiya. 1956c. *Japanese music and drama in the Meiji Era*, 368, 388, and 396. Tokyo: Obunsha.

Part III
Max Planck

Chapter 16
Berlin: Capital of the New German Reich

Paris was the center of scientific research and a fertile breeding ground for mathematical physics and the exact sciences for more than half of the nineteenth century. Its two great schools, the *École Normale Supérieure* and the *École Polytechnique*, claim a record of celebrated scientists and mathematicians unmatched by any similar institutions in Europe. To a large extent, it was the Revolutionary Government of the 1790s that contributed to elevating nineteenth-century science in France to a position of eminence and promoting its commanding social power and national prestige. In contrast, German universities, many of which had been established in the fifteenth and sixteenth centuries under the influence of the humanistic sciences, did not, until well into the century, achieve a comparable level of excellence in the sciences, at least not in the sense in which the word "science" is used in France and England. The German word *Wissenschaft* has a much wider meaning than the French word "*science*" or the English word "science." According to the German ideal, only those disciplines that employ exact methods deserve to be referred to as *Wissenschaft*. Long before the spirit and methods of the exact sciences were introduced into Germany with their French and English meanings, the Germans had cultivated *Wissenschaft* in their universities and had established research departments and institutes in disciplines such as philosophy, history, linguistics, jurisprudence, classics, biblical criticism, and music.[1]

In contrast to the scientific spirit that had characterized the exact sciences and their mathematical and analytical methods in early nineteenth-century France and the *Académie des Sciences*, little was done during this period to further directly the development of the exact sciences in Germany. Indirectly, nevertheless, it was the founding of German institutions of higher learning and culture and the system of learning at the German universities that set the stage for the development of the

[1] For a full range of historical and national meanings of the words "Wissenschaft," "Science," and other words used in the text see Joachim Ritter, Karlfried Gründer et al. (eds.), *Historisches Wörterbuch der Philosophie*, 12 volumes and Index (vol. 13), Basel (1971–2007). For "Science," *8* (1992) 1498–1503. For "Wissenschaft," *12* (2004) 902–948.

E. Hiebert, *The Helmholtz Legacy in Physiological Acoustics*, Archimedes 39,
DOI 10.1007/978-3-319-06602-8_16, © Springer International Publishing Switzerland 2014

natural sciences with specific emphasis on research, criticism, and orderly thoroughness (*ordentliche Gründlichkeit*). As John Theodore Merz stated it: "Before the methods of exact science were introduced into Germany under English and French influences, the Germans possessed many scientific methods. There was the science of philosophical criticism, established by Kant; the science of historical criticism, of Biblical criticism; the science of philology: all these professed to have methods as definite, aims as lofty, and a style as pure, as the exact sciences brought with them."[2]

The University of Berlin, essentially the creation of linguist, Prussian statesman, and minister of education Wilhelm von Humboldt (1767–1835), was founded in 1809, the year referred to by its founder as the pinnacle of "Prussia's greatest misery." It was named the Humboldt University in 1828. From then on the pace of development in the exact sciences was such that already before the Franco-Prussian war of 1870/71 the growing interest in the physical sciences had led so visibly to an increase in the number of physics students at the University of Berlin that its laboratories had become totally outdated. When Hermann von Helmholtz (1821–1894) became successor to Gustav Magnus (1802–1870) in the spring of 1871 he accepted the professorship on condition that a new institute, "the greatest to be desired in the German Empire," would be built to house the discipline of physics. Over a million marks were set aside for the construction of the Helmholtz physical institute. Completed in 1876, it housed a lecture room with 300 seats and an official residence (*Dienstwohnung*) for the family of the *Reichskanzler*, as Helmholtz came to be called. The physical institute became the meeting place for the regular 2-week sessions of the *Physikalische Gesellschaft* for which the master of the house functioned as chair for decades. With the development of the universities and institutes that followed German unification and the establishment of Berlin as the capital of the new German *Reich* came the vision of Berlin as a world center for scientific research and learning alongside Paris and Vienna.

In the spring of 1889, on the recommendation of the philosophical faculty of the University of Berlin, Max Planck (1858–1947) was appointed successor to Robert Kirchhoff (1824–1887), first as *Extraordinarius* and director of an Institute for Theoretical Physics that explicitly had been created for him, and then, in 1892, as *Ordinarius*. At the same time August Kundt (1839–1894), one of Germany's leading experimental physicists, was appointed to the physics chair that had been vacated when Helmholtz became director of the newly founded *Physikalisch-technische-Reichsantalt* (PTR) in Berlin-Charlottenburg. Positions in theoretical physics in Germany were few at the time and did not have the high distinction attributed to experimental physics. For Planck, nevertheless, conditions in Berlin were optimum for exploring the theoretical domain of thermodynamics, a subject

[2] John Theodore Merz, *A History of European Thought in the Nineteenth Century*, 4 vols., London 1896–1914 and new York, 1965. Vol. 1, p. 173. Written during the waning years of the century, this monumental comparative and well-documented study provides a first-hand view of the origins and influences of new modes of thought, schools, ideas, and systems in nineteenth-century Germany, France, and England.

in which he had become an essential expert by age 40. As a branch of science that in addition to mechanics and electrodynamics embodies the theoretical foundations of nineteenth century physics, thermodynamics had always been Planck's *Lieblingsthema* (favorite theme). Having completed the Ph.D. degree in 1879 with a thesis on the second law of thermodynamics that notably contained important criticisms of Clausius's definition of irreversibility, he published a series of fundamental papers on thermodynamics that were to become the celebrated *Vorlesungen über Thermodynamik* (1897).

When Planck left his teaching post at the University of Kiel and arrived in Berlin, he most likely was looked upon as a half-baked theoretical physicist among colleagues whose average age was well above 60. He nevertheless had the good fortune to have the thermodynamics-famed Helmholtz as a close friend and colleague – one who more than anyone else would have been apprised of the status of thermodynamics and the residual open problems in thermodynamics that Planck was struggling with at the time. Helmholtz, who had been professor of physiology at Heidelberg for 13 years and professor of theoretical physics in Berlin for 18 years, was clearing the way for Planck to take his place. At the time of Planck's arrival in Berlin Helmholtz was beginning to be recognized as the leading personality in Germany's physics community. It must have been a tremendous inspiration for the almost 40 years younger Planck to work under Helmholtz's patronage just as Berlin was moving in the direction of becoming the European metropolis for the sciences and technology. The situation engendered in Planck the impression that the university, working alongside Helmholtz and his well-established institutes, had taken the lead in physics research; that this preeminence in the sciences exemplified Germany's image as prime mover in the world of science and technology. In an autobiographical statement, reminiscing about his 40 years in Berlin, and written shortly before he died, Planck spoke of his close personal friendship with Helmholtz:

> These were the years during which I perhaps experienced the strongest expansion of my entire scientific way of thinking. Now for the first time I came in close contact with men who at the time held the world's leadership in scientific research. Above all with Helmholtz. But I also learned to know him and admire him as much from his human side as I always had for his science. For in his entire personality, in his unerring judgment, and in his simple nature were embodied the dignity and veracity of his science. To this was joined a human kindness that went straight to my heart. When he looked at me with his calm and penetrating but basically benevolent eyes I was overcome with a feeling of boundless childlike devotion so that without any reserve I could have entrusted him, in confidence, with all that was in my heart and would find him to be a fair and gentle judge; and that an appreciative or even praiseworthy word from him would please me more than any external achievement.[3]

When Helmholtz died in 1894 the majority of leading theoretical physicists in Germany took it for granted, rather unpretentiously it seems, that Planck would be

---

[3] Max Planck "Wissenschaftliche Selbstbiographie," (1948), *Physikalische Abhandlungen und Vorträge,* III, 1958, 381–822.

looked upon as the major spokesman for German science. It was a responsibility that for Planck was to entail a strong sense of Prussian duty. With Planck's discovery of energy quanta in 1900 and its recognition in 1918 by the Nobel Foundation at a time of Germany's military defeat, Planck was well placed in Berlin to take on the position of scientific *Reichskanzler* that Helmholtz had occupied. Helmholtz's move from the University of Heidelberg to the University of Berlin had represented a return to the locus, the culture, and the intellectual environment in which he had been born and in which he had lived for some 30 years alternatively at home and while attending the Volksschule, the Gymnasium, and the University and its environs. Planck arrived in Berlin almost two decades after Helmholtz had been teaching there. Apart from the challenging professorship in theoretical physics and its institute, the greatest attraction for the young Planck unquestionably was the opportunity to work for a number of years alongside Helmholtz. Within 5 years of his arrival in Berlin Planck was elected to the German Physical Society (*Deutsche Physikalische Gesellschaft*) – a post which, right from the start, brought recognition in Berlin's academic quarters and its social and government-sponsored institutions.[4] Election to the *Gesellschaft* in 1894 was for the Prussianophilic Planck a great honor since membership was taken to signify a type of knighthood into the temple of learning made up of Germany's outstanding intellectuals at the universities and in other scientific institutions. It was at meetings of the *Gesellschaft* that the young Planck first had the opportunity to meet eminent older men such as Gustav Kirchhoff and Ludwig Boltzmann. Occasions to establish close relationships with younger German scientists such as Willy Wien, Albert Einstein, and Max von Laue were provided by the various meetings of the *Gesellschaft Deutscher Naturforscher und Ärzte*.[5] As had been the case with Helmholtz, Planck's circle of associates in Berlin came, in the course of time, to include not only academics in the scientific societies and at the university but persons in the humanities, the arts, industrialists, and persons of power and influence in the

[4] The *Physikalische Gesellschaft* was founded in January 1845 by three physicists, two physiologists, and a chemist. Motivated by discussions that had taken place in the *Physikalisches Kolloquium* of their teacher and Helmholtz's predecessor Gustav Magnus (1802–1870), they formed their own small study group and at first met in each other's homes without the professor. The original group of six included Gustav Karsten, Hermann Knoblauch, Wilhelm Beetz (who became one of Helmholtz's teachers in Munich), Ernst Brücke and Emil Du Bois Reymond (Helmholtz's two closest colleagues in the Johannes Müller circle of physiologists), and Wilhelm Heintz. By the end of 1845 the *Gesellschaft* had 53 members one of whom was Helmholtz. Two years later Helmholtz presented the young *Gesellschaft* with a paper on the conservation of energy. In 1899 the *Gesellschaft* was restructured to become today's *Deutsche Physikalische Gesellschaft (DPG)*. At a meeting of the *DPG* on 14 December 1900 Planck presented his classic radiation law and quantum hypothesis that led to quantum theory.

[5] For a concise biographical sketch of Planck and his interests in physics, physiology, the humanities and the arts: Hans Kangro "Max Planck," *Dictionary of Scientific Biography* (*DSB*, *XI* (1975) 7–17. The late Hans Kangro, personal friend to many of us, was a Hamburg historian of science who was acquainted with many of the persons here mentioned.

Prussian government.[6] The contrastive paths and contextual circumstances by means of which Planck's contacts with colleagues and close interactions with them and their experimental investigations and ideas entered into formulation of the theory for which Planck achieved worldwide and lasting fame are manifold. From all accounts it is apparent that Helmholtz took no direct part in these developments except to provide institutional support for the investigations at the PTR that led to the quantum theory. The quantum story provides a paradigm case of creative investigations and collaborations at the frontier of experimental and theoretical physics and has been examined *in extenso* and with considerable esprit by numerous historians of science and almost as many scientists and philosophers of science.[7]

[6] Planck's closest associates comprised a motley crew and included: the electrophysiologist Emil Du Bois Reymond who had been Helmholtz's first and closest colleague in the Berlin circle of Johannes Müller; the affluent and influential Werner von Siemens who became the prime mover for the *PTR*; the physicist August Kundt who inherited Helmholtz's post at the University of Berlin when he became president of the *PTR*; the Protestant theologian Adolf Harnack who became President of the *Kaiser-Wilhelm-Gesellschaft zur Förderung der Wissenschaften* (renamed *Max-Planck Gesellschaft* in 1948); the politically liberal historian of ancient history at the University of Berlin, Theodore Mommsen, who became Secretary of the *Prussian Academy of Sciences* and was cited in 1902 (Nobel Prize for Literature) as "the greatest living master of the art of historical writing, with special reference to his monumental work, *A History of Rome*"; and the strong nationalist and philologist Wilhelm Scherer who was the first to maintain a phonetic development of the alphabet.

[7] We may single out, in particular, a number of works that examine Planck's life and scientific career from points of view that encompass important facets of the intellectual, social, political, and institutional contexts of early twentieth-century Berlin: Max Born, "Max Karl Ernst Ludwig Planck," *Obituary Notices of Fellows of the Royal Society, 6* (1948) 161–188; Martin Klein, "Max Planck and the Beginnings of Quantum Theory," *Archive for History of Exact Sciences, 1* (1961) 460–479; *Ibid.*, "Thermodynamics and Quanta in Planck's Work," *Physics Today, 19* (1966) 23–32; Hans Kangro, *Vorgeschichte des Plankschen Strahlungs-gesetzes*, Wiesbaden, 1970; Armin Hermann, *Max Planck. In Selstzeugnissen und Bilddokumenten*, Hamburg, 1973; *Ibid.*, "Max Planck," in Wilhelm Treue and Gerhard Hildebrandt (eds.) *Berlinische Lebensbilder*, Berlin, 1987, pp. 115–132; "Max Planck," in Karl von Meÿenn (ed.) *Die Grossen Physiker* II (1997) 143–156; John Heilbron, *Dilemmas of an Upright Man. Max Planck as Spokesman for German Science*, Berkeley, 1986; Annette Vogt, "Berliner Wissenschaft im Abgesang des Wilhelminischen Reiches," pp. 304–395 in Hubert Laitko (ed.), *Wissenschaft in Berlin*, Berlin, 1987. The essays in the Laitko volume provide information and unique insights by historians of science living in the German Democratic Republic under the political constraints of the system at the time. Max-Planck-Gesellschaft, Munich (hrsg.), *Max Planck (1858–1947). Zum Gedenken an seinen 50. Todestag am 4. Oktober 1997.* Max-Planck-Gesellschaft Berichte und Mitteilungen Heft 3/97, Munich, 1997; *Planck-Bibliographie*. Berichte und Mitteilungen Heft 4/97, Munich, 1997; Fritz Stern, Max Planck: "Grösse des Menschen und Gewalt der Geschichte," in *Max Planck. Vorträge und Ausstellung zum 50. Todestag*, MPG (hrsg.) Munich, 1997, pp. 34–53. Archiv der Max-Planck-Gesellschaft, hrsg. Lorenz Friedrich Beck, *Max Planck und die Max-Planck-Gesellschaft*, Berlin, 2008; Dieter Hoffmann, *Max Planck. Die Enstehung der modernen Physik*. Berlin, 2008.

References

Born, Max. 1948. Max Karl Ernst Ludwig Planck. *Obituary Notices of Fellows of the Royal Society* 6: 161–188.

Heilbron, John. 1986. *Dilemmas of an upright man. Max Planck as spokesman for German science*. Berkeley: University of California.

Hermann, Armin. 1973. Max Planck. In *Selstzeugnissen und Bilddokumenten*. Hamburg: Rowohlt.

Hermann, Armin. 1987. Max Planck. In *Berlinische Lebensbilder*, ed. Wilhelm Treue and Gerhard Hildebrandt, 115–132. Berlin: Colloquium Verlag.

Hermann, Armin. 1997. Max Planck. In *Die Grossen Physiker*, vol. II, ed. Karl von Meÿenn, 143–156. München: Beck.

Hoffmann, Dieter. 2008. *Max Planck. Die Enstehung der modernen Physik*. Berlin: Springer.

Kangro, Hans. 1970. *Vorgeschichte des Plankschen Strahlungs-gesetzes*. Wiesbaden: Franz Steiner.

Kangro, Hans. 1975. Max Planck. *DSB* XI: 7–17.

Klein, Martin. 1961. Max Planck and the beginnings of quantum theory. *Archive for History of Exact Sciences* 1: 460–479.

Klein, Martin. 1966. Thermodynamics and quanta in Planck's work. *Physics Today* 19: 23–32.

Merz, John Theodore. 1896–1914. *A history of European thought in the nineteenth century*, 4 vols. London: Blackwood.

Merz, John Theodore. 1965. *A history of European thought in the nineteenth century*, vol. 1, 173. New York: Dover Publications.

Planck, Max. 1958. Wissenschaftliche Selbstbiographie (1948). In *Physikalische Abhandlungen und Vorträge*, vol. III, 381–822. Braunschweig: F. Vieweg.

Vogt, Annette. 1987. Berliner Wissenschaft im Abgesang des Wilhelminischen Reiches. In *Wissenschaft in Berlin*, ed. Hubert Laitko, 304–395. Berlin: Dietz Verlag.

Chapter 17
From Thermodynamics to the Quantum of Action

Thermodynamics was one of the most actively researched areas in physics after the middle of the nineteenth century. In 1859 Gustav Kirchhoff, then in Heidelberg, recognized that the spectrum of the light emitted by a glowing body depends solely on the temperature of the glowing body. He reasoned that there ought to be a law of nature that describes this radiation for all bodies irrespective of their composition. Based on the observation that every body absorbs those frequencies in the spectrum that it emits, Kirchhoff in the 1860s enunciated what came to be known as Kirchhoff's law; namely, that there is a constant relationship between wavelength and temperature. This law holds the key to the relationship between radiation and thermodynamics and led, by way of Planck's reasoning, to the world of the quantum. In 1896 Planck provided a theoretical explanation of the Kirchhoff black-body law. He found, using the proper frequencies of oscillation, that the resulting exchange of energy in emission and absorption gives the required equilibrium. By interpolating the Wien displacement law, which is valid at high frequencies and low temperature, and the Rayleigh-Jeans formula, which is valid at low frequencies and high temperature, Planck obtained excellent agreement with the PTR experiments.[1]

In a lecture at the Johns Hopkins University in Baltimore in April 1900 the British physicist Lord Kelvin spoke of the triumphs of physics and questioned whether Newton's theory of motion might be extended to embrace the phenomena of heat and light. He noted that in "the clear blue sky of physics" there remained on the horizon "just two small clouds of incomprehension" that obscured the "beauty and clearness" of the theory. The first involved the way light travels through space, the second was the problem of distributing energy equally among vibrating molecules as exhibited in the case of the colors of glowing bodies. While the

[1] According to the Wien displacement law the wavelength at which maximum energy is radiated is inversely proportional to the absolute temperature of the body. The Rayleigh-Jeans formula gives the distribution of energy in the spectrum of black-body radiation as a function of frequency and temperature.

E. Hiebert, *The Helmholtz Legacy in Physiological Acoustics*, Archimedes 39,
DOI 10.1007/978-3-319-06602-8_17, © Springer International Publishing Switzerland 2014

solution that Kelvin came up was farfetched, the two clouds of incomprehension that he had referred to turned out to be connected with the two most revolutionary accomplishments of twentieth-century physics – relativity and quantum theory. The experimental underpinnings to the latter relate directly to investigations on black body radiation undertaken at the PTR in Berlin at the end of the nineteenth century. At a meeting of the *German Physical Society* in 1900 important new information on wavelength as a function of temperature in the long wavelength range became available. Using the PTR data Planck was able almost overnight to formulate an equation that expressed radiation as a function of wavelength and temperature, and the equation was in remarkable agreement with the PTR studies.

Eight weeks later, at the 14th of December 1900 meeting of the *German Physical Society* Planck presented a lecture on "the theory of the law of energy distribution in the normal spectrum" in which he introduced the constant "h" as the quantum of action [*Wirkungsquantum*]. Although the idea was presented without fanfare and did not create much of a sensation, the event may best be designated as the point at which an inconspicuous revolution had been launched. In a letter to experimental physicist Robert W. Wood at the Johns Hopkins University Planck wrote in 1931: "In summary I may designate the entire thing as an act of desperation [*ein Akt der Verzweiflung*], for by nature I am peaceful and averse to dubious adventures... .but a theoretical interpretation had to be found. no matter what the cost." Planck remarked on another occasion that his quantum theory had been established as "an interpolation formula based on a fortunate conjecture [*eine glücklich erratene Interpolationsformel*]" that "I really did not think much about [*ich machte mir nicht viel dabei*]." Planck unintentionally had become a revolutionary – *ein Revolutionär wider Willlen* (a revolutionary contrary to his own will), as his colleagues referred to the conjecture. Planck was confident that the outcome of the deliberations that had led to the quantum of action had been sound on both experimental and theoretical grounds; but it embodied revolutionary implications connected to the quantum of action that he was not ready to accept. The search for ways to explain thermal phenomena on the basis of the laws of thermodynamics had been a subject of Planck's deliberations since the days of his studies in Munich. Decades later, in 1943, Planck published a retroflective essay on "The History of the Discovery of the Physical Quantum of Action" in which he outlined the line of thought that had taken him from his interests in the second law of thermodynamics and radiation theory to the idea of the quantum of action. His incentive for writing the 1943 essay was clear-cut. "Since the emergence of the elementary quantum of action has initiated a new epoch into physical science I feel the need and the obligation to provide a summary account, for the later generation, of what is mirrored in my memory and explain, as best I can, the many entangled paths on which I came to determine this universal constant."[2] As Planck remarked in the essay – one that is largely given over to a detailed analysis of how Clausius and

[2] Max Planck, Zur Geschichte der Auffindung des physikalischen Wirkungsquantum. Fassung letzter Hand (1943), *Physikalische Abhandlungen und Vorträge*, *3* (1958) p. 255.

the second law, Kirchhoff's law, and the experiments of Lummer, Pringsheim, and Paschen at the PTR entered into his deliberations – it already was in his student years at the university that he came to recognize the prime importance of "large general laws that have significance for all process of nature independent of the properties of the bodies involved in the processes." With this ambitious and overarching idea in mind, Planck's deliberations about thermodynamics first led to investigations connected with physical and chemical equilibrium changes in the various states of aggregation for gaseous mixtures, solutions, and dissociation theory – an area in which the American chemist Josiah Willard Gibbs already had pioneered in the *Transactions of the Connecticut Academy of Arts and Sciences* in the 1870s. As Planck remarked, his own deliberations along similar lines were noteworthy but granted him "no special achievement [*keine besonderen Erfolge*]." By contrast to the undistinguished chemical equilibrium studies, Planck's activities in the area of thermal radiation thrust him "into virgin territory [*Neuland*]" and led him in increments, by way of the Kirchhoff law, the black-body radiation data at the PTR, and Boltzmann's statistical conception of entropy, to the risky game (*gewagtes Spiel*) of dealing with quantized units of energy, namely, quanta of action and the universal constant h. "But then there arose the most difficult theoretical problem, namely, to bestow a physical meaning on this strange constant [*diesen sonderbaren Konstant*], since its adoption denoted a break with classical theory that is more radical than I ever suspected at first." The nature of the energy element h remained unclear.

> For several years I made repeated attempts to somehow or other incorporate the quantum of action into the system of classical physics, but I did not succeed. Rather, the shaping [*Ausgestaltung*] of quantum physics remained, as is well known, for younger persons [*jüngere Kräfte*] among whom I here list in chronological order only the names of A. Einstein, N. Bohr, P. Jordan, W. Heisenberg, L. de Broglie, E. Schrödinger, and P. Dirac. Among German physicists who deserve much credit there is in first place A. Sommerfeld for the mathematical construction of the theory and Cl. Schaefer for the promotion of physical comprehension.[3]

The difficult "theoretical problem" of giving physical meaning to the "strange constant" that Planck had handed over to the younger generation was approached head-on by the 47-year younger Werner Heisenberg (1901–1976). Since the 1920s he had been actively working on problems in quantum theory that bear on issues referred to by Planck in his 1943 *Geschichte*. In an essay of 1958, entitled "Planck's Discovery of the Philosophical Problems of Atomic Theory," Heisenberg took the position that coming to terms with the issues brought on by the developments in quantum physics (when seen against the background and within the context of classical physics) was more a matter of raising general philosophical questions and having specific prerequisites or basic requirements than focusing on problems encountered in specific scientific disciplines.[4]

[3] Max Planck, Zur Geschichte. . . ., (1943), p. 267.

[4] Werner Heisenberg, Die Plancksche Entdeckung und die philosophischen Grundfragen des Atomlehre (1958) reprinted in Max-Planck Gesellschaft, München (hrsg.) Heft 3 (1997) pp. 40–54.

Intent on demonstrating the advantages of exploring fundamental questions in physics by conceiving them from a philosophical point of view, Heisenberg referred to Newtonian mechanics as the archetypical example of the way in which questions about nature are formulated. What, it was asked, is actually meant by the "understanding" [*Verständnis*] or "explanation" of nature [*Erklärung der Natur*]. The extraordinary influence of Newton's *Principia* on the thinking of the century that followed rested, as Heisenberg understood the situation, not on the special axioms or the result of Newtonian mechanics – say the well-known formula force = mass x acceleration – but on the fact that for the first time natural processes could be described mathematically as a function of time; that, in fact, such a mathematical description of nature was at all possible in principle. According to Heisenberg, "the interest of the natural scientist in philosophical thinking... .resides above all in the formulation of questions and only secondarily in the answers." The former are valuable to the extent that the development of human thinking on philosophical terms becomes fruitful. By contrast the answers in most cases may be the outcome of prevailing circumstances and may lose their meaning as factual knowledge expands. In general, Heisenberg explained, it would be contrary to the spirit of the natural sciences if one would elevate the answers to the status of dogma. Rather, one must attempt, without prejudice, to learn as much as possible from both the old and the new question formulation.[5]

On Heisenberg's models of formulating questions and answers to question, the in-depth meaning of Planck's discovery arose as a consequence of posing a general philosophical question to explain a very special problem connected with knowledge about thermal radiation. The question was:

What can the Planck $\rho_\nu = \frac{8\pi\nu^2}{c^3}\frac{h\nu}{\rho^{\frac{h\nu}{kT}} - 1}$ formula mean for philosophy?

One probably can best clarify the element of novelty that Planck introduced into modern natural science by referring once more to the problem that Plato and Democritus had struggled with two and a half millennia ago – a struggle that signified the decisive differences of opinion [*Meinungsverschiedenheiten*] between these two philosophers.[6]

According to Heisenberg the systematic thinking in Greek natural philosophy from Thales to Democritus led ultimately to questions about the smallest constituents of matter. Atoms were taken to be the absolute ultimate "givens" of matter – indivisible and unchangeable essential existences from which all else, all natural phenomena, but not their own existence, needed to be explained. On the other hand, for Plato elementary particles were not to be taken as the ultimate units of existence;

[5] Heisenberg, Die Plancksche Entdeckung... (1958) 40–41.

[6] Heisenberg, Die Plancksche Entdeckung... (1958) 41.

like everything else in the natural world the atoms needed to be explained mathematically. The Greek debate about primacy of form, picture, image or idea, on the one hand, and matter, material existence on the other influenced and guided much of the thinking among natural philosophers in the Western world into modern times. Plato felt that the idealism-materialism debate was so central to natural philosophy as to suggest that the atomistically oriented books of Democritus be burned.

But what did the Planck discovery have to do with these old questions? It revealed that the atom, the smallest units of matter that had been the backbone of nineteenth-century chemistry, could no longer be taken as the only object of physical research. This discovery made it clear that the notion of discontinuity that had made itself known so successfully in connection with problems in the structure of matter also was revealing itself, but in a very different manner, in areas of investigation such as thermal radiation. According to Heisenberg

> Planck's discovery suggested that the tug toward the notion of discontinuity [*Unstetigkeit*] that had been revealed independently both in the existence of atoms and in thermal radiation was to be understood as the result of a much more general law of nature. Thereby Plato's idea was brought anew into the natural sciences; for in the last analysis the atomic structure of matter is founded on mathematical law or mathematical symmetry – the existence of atoms or elementary particles as an expression of mathematical structure – that was the new element of chance that Planck demonstrated with his discovery. He had raised basic philosophical questions.[7]

It took a quarter of a century to establish a mathematical formulation of the structure of the atom based on the Bohr theory that was free from contradictions. Even so it took another two decades to reach a firm understanding of the structure of matter. In any case it was Planck's discovery that introduced the possibility of recognizing a new type of natural law that can be manipulated to ask questions about processes that are observed experimentally and analyzed mathematically but for which it is not possible to form a physical picture, a visual image. "In the last analysis the nonvisualizeable character [*unanschauliche Character*] of modern atomic physics rests on Planck's quantum of action – on the existence of a criterion for the notion of atomic smallness in the natural laws." According to Heisenberg, quantum theory and relativity theory brought drastic changes to our view of the world because they have made it clear that the visualizable conceptions (*anschauliche Vorstellungen*) that are grasped in daily experience are valid only for a restricted scope of experience; that they by no means belong to the irrefutable preconceptions of the natural sciences.[8]

In conclusion Heisenberg remarked:

> The Planck discovery, whose individual stages of development over the past half century I have attempted to depict, has led to a state in which the goal, namely the understanding of the atomic structure in terms of simple mathematical symmetry properties, is clearly achieved in outline. Even if one regards with skepticism the developments of the latter years that I have spoken of, developments that constitute the uppermost responsibilities of

[7] Heisenberg, Die Plancksche Entdeckung... (1958) 42.

[8] Heisenberg, Die Plancksche Entdeckung... (1958) 42–43 and 45.

> the natural scientist, one nevertheless can assert that one is confronted by structures of extraordinary simplicity, unitary construction, and beauty. They are of interest, in particular, because they no longer relate only to a special demand in physics but to the world as a whole.
>
> Max Planck's 100th birthday comes at a time which, when compared to earlier epochs, leaves one with a very chaotic impression in regard to areas such as politics, art, and standards of value. If one therefore thinks of so harmonious a personality as Max Planck it is reassuring that at least in the one domain to which Planck devoted his life nothing chaotic can be found. Here simplicity and transparent clarity are as determinative as they were at the time of Plato, Kepler, and Newton.[9]

Planck's 1943 account of the history of the discovery that he had made in 1900, and Heisenberg's reflections on that discovery and on the quantum's status in the physical sciences some 60 years later, present us with a contrast in the philosophical perspectives of two major contributors to quantum physics that is truly astounding. Inspired by the research outlook and teaching of Arnold Sommerfeld (1868–1951) in Munich, Werner Heisenberg began his studies in theoretical physics in 1920. He advanced rapidly in his comprehension of the new quantum ideas. In close collaboration with Niels Bohr (1885–1962) in Copenhagen he developed the physical and philosophical formalism that led to the enunciation of the uncertainty (or indeterminacy) principle. According to this principle the act of measuring one of a pair of conjugate physical quantities characterizing a micro-system (such as position-momentum or energy-time) destroys the possibility of measuring the other variable to arbitrary accuracy. After 1927 the principle came to be embodied in the Copenhagen interpretation of "quantum mechanics." For its creation Heisenberg was awarded the Nobel Prize in physics in 1932.

Although Planck had a comprehensive grasp of the developments that led to quantum mechanics in the writings of scientists such as Bohr, Sommerfeld, Born, Jordan, and Heisenberg, he was hard pressed to accept the truth status of quantum mechanics because of philosophical implications connected to the uncertainty principle, wave-particle duality, and the quantum's statistical laws. Although the quantum idea had given Planck something close to a satisfactory physical explanation for thermal radiation he was unable to come to grips, philosophically, with an acausal mechanics at the micro-level. Like Einstein, who had made major contributions to the formulation of quantum mechanics, Planck was a realist and could not rid himself of the conviction that quantum theoretic ideas represent no more than a transitory stage in the development of a more satisfactory theory – one that is ultimately commensurate with traditional classical theory.

The fact that scientists are not always able or are reluctant to fit a newly discovered scientific truth into their own world view was well known to Planck. A case in point is the running debate concerning heat as a form of energy that was set in motion at the 1895 Lübeck meeting of the German Society for Scientists and Physicians – a debate between Ludwig Boltzmann, who was not at the meeting, and Wilhelm Ostwald and his fellow-energeticist Georg Helm. It was a debate in

[9] Heisenberg, Die Plancksche Entdeckung... (1958) 54.

which Boltzmann recognized, and the energeticists did not, that there is a thermodynamically-significant difference between mechanical and thermal processes.[10] In commenting on the Lübeck controversy in a lecture to the society of German engineers in Berlin, but now applying it to himself as one whose quantum theory was being widely accepted by the upcoming younger generation of scientists, Planck remarked: "A new big scientific idea tends not to make its way by gradually convincing and converting its opponents – that a Saul becomes a Paul is a rarity – but rather by having its opponents gradually die out while the new generation becomes familiar with the idea from the outset. Here again we learn that whoever has the youth (in hand) has the future. For this reason a proper programming of school instruction belongs to the most important conditions for scientific progress."[11] Planck was intent on having his colleagues comprehend the depth of sentiment inherent in the radical break with the past that his discovery in 1900 had set in motion because he brought up the same topic at the very end of his life when he remarked: "This experience provided me with the opportunity to establish what I believe to be a remarkable fact. A new scientific truth [*eine neue wissenschaftliche Wahrheit*] tends to gain acceptance not so much by having its opponents [*Gegner*] convinced and declare themselves. But rather much more because the opponents gradually die out and the upcoming generation [*heranwachsende Generation*] makes itself familiar with the truth from the start."[12] This phenomenon has come to be referred to as "Planck's principle."[13]

In order fully to comprehend the origins and depth of the philosophical problems that Planck was confronted with in the 1920s it is imperative to explore the trajectory that the discipline of theoretical physics took in his mind shortly after his arrival in Berlin. While experimental investigations on black-body radiation and theoretical deliberations concerning their meaning were the hot topic of discussion in turn-of-the-century Berlin, Planck, shortly after arrival at his new post, had become deeply engrossed not only in the thermodynamics of heat radiation but in music. A vigorous enthusiasm for music, its enjoyment and performance had been cultivated in Planck since his youth. In Berlin he now would be working within earshot of the author of *Tonempfindungen* who almost three decades earlier

[10] For an analysis of the energetics debate, Erwin Hiebert, The Energetics Controversy and the New Thermodynamics, in Duane Roller, *Perspectives in the History of Science and Technology*, Norman, Okla., 1971, 329–348.

[11] Planck, "Ursprung und Auswirkung wissenschaftlicher Ideen" (1933), *Physikalische Abhandlungen und Vorträge*, vol. 3 (1958) 245.

[12] Planck, "Wissenschaftliche Selbstbiographie" (1948), *Physikalische Abhandlungen und Vorträge*, vol. 3 (1958) 388–389.

[13] For more on the topic: Geoffrey Gorham, "Planck's Principle and Jean's Conversion," *Studies in the History and Philosophy of Science*, *23* (1991) 471–497. For more on the topic: Geoffrey Gorham, "Planck's Principle and FoJean's Conversion," *Studies in the History and Philosophy of Science*, *23* (1991) 471–497.

had demonstrated that an enharmonic harmonium, such as the one that was housed next door to Planck's own office in the physics department, was available for engaging in research on the theory of music. Common interests in music theory between the two theoretical physicists was all it took to lure Planck into the laboratory; not to pursue thermal physics as black-body radiation enthusiasts might have expected, but to study the quality of aural pitch perception of tones produced by the Helmholtz-commissioned Eitz harmonium that had been placed in custody of August Kundt when Helmholtz moved to the PTR. Except for the few experiments that Planck was engaged in as a student in Munich in connection with the lectures of Wilhelm Beetz, the experiments with the Eitz harmonium most likely were the only experiments that Planck was to become involved in.[14]

Before examining, in specifics, the nature and significance of Planck's experimental ventures and writings on music, it is essential to identify some of the cultural and political trends dominant in the city of his birth and training. Munich, an important center for the arts, commerce, and culture, was strong at mid-century in Prussian-German traditions. Planck sang oratorios in the boys' choir, played the organ in church services, and was at one with the family in affirming the ever growing middle-class sentiments for unification of the fatherland. As a 13-year-old at the humanistically structured *Maximiliansgymnasium* in Munich he experienced, first, the war of 1870–1871 in which his brother was killed, and then the establishment of the German *Kaiserreich.* Politically conservative, loyal to the Prussian conception of state, the Planck family maintained its fidelity to Wilhelm II during World War I. On at least one occasion before the war Planck expressed his devotion to the cause of a Germany united in battle. Reminiscing about the days just before the outbreak of World War I, Planck referred to what it meant to be of one mind with the "heroes who had sealed the truthfulness of their love for the fatherland with their heart's blood [*Herzblut*]."[15] In spite of the experience of having lost his oldest son in the first world war, Planck in an address on Leibniz day at the Berlin academy remarked in 1919: "The most terrible war that the world has seen is over but what burns more deeply than all its horror and suffering is the shame that the peace declaration has thrust upon us by our enemies."[16] In this respect Planck was not the exception but rather the prototype of the established Berlin scientist. Politics did not belong to the hallowed corridors of the sciences.

A prominent historian of modern physics and author of important publications on the life and work of Max Planck, Armin Hermann, has pointed out that already as a student Planck had no difficulty accepting the authority of the state, the school,

[14] Planck's *Physikalische Abhandlungen und Vorträge* lists 160 entries all of which, with the exception of the philosophical, historical, and biographical lectures and essays, are theoretical in nature. The single experimental paper deals with the Eitz harmonium.

[15] Max Planck, "Neue Bahnen der physikalischen Erkenntnis." Rede gehalten beim Antritt des Rektorats der Friedrich-Wilhelms-Universität, Berlin, 5. Oktober, 1913. *Vorträge und Erinnerungen*, Darmstadt, 1970, 80.

[16] Max Planck in seinen Akademie-Ansprachen. *Erinnerungsschrift der Deutschen Akademie der Wissenschaften zu Berlin*, Berlin, 1948, 29.

and the church. And yet in spite of this authoritarianism – that also dominated his physics – his thinking led to an intellectual revolution in physics. "He accepted the authority of the school in the same way that he later accepted the authority of the overall structure of physics. Totally contrary to his will, this thinking led to an intellectual revolution at the turn of the century."[17] Planck's lifelong conservatism had much in common with attributes common among members of the European bourgeoisie – an attitude that seems to have held sway not only in regard to order at the state level but in regard to accomplishments in the physical sciences. As Hermann shows, it was in this climate of thinking that Planck formulated his opinions. "An enormous conflict, a huge inner self-debate led him to revise his convictions in later life. Painfully encountered experiences taught him that neither the structure of science nor the structure of the state should be regarded as sacrosanct."[18]

References

Gorham, Geoffrey. 1991. Planck's principle and Jean's conversion. *Studies in the History and Philosophy of Science* 23: 471–497.

Hermann, Armin. 1973. *Max Planck in Selbstzeugnissen und Bilddokumenten*, 9. Hamburg: Rowohlt.

Planck, Max. 1958a. Zur Geschichte der Auffindung des physikalischen Wirkungsquantum. Fassung letzter Hand (1943). In *Physikalische Abhandlungen und Vorträge*, vol. 3, 255. Braunschweig: F. Vieweg.

Planck, Max. 1958b. Ursprung und Auswirkung wissenschaftlicher Ideen (1933). In *Physikalische Abhandlungen und Vorträge*, vol. 3, 245. Braunschweig: F. Vieweg.

Planck, Max. 1958c. Wissenschaftliche Selbstbiographie (1948). In *Physikalische Abhandlungen und Vorträge*, vol. 3, 388–389. Braunschweig: F. Vieweg.

[17] Armin Hermann, *Max Planck in Selbstzeugnissen und Bilddokumenten*, Hamburg, 1973, 9.

[18] Hermann, *Max Planck* (1973), 10.

Chapter 18
Objective Laws as Stepping Stones to the Deity of Creation

Planck was a devoutly religious person and believed in an almighty, omniscient, and benevolent God – not in a personal sense but as a deity [*Gottheit*] identical in character with the power of natural law. He was convinced that religion and science, properly conceived, could not be in conflict; that both had the absolute as their goal and provided shelter for the believer. In Berlin Grunewald Planck served for almost three decades as a churchwarden. His views on science and religion are expressed most unreservedly and most comprehensively in a lecture of 1937 on *Religion and the Natural Sciences*. In this lecture he sought to clarify the question whether knowledge acquired in the sciences is compatible with the religious sentiments for a scientist brought up in the spirit of the exact sciences. More concisely, "whether a scholar can be a scientist and be genuinely religious at the same time." Planck chose to examine these general questions at three different levels: man's personal relationship with God, religion as an organized societal phenomenon, and the extent to which the natural sciences enter into man's religious belief system.[1]

The foundational premise on which Planck forged his religious belief system was one that recognized "religion as the link that binds man to God." It was a life anchored in a respectful attitude of reverence toward the supernatural power that maintains the well-being to which human life is subject. The highest aim of the religious person, Planck believed, was to be in harmony with this power and uphold it with well-meaning as a lasting endeavor. Only in such a relationship would man achieve the inner peace of mind (*Seelenfrieden*) that binds man to God and affords unconditional belief and trust in his omnipotence and in the readiness to help (*Hilfsbereitschaft*). To this extent religion was seen to be rooted in the consciousness of the individual person.[2]

---

[1] Max Planck, *Religion und Naturwissenschaft*. Vortrag gehalten im Baltikum (1937), sixth unaltered edition Leipzig 1938, 8.

[2] Planck, *Religion* (1938), 9.

E. Hiebert, *The Helmholtz Legacy in Physiological Acoustics*, Archimedes 39, DOI 10.1007/978-3-319-06602-8_18, © Springer International Publishing Switzerland 2014

The personal link of the individual to God is but one element in Planck's views on the relationship between religion and the natural sciences. For "in the realm of nature as in the realm of the spirit there is no area that God does not omnipresently permeate." Common to all religions, as Planck pointed out, is the practice of casting God in humanlike form and designating a set of symbols that bestow on religion a vivid graphic quality by which the sanctity of the incomprehensible deity [*die Gottheit*] is carried over into the sanctity of comprehensible symbols. From these symbols arise the strongest motives for the arts to put themselves in the service of religion. While a work of art has its own meaning and does not follow any rules it nevertheless lends itself to special interpretation. "This symbolism is most distinctly perceived in the most abstract of all the arts, in music."[3]

Beyond the recognition of the individual's responsibility to God and religion as an organized social and institutional phenomenon that seeks God by way of religious symbols, Planck claimed that the fundamental religious question lies elsewhere: "Does the higher might that stands behind religious symbols and gives religion its essential meaning have its seat solely in the mind of man and cease to exist at death or does it represent something more? In other words: does God live only in the souls of believers or does he rule the world independent of whether one believes in him or not? That is the point at which minds differ basically and definitively. From a scientific point of view this can never be clarified with logical arguments based on facts. Rather the answer to the question is entirely a matter of religious belief." According to Planck the truly religious-minded person (*der wahrhaft religiös gesinnter Mensch*) knows that God is eternal, has existed before the earth and man existed, and will exist long after everything lies in ruins.[4]

With these remarks about the question of God's existence and all that is related thereto on the part of the individual religious believer, Planck, the theoretical physicist, finally takes hold of issues that more incisively lie at the heart of his conception of the natural sciences and their meaning for religion. The key question for him is whether over the long haul of history scientists have been able to develop a theoretical system of the world of natural phenomena that is true – whether it is real and exists as such, and is not merely a mental construction on the part of the scientists who have conceived it. Planck's answer to this question was in the affirmative. He believed that the so-called "universal constants" – the invariable building blocks that hold together the doctrines of theoretical physics – are real and independent of human intelligence. They provide the evidence that is necessary to confer reality on the system of doctrines established in theoretical physics. "The existence of these constants is tangible proof for the existence of a reality in nature that is independent of all human measurements." In summary: "physical science requires that we adopt [the existence of] a real world that is independent of us – a world, nevertheless, that we can never know directly but only indirectly by investigations using the spectacles of our sensations."[5] For Planck, belief in the real

[3] Planck, *Religion* (1938), 9–11.

[4] Planck, *Religion* (1938), 15–16.

[5] Planck, *Religion* (1938), 20–21.

existence of the physical world that theoretical physics explores and describes should give rise in every unbiased person to the impression that nature is governed by an intelligent will conscious of purpose.

Having laid out the criteria that religion and the natural sciences have independently set for themselves in formulating and resolving meaningful questions addressed to each other, Planck finally reflects on the extent to which they can be in agreement with each other. He writes: "Nothing hinders us from identifying the scientific system of the world [*die einheitliche Weltanschauung*] with the God of religion [*den Gott der Religion*] – two powers [*zwei Mächte*] that are effective and yet mysterious [*wirksam und doch geheimnisvoll*]. Accordingly, the religious person seeks to draw close to the deity [*Gottheit*] by means of graphically perceived symbols that to a degree are made known to the probing person by means of sensations."

Given this congruence between the "powers" of the deity and the powers of the scientific world view, Planck nevertheless recognized a fundamental difference in the way they are received by man. "For the religious person God is given directly and with elemental primacy. From him and his almighty will pours forth all that occurs in the corporeal and spiritual worlds, and although God is not knowable by the intellect he is understood in religious symbols and conveys his sacred message to the souls of all who faithfully trust him. By contrast, the only elemental givens for the natural scientist are the sensations and the measurements that can be derived from them." Beyond sensations the natural scientist seeks the forever unreachable goal of God and his world order to the extent that this is possible. This denotes that "if both religion and natural science need belief in God for their existence then God stands at the beginning for the one and at the end for the other in regard to all thinking."[6]

It is evident from our analysis of Planck's lecture on religion and the natural sciences that he was a strong proponent of realism from points of view that encompass both religious belief and scientific knowledge. This realism shows up elsewhere, especially in his devastating critique of Ernst Mach's sensationalism, but also in his critique of other scientist-philosophers who have been identified as belonging to the camps of positivism and subjective idealism. One may well puzzle about the conceptual and inferential order in which Planck reached his views about the natural sciences and the God of the natural sciences in relation to his views on religion and the God of religion. Did Planck formulate his views on religion and the God of religion by way of his developing ideas on the natural sciences, or were his views on the natural sciences and the God of the natural sciences engendered on the basis of his developing views on religion and the God of religion? In either case, religion and the natural sciences, as conceived and formulated by Planck, are not in conflict. Rather, they advance in parallel, strengthen each other, stand in need of each other, and in a unique Planckian way are integrated into each other's conceptual frameworks and forms of expression. Three overlapping special interests ruled Planck's life and thinking: theoretical physics, religion, and music. An effort is made here and elsewhere in the text to illuminate their mutual interdependencies. In matters of religion we may give Planck the last word: "Religion and the natural sciences have in common the steadily advancing never-flagging struggle against

[6] Planck, *Religion* (1938), 29–30.

skepticism, against dogmatism, and against superstition. The guiding watchword [das *richtungweisende Losungswort*] to this struggle from time immemorial and for all time to come is: Onward to God [*hin zu Gott*]."[7]

Planck's concept of God was closely linked to his belief in the objective existence of natural law and the human mind's ability to discover the existence of natural laws in the outer world by "pure thinking." It has been argued that Planck's belief in the existence of objective laws was a steppingstone to his religious beliefs.[8] Given that Planck's religious beliefs probably were acquired in his early years, prior to the time he formulated his scientific world view, it is equally plausible to argue that his belief in an omnipotent God served as steppingstone to his belief in the objective existence of the laws of nature. Planck's somewhat cryptic remarks about religion in relation to science gave rise after his death to a range of interpretations, many of which were slanted in support of a particular religious or sectarian persuasion. He was claimed by Protestant and Catholic Christians as one who courageously had broken with nineteenth-century skepticism and positivism. He also was championed by some sectarian free thinkers who sided with him on views about God that were not specifically tied to Christian doctrine. In the German Democratic Republic his strong realist position and unremitting critique of scientific positivism were seen by Marxists to give strong support to dialectical materialism.[9]

As the historian of physics Dieter Hoffmann at the Max Planck Institute for the History of Science in Berlin has demonstrated persuasively in a recent volume, Planck was a genuine child of the *Wilhelmian Kaiserreich* and sought to live the life of Prussian virtues – duty, obedience, and modesty.[10] Planck's Protestantism was laced with a characteristic piousness that seldom questioned the state and its

---

[7] Planck, *Religion* (1938), 32.

[8] Paul Feyerabend, Max Planck, *The Encyclopedia of Philosophy*, *6* (1967) 313.

[9] Erich Dinkler, "Max Planck und die Religion," *Zeitscrift für Theologie und Kirche*, Göttingen, *56* (1959) 201–223. For a cogently argued analysis of Planck's philosophical views about physics correlated with his writings that touch on religion L. Jánossy, "Plancks philosophistische Ansichten in der Physik," 389–407 in *Max-Planck-Festschrift*, Berlin, 1958.

[10] Dieter Hoffmann, *Max Planck. Die Enstehung der modernen Physik*, Munich, 2008. In chapter 5 on "Planck as the representative of German Science" (67–83) Author Hoffmann, who was born and brought up in the German Democratic Republic and was exposed to all its crisscrossing political and cultural trajectories, has shown persuasively and in a well-nuanced and document-rich account how Planck's sense of duty, devotion to state, patriotism, German Protestantism, and loyalty to the crown meshed with and entered into his social, political, and scientific outlook on life. The author has known and discussed history of science matters with Dieter Hoffman since about 1984; that is, prior to and after the so-called *Wende* that unified the German Democratic Republic and Western Germany in 1989. Unfortunately, for scholars pursuing the personal lifestyle of Planck there essentially is no home-based *Planck-Nachlass*. Planck's home on the *Wangenheimstrass* in Berlin-Dahlem, along with his correspondence, diaries, and scientific records, was all but totally destroyed in an aerial bomb attack in February 1944. The most important extant Planck documents include the archives of the institutions that were connected with Planck's life: the University, the Academy, the Max Planck Society in Berlin, and Planck's letters to his son Erwin and to his assistant Lise Meitner. Albert Einstein whose *Collected Papers* have been in the process of organization and publication since 1987, provides crucial documentary information for examining the life and work of Planck.

activities. When problems arose within the state Planck was disposed to adjust to them rather than oppose them. Believing that the search for truth was his only major obligation, he felt that he was absolved from responsibility for the development of political adversity. When the first world war broke out in 1914 Planck aligned himself with German patriots who supported those on the nationalistic military front who were ready to go to war for the love of country. Planck was one of the 93 German intellectuals who signed the proclamation (*Aufruf*) entitled "To the Civilized World [*An die Kulturwelt*]," directed to French and British intellectuals. The proclamation denounced as war propaganda the information that was being reported concerning German activities in Belgium. Einstein, who was with Planck in Berlin at the time, opposed the *Aufruf* and wrote to his friend Paul Ehrenfest in Holland: "It is unbelievable how the European mania has taken hold. At such times one recognizes to what kind of sad animal species [*Viehgattung*] we belong. I calmly drowse away in my peaceful ruminations and sense only a mixture of pity and abhorrence."[11]

Military collapse, resignation of the Kaiser, and the proclamation of the Republic were for Planck "days of national misfortune." He nevertheless found solace in the reflection that what had been lost in material welfare could be replaced with intellectual forces. He was convinced that in spite of a military catastrophe German science and research would ever remain for him and his colleagues the last refuge. Planck's belief in the power of science was expressed in 1918 at a plenum session of the Academy: "Where enemies have taken weapons and might from our fatherland, where heavy inner crises have broken in and perhaps still may be more severe in future, what no outer enemy has yet taken from us is the position which German science holds in the world."[12]

During the Weimar Republic Planck came to occupy a position equivalent to that of "science advisor to the state." His scientific research and official functions gave him the status of being the main representative and spokesman for German science. In 1923 he received the order *Pour le merité für Wissenschaft und Künste (Friedensklasse)*.[13] This high Prussian honor, and the position that Planck served as deputy and mediator of German science during the Wilhelmian period – the *Golden Age* of German science – brought its own problems. Planck could not conceivably have anticipated what was to come in the 1930s and beyond. In order to defend and sustain the elevated position he had attained in the eyes of the government and among his colleagues and countrymen, Planck was prepared to compromise with National Socialism in its initial stages and tolerate what to his utter disillusionment developed into the steady encroachment of power at all levels including the natural sciences.

[11] Einstein to Paul Ehrenfest, 19. viii (1914), *The Collected Papers of Albert Einstein, 84* (1998) Doc. 84, 56.

[12] Dieter Hoffmann, *Die Entstehung....* (2008), 76–77.

[13] The Prussian order of *Pour le merité* was established in 1740 for high-ranking military and civil personnel by Frederick the Great and was extended by Frederick William IV of Prussia to include science and the arts in 1842.

Although none of Planck's biographers have maintained that he was sympathetic to National Socialism, several have mentioned that his posture as a leader in science inevitably represented a conflicting situation in which his deportment by no means remained free from ambivalence and questionable judgment. This was the case elsewhere, notably in the early phases of the *Third Reich*, when many others among Germany's intellectual elite silently ignored what was going on and misjudged the depth and durability of National Socialism. Eventually, but much too late, they were compelled to understand that the authoritarian power and the persecution and anti-semitic expulsionism that the regime had tolerated was irreversible. By the mid-1930s Planck had distanced himself definitively from the Nazi dictatorship. As president of the Kaiser Wilhelm Society Planck prevented the Nazis from canceling the commemoration services for Fritz Haber (1868–1934) who had died in exile in Switzerland after having been forced to leave Germany in 1933. The service was held at the *Harnack-Haus* with Planck and Otto Hahn presenting commemoration addresses. The discrimination procedures and persecution of highly trained Jewish scientists became for Planck a heavy burden and created in him the feeling of abandoned despair. At some point he came to the conclusion that all he could do was to shelter German science and its institutions to the extent that when the war was over the preeminence that German science had achieved after the turn of the century would not be totally obliterated and forgotten.[14]

With Albert Einstein Planck developed a close, lasting, and rewarding friendship that flourished in a unique way during their years together at the University of Berlin (1914–1933). Even before they met Planck had enthusiastically supported Einstein's work in relativity theory. In turn Einstein not only supported Planck in his work on quantum theory but made significant contributions to the field himself. In fact in the early 1930s Einstein recognized what few others did at the time; namely that Planck's quantum theory was truly revolutionary and would radically change physics at every level of the discipline. Planck was one of the very first physicists to recognize the fundamental significance of Einstein's first paper on relativity – the 1905 paper on the electrodynamics of moving bodies. In that paper Einstein had put forward a simple and consistent theory of the electrodynamics of moving bodies based on Maxwell's theory of stationary bodies. In 1906 Planck delivered a lecture on the principle of relativity and the fundamental equations of mechanics that precisely and convincingly supported the Einstein results. Shortly thereafter, at a meeting of the German scientists and physicians at Stuttgart, Planck defended Einstein's theory in the face of criticism. In 1909 in his lectures at Columbia University Planck referred to Einstein's revision of the Newtonian

[14] Dieter Hoffmann, *Die Entstehung....* (2008) chapter 6. Zwischen Anpassung und Auflehnung: Das Dritte Reich, 84–103. John L. Heilbron, *The Dilemmas of an Upright Man. Max Planck as Spokesman for German Science*, Berkeley, 1986.

concept of time and the enormous challenge that relativity theory had presented for modern physics. He remarked on that occasion:

> It scarcely needs to be emphasized that this conception of time makes a very high demand on the degree of abstraction and the imaginative power of the physicist. It probably surpasses in boldness all that until now has been accomplished in the speculative study of natural science. In comparison, non-euclidean geometry is child's play. And yet, in contrast to non-euclidean geometry, the relativity principle lays claim, and with good reason, to genuine physical meaning. In regard to the breadth and depth of the vertical changes [*Umwälzung*] in the physical world view, the relativity principle probably is qualified to be compared with the adoption of the Copernican system of the world.[15]

It is unquestionably the case that support for the theory of relativity on the part of the well-established Planck contributed significantly to putting the 26-year-old and 21-year younger Swiss patent office examiner in Bern on the map of theoretical physics. Dieter Hoffmann has offered the comment that all these moves in behalf of Einstein show that "Planck was the discoverer of Einstein; that they led to the establishment of trust and friendship between two scholars [whose] character and political views hardly could have been farther apart."[16]

At this point in our discussion it is essential to search out the deep rationale – deeper than agreement about the merits of relativity theory – to elucidate the nature of the friendship and trust that Planck and Einstein shared over a period of almost three decades. Both men were self-assured about what they were doing and became luminaries whose personal lives and political ideals and allegiances nevertheless were worlds apart. What brought them together, as we will want to show, were mainly two theory-prone components – one in epistemology and the other in aesthetics. First, as to epistemology: both men were devoted to a life in theoretical physics and held firmly to the conviction that it is possible on the basis of pure thinking and mathematical reasoning to discover laws and formulate theories about the external world and its ways that are absolute and independent of the means by which they were acquired. Both men were dyed-in-the-wool realists who had no use for any shade of phenomenalism, Machian sensationalism, Vienna Circle positivism, or Berkeleyian subjective idealism. This realism made itself known and expressed most decisively, as they themselves were anxious to assert, in the areas of their own major contributions to relativity and quantum theory. In essence they both ruled out of court the idea that it is possible to construct a sound and true science of the external physical world that challenges scientific determinism and classical causality.

Although both men recognized that the development of quantum physics had opened up fertile fields of investigation and provided crucial new information about the world of atoms and molecules they rejected the quantum mechanics of 1925, and its Copenhagen interpretation, and looked upon it as a temporary and

---

[15] Max Planck, *Acht Vorlesungen über theoretische Physik*, gehalten an der Columbia University in the City of New York im Frühjahr 1909, Leipzig, 1910.

[16] Dieter Hoffmann,. *Max Planck. Die Enstehung....* (2008), 48.

unfinished science in the ongoing development of quantum-related ideas that eventually would take physics in the direction of a more unitary picture rooted in determinism and causality. They both were optimistic about the physics of the future and maintained that relativity and quantum theory embody important refinements in the notions of space, time, and the microstructure of matter; that the monumentally successful domains of mechanics, thermodynamics, and electrodynamics theory nevertheless had set the stage for a future in physics that avoids acausality and the implications of indeterminacy.

Notoriously, almost scandalously, successful as post-1925 quantum mechanics has proven to be (it remains, thus far, uncontradicted by any experiment), it nevertheless, in the transfigured form of quantum electrodynamics and quantum entanglement continues to be an outrageously strange theory. The struggle to fully understand and predict the behavior of matter at both the atomic and the sub-atomic levels defies commonsense understanding. Indeed, no one could have anticipated the extent to which quantum mechanics has transformed the scientist's sense of reality. Science continues to advance by drastically simplifying what transpires in the world of nature by cutting out everything that cannot be correlated conceptually and mathematically/logically within the broad spectrum of universality. Still a nagging question remains. May reality, actually, be not unitary but plural – one reality for the causal and deterministic commonsense world of classical physics, and another reality for the microworld of neutrons, quarks, and strings? If the unitary world view be entertained as overall goal then scientists in the Planck-Einstein-realist camp should want to urge physicists to keep on searching theoretically, mathematically, for the unitary theory that hopefully, at least for the world of forces, fields, and superstrings, leads to a theory of everything. If the latter view of a pluralism of reality be entertained, scientists in the camp of the Niels-Bohr Copenhageners of quantum mechanics should want to urge physicists to shift their emphasis in the direction of experimentation and leave string theory to the mathematicians.

On the premise that the historian of science is able, on the one hand, to become knowledgeable about and sufficiently skilled in some branch of philosophy of science, and on the other hand, is convinced that the philosophy of science is an advantageous and unique mediator for identifying and exploring important issues in the history of science, it would seem plausible and defensible to maintain an open-minded disposition – a pluralistic outlook on the direction and the course that theoretical physics might take in the future. To remain open to pluralism and to do so persuasively and with humility using the tools and resources of the human mind would seem in the long run to merit the most serious concern; for to entertain a philosophy of pluralism is to recognize that the overall object of scientific research is linked to an incredibly complex world. It may well be the case that even where unification is the ideal – as it invariably should be – the growth of knowledge, as pragmatic goal, should assume priority over conceptual schemes designed to achieve the growth of knowledge. The history of science is a discipline that claims close proximity not only to the natural sciences and their technologies, but to the humanities (foremost), the social sciences, and the practical and fine arts. One of the objectives here, in fact, has been to examine what transpires at the common frontier of the sciences and the humanities. As for quantum mechanics, the historian of science may well afford to leave the problem hanging and blithely opt for close cooperation between the theoretical physicist, the pure thinker so-called, and the experimentalist. It is high time at this point to move on to a consideration of music – a reality and a domain of knowledge that is not as far removed from the natural sciences and mathematics as one might assume.

Thus far an effort has been made to explore how, why, and to what end it came about that two of the most famous mental giants of early twentieth-century physics, one working at the level of atomic entities and the other at the level of astronomical entities, developed a scientifically- and philosophically-rewarding affiliation that has changed man's conception of the world at both ends of the cosmic spectrum. We have seen that both Planck and Einstein held, or came in their work to hold, very firm epistemological views. Both men believed that the human mind is able, on the basis of pure thinking and mathematical/logical reasoning, to discover (confer reality on) laws of nature whose truth status is confirmed by experimental investigation. On these premises both men rejected the Copenhagen interpretation of quantum mechanics which of necessity stipulates that the laws which govern the micro-world violate the time-honored principle of classical causality and opens the door to a pluralistic conception of reality.

Whereas Planck and Einstein may have been of one mind in their critical appraisals of the development of quantum mechanics, in other matters such as political loyalty, allegiance to the state, and affiliation with religious institutions, the 20-year-younger Einstein was the confrontational and obstreperous contra-figure to the socially- and politically-adjusted Planck. They nevertheless respected and even took satisfaction in each other's characteristic idiosyncrasies. Almost immediately after meeting they discovered that they were on the same track in their passion for music and the making of music. During the years of their association in Berlin they regularly played together and jointly fostered the playing of music in home, family, and musical ensemble. In a recent volume entitled *Einstein for the 21st Century* the musician, conductor and educator Leon Botstein has given a splendid appraisal of Einstein and music and the cultural environment in which his musical image was fostered and embroidered by his biographers. According to Botstein:

> Few aspects of Einstein's life and personality are cited with such regularity, in nearly identical phrases and commentary, as his devotion to and love of music. Every biographical account tells the same story. His mother, a pianist of reasonable amateur proficiency, wanted her son to play the violin. It is speculated that she wished to have a partner in the family for *Hausmusik*. Musical culture in the home, which meant chamber music, usually with piano, was a highly prized symbol of successful middle-class acculturation in late 19th-century, German-speaking Europe. It signaled *Bildung*, a sign of status and achievement particularly prized by assimilated urban Jews.[17]

[17] Leon Botstein, "Einstein and Music," 161–175 and 331 in *Einstein for the 21st Century: His Legacy in Science, Art, and Modern Culture*, Princeton (2008), ed. by Peter Galison, Gerald Holton, and Silvan S. Schweber.

In many ways Planck benefited, as Einstein did, from the importance that urban German intellectuals placed on *Hausmusik*. This means that Planck's music-making life is well presented. On the other hand Planck's biographers – physicists as well as others – have all but ignored his own published writings in the area of his involvement in music listening and music theory. In what follows the objective has been to document and provide commentary to when, where, and in what contexts Planck studied and cultivated music and became involved in music-making as keyboard player (piano and organ) and as vocalist and organizer-conductor of chamber and choral music sessions and concerts. Planck's early music efforts in the parental home and as student in Munich were taken up anew when he moved to Berlin and settled in his home in the *Villenkolonie* Berlin Grunewald. It was in the latter environment, in the early decades of the twentieth century that Planck more than any other scientist at the university provided the inspiration for music sessions with friends, academic colleagues, and visiting musicians. Most of the time the groups met in Planck's home on the Wangenheim Strasse no. 21 in the *Villenkolonie* in Grunewald; large groups met at the Sing-Akademie.[18]

In what follows about the music-making activities in Berlin's *Villenkolonie* we will examine the findings and the significance of Planck's experiments and writings that pertain to his music-theoretic investigations on temperament in connection with vocal music using an enharmonic harmonium that had been placed in the physics institute at the University of Berlin by Helmholtz. How these investigations were seen by Planck to relate to his activities at the university and what his colleagues made of the work of a theoretical physicist who was dabbling in music on the side will also be touched upon. Our topic is Planck and music, or more succinctly, "music theory." Einstein and his music story will come into the picture only to the extent that it impinges on Planck's musical activities.

When Planck graduated from Munich's classical Maximilian-Gymnasium in 1874, he had made no decision about the area in which he wanted to continue his university studies. Hans Kangro relates:

> As a child he already had displayed considerable ability in music; (and) he was an excellent performer on the piano and the organ. His home and early school background, with emphasis on *Bildung* in the classical humanistic sense of a liberal education, could well

[18] The history of the *Sing-Akademie* in Berlin is one of the most fascinating chapters in Berlin's musical culture. It was established in 1791 on the private initiative of a community of singers and instrumentalists who over the years performed most of the important choral works of the eighteenth and nineteenth centuries. This included virtually all of the major choral works of Bach, Mozart, Händel, Haydn, Beethoven, Brahms, and Mendelssohn – several as premiere performances conducted by the composer. Planck was able, in his many *Hausmusik* encounters, both to draw on the resources of the *Sing-Akademie* and its musicians and participate with friends in their concerts. When the young Beethoven visited the *Sing-Akademie* in 1796 he was privileged to improvise for his audience. During the years that Planck was in Berlin, the *Sing-Akademie* offerings included, in addition to the works of the above-mentioned classical composers, the music of composers such as César Franck, Edward Elgar, Giuseppe Verdi, Max Reger, Hector Berlioz, and Richard Strauss. The indispensable history of the *Sing-Akademie* is recorded in the Staaatliches Institut für Musikforschung, Preussischer Kulturbesitz, *Die Sing-Akademie und ihre Direktoren*, ed. by Gottfried Eberle and Michael Rautenberg, Berlin, 1998.

> have prepared him to pursue mathematics, music, classical philology, or religion. On seeking advice from a professional musician he was told: "If you have to *ask*, you'd better study something else."[19]

As recorded in his *Gymnasium* report card, the young Planck was a clearheaded logical thinker who demonstrated remarkable aptitude for music and philology but exhibited no particular strength in mathematics and the sciences. Almost every year he received the school prize in religious instruction. On Sundays and at festive church services he made a name for himself by playing the organ and on occasion stood out conspicuously as soloist with his high soprano voice in church and school choirs. The families of the established bourgeoisie, that included Planck and his friends and schoolmates, made full use of *Hausmusik* and theatrical play-acting. At such times Planck not only took over the role of theatrical organizer and producer but enriched the sessions with his own vocal compositions. He even composed an operetta, *Die Liebe im Walde*, that has not survived. In 1874 when Planck had completed his *Abitur* he experienced difficulty in choosing a subject for further study. With keen affection for music and being endowed, as it is said, with absolute pitch, he was led earnestly to consider the life of a concert pianist with classical philology as a strong alternative.[20]

As it turned out, Planck chose physics and mathematics. It was a choice that he later, as an 80-year-old, attributed to the inspirational guidance of his mathematics teacher at the *Gymnasium* Herman Müller. "This middle-aged shrewd and witty man understood how to explain… the significance of physical laws by means of drastic examples. Thus it was that the principle of conservation of energy, a principle that has absolute validity independent of man, came to me as a message of salvation [*Heilsbotschaft*]." The way in which Müller described the conservation principle to the students was to envision a mason laboriously dragging bricks to the roof of a house. His work is not lost. It remains intact and is stored for years until some day, perchance, it comes loose and falls on someone's head below. "This description," Planck remarked, "is still unforgettable for me."[21]

On completion of the qualifying examinations for entrance to the University of Berlin, Planck studied physics and mathematics: first in Munich and then with Kirchhoff and Helmholtz in Berlin. His doctoral dissertation of 1879 in Munich was on the second law of thermodynamics. Following *Habilitation* in Munich in 1880 he became *Privatdozent* in Munich 1880–1885, then a.o. professor for theoretical physics at the University of Kiel (1885–1889), and finally in 1889, on the death of Kirchhoff, a.o. professor and director of the Institute for Theoretical Physics at the University of Berlin. In retrospect Planck's decision to pursue theoretical physics is surprising in light of what his teacher, the experimental physicist Philipp von Jolly (1809–1884) at the University of Munich had communicated to him when he

---

[19] Hans Kangro, Max Planck *DSB*, *11* (1975) 7.

[20] Dieter Hoffmann, *Max Planck. Die Enstehung….* (2008), 11–12.

[21] Planck, Wissenschaftlische Selbstbiographie, *Physikalische Abhandlungen und Reden*, *3* (1958) 374–75.

completed his *Gymnasium* studies. In a lecture delivered at the same university 50 years later, Planck related that von Jolly had told him that with the discovery of the principle of conservation of energy, physics had more or less reached its stable form; that the "system of physics as a whole generally was secure, and that theoretical physics noticeably was reaching the point of completion that geometry already had occupied for centuries."[22] Physics, at the time and until the end of the century, was considered to be no more than a branch of the discipline that had served as mathematical support for experimental findings that had given the discipline a unified structure under the heading of mechanics, thermodynamics, and electrodynamics. Experimental physicists occupied the leading teaching positions in the physical institutes at the universities. Planck may be looked upon in this environment as one of the few physicists of his generation whose ambition it was to take physics in a new direction. He was prepared to teach physics but to engage in experimental physics with an eye to becoming the director of a physics institute that included the teaching of theoretical physics. The persons who would have been counted as full-fledged theoretical physicists at the end of the century would include Helmholtz, Planck, Einstein, Kirchhoff, Boltzmann, Lorentz, and Gibbs. After the turn of the century it was, above all, relativity theory and quantum theory that put theoretical physics on the map. Every major university thereafter was on the lookout for a theoretical physicist. In spite of the fact that Planck gave up the idea of music as a profession, he cultivated music and its performance with unprecedented intensity all his life – a component of Planck's career that physicists have taken scant notice of.

In 1875, Planck and three close friends from Munich's Academic Choral Society (*Akademischer Gesangverein München*), began a four-way round letter that was kept alive until one of the group died in 1927.[23] The letters, collected in a *Brieftagebuch*, contain passing remarks that reveal an intimacy in listening to and performing music that is entirely consistent with what has been written elsewhere about Planck's infatuation with music. The *Brieftagebuch* notably documents Planck's exposure to music in family settings and in the company of his very close friend the violinist Joseph Joachim and the string trios and quartets that Joachim and his students organized in Berlin with members in the Brahms and Schumann circles.[24] Armin Hermann mentions that during his student days in Munich Planck served the above mentioned Academic Choral Society as second choirmaster and tenor soloist. With his high voice he was able in rehearsals to double for the female voice in operas and operettas. According to a retrospective account in the Choral Society's journal, music for Planck was not a matter of

---

[22] Planck, "Vom Relativen zum Absoluten," *Physikalische Abhandlungen und Vorträge*, *3* (1958) 145.

[23] The group included Planck, the applied mathematician Carl Runge, the physicist Bernard Karsten (instigator of the *Physikalische Gesellschaft* that had been founded by six students at the University of Berlin in 1845), and the lawyer Adolf Leopold.

[24] Klaus Hentschel and Renate Tobies, *Brieftagebuch zwischen Max Planck, Carl Runge, Bernhard Karsten und Adolf Leopold*, Berlin, 1999.

dilettantism or youthful antics. "His contributions as a composer, singer, pianist, and organ-player were highly valued expressions of one who was endowed with an unusual musical aptitude, absolute pitch, and considerable technical knowledge about music." When Planck moved to Berlin in 1889 he developed a friendship with Joachim and became attached to the *Hochschule für Musik*, where Joachim was the director. For a period of 50 years Planck became a part of the best that Berlin had to offer in music. His home on occasion became the locus for trios: Planck at the piano, Einstein on the violin, and Planck's son Erwin on the cello.[25]

During the years that Planck was at the University of Munich as *Privatdozent* (1880–1885) and served as leader of the *Akademischer Gesangverein*, he decided to study harmony and composition with the composer, organist, and music teacher Joseph Rheinberger (1839–1901), professor for composition and organ instruction at the newly founded Royal School of Music in Munich.[26] From all that has been written about Rheinberger it is evident that his students were exposed, as Planck was for 5 years, to a sound and comprehensive knowledge of counterpoint, fugue, and score-reading. Rheinberger had been trained in the music of Bach and the Viennese Classical School of Haydn, Mozart, Beethoven, and Schubert, and had used them as his model. As a composer whose music and teaching were firmly rooted in the classical tradition, Rheinberger attracted top-notch students notably from Austria, Italy, and New England.[27] Planck began the serious study of music theory in 1880 at a time when the general tendency was one that fostered greater freedom in musical expression. Prudent in his approach to teaching, Rheinberger sought, in the mercurial Munich environment, to stress the critical acquisition of well-established techniques of harmony and counterpoint in conjunction with the classical music tradition. In the music and spirit of high romanticism, and notably in the works of Richard Wagner, Rheinberger found nothing of value for his own artistic expression. As one who at the age of 25 had witnessed the Munich *première* and subsequent quarrels surrounding *Tristan und Isolde*, Rheinberger decided,

---

[25] Armin Hermann, "Max Planck," in Karl von Meÿenn (ed.), *Die Grossen Physiker*, vol. 2, Munich, 1997, 143–156.

[26] For information on Rheinberger: Anton Würz, "Joseph Gabriel Rheinberger," *MGG*, *11* (1963) col. 377–381; Anton Würz and Siegfried Gmeinwieser, "Rheinberger," *NG*, *21* (2001), 257–258; Wolfgang Hochstein, *MGG*, Personenteil *13* (2005), col. 1615–1621.

[27] Rheinberger's students, apart from Planck, included: Engelbert Humperdinck, composer of *Hänsel und Gretl* (1890–1893, première Weimar under Richard Strauss); the Italian opera composer Ernanno Wolf-Ferrari; the Austrian composer Ludwig Thuille, professor of theory and composition at the Royal School of Music in Munich; Wilhelm Furtwängler (1886–1954), the leading German orchestra conductor of the Leipzig Gewandhaus Orchestra in the 1920s and then the Berlin Philharmonic until his resignation in 1934; and two prominent Boston composers, Horatio Parker, professor of music at Yale, and G. W. Chadwick, director of the New England Conservatory in Boston. As recently as December 2007 one of Rheinberger's best known cantatas, *the Star of Bethlehem*, Op. 164 (scored for soli, chorus, orchestra, and organ, and premiered in 1892 in the Dresden Kreuzkirche) was performed by the Dedham Choral Society (Boston environs) and orchestra.

forever, to take a direction in music contrary to the ascending new directions in music.[28] It was said that "although he disliked the works of Wagner and Liszt and was no partisan of the *New German School* he never tried to influence young artists in his care with his personal views." His strength rather was "in the indisputable mastery and planned coherence of a compositional style that was imbued with the spirit of polyphonic thinking rather than with compelling inventiveness or vivid conception." The pianist, conductor, and uncompromising music critic Hans von Bülow, who enjoyed the reputation of being referred to as Germany's great *Klavierpedagogue*, spoke of Rheinberger as Germany's number one contrapuntist and music teacher, "a truly ideal teacher of composition unrivaled in the whole of Germany and beyond, in skill, refinement and devotion to his subject; in short one of the worthiest musicians and human beings in the world."[29]

As the most prominent personality in the so-called Munich School of the second half of the nineteenth century, Rheinberger directed his efforts mainly to the composition, performance, and teaching of scared and secular choral and organ music. Rheinberger's emphasis on the vocal performance of music discernibly attracted the attention of the enterprising young Dozent Planck in Munich was to remain intact for him during the four decades of his musical life in Berlin. It took the form of active participation in choral groups and the writing of an idea-rich research article on natural intonation in modern vocal music (examined below). The characteristic benchmarks of Rheinberger's approach to vocal music that would seem to find support in Planck's musical activities and writings in Berlin are just those that a prominent musicologist has put forward in the recently updated article on Rheinberger in *Die Musik in Geschichte und Gegenwart (MGG).*[30]

Characteristic in Rheinberger's music is his "combination of a song-like, often Gregorian-inspired melody that demonstrates absolute mastery of counterpoint and stylization." His works are distinguished by "an attitude not structured on superficial effects but on well-shaped lyrical features and an unmistakable striving toward tonal beauty." According to one of Rheinberger's students, Wilhelm Furtwängler, "a naturalness in the shaping of voice-leading [*Stimmführung der Formgebung*]" was for Rheinberger "the supreme law in music-making." His academic-classic position (he was referred to as the "*German Cherubini*") led him to distance himself from the *New German School* (*neudeutsche Schule* associated with Liszt, Berlioz, and Wagner) and brought his artistic outlook "to stand lonely in the stream of modern music." In spite of his self-imposed traditional sentiments, it is apparent that Rheinberger took advantage of the harmonic achievements of the time and actually experimented with the *Leitmotiv*. Dominant seventh and ninth chords, intensified altered fourths, free dissonances, and other methods typical of the time were never used in excess or as ends in themselves. "In spite of

---

[28] Würz, "Josef Rheinberger" *MGG* (1963), col. 380–381.

[29] Würz and Gmeinwieser, "Rheinberger" *NG* (2001), 257–258.

[30] The information that follows is taken from Wolfgang Hochstein, *MGG* Personenteil *13* (2005), col. 1619–1620.

all their intensities and expressiveness Rheinberger's harmonies never became brusque or flashy. His modulations were elegant and compelling without being anticipated in their development." In liturgical compositions Rheinberger refused to join those trends in the *Cecilian movement* that sought the reform of Catholic church music by promoting a sober style of unaccompanied choral singing. Rather, he pressed for a balance between restrictive tendencies in church music and the art of making music in its own right.

The artistic outlook of Rheinberger and his Munich School, the striving for tonal purity, the strong personal commitment to sacred (Protestant) and secular choral music, and the participation in various choral ensembles, taken together are fully consistent with the path that Planck's interest in music took in Berlin. His emphasis on *Hausmusik* sessions in his own home in Grunewald and his experiment-based writings on music theory are all connected with vocal music.

The well-known Swiss ophthalmologist Ernst Wölfflin (1873–1960), who studied medicine at the University of Munich, has written about his personal encounters with Planck when he was in Munich as *Privatdozent*. Wölfflin wrote:

> I recall that I first heard the name Max Planck when my parents, from time to time, invited the young *Privatdozent* to our house in Munich for social occasions. It remains distinct in my memory how my father became engaged with him in lively discussions dealing with musical problems. He was taking piano lessons from Professor Rheinberger who, besides, instructed him in harmony and composition. As a student Planck was leader of the *Akademischer Gesangverein*. He also tried his skills at composition. Unfortunately most of his works burned in the air raids. Only a romance for piano and violin has survived.[31]

Later encounters with Planck led Wölfflin to believe that apart from an extraordinarily active life in physics, Planck gave himself to music-making so consistently and intently because he experienced in music not only "a gratifying relaxation from hard work but a revelation from on high [*eine Offenbarung aus höheren Sphären*] that gave him the fortitude to carry the inner strains of life with philosophical repose." In spite of the many honors, awards, and decorations, that included the Nobel Prize for physics in 1918, Planck remained a simple person; but "whoever had the good fortune to know Planck more intimately knew that he possessed a nature that hermetically isolates its inner self from the outer world. His feelings and moods were not easy to fathom."

Planck admired Helmholtz not only as a physiologist and theoretical physicist and for his fundamental contributions to optics and acoustics, he was an enthusiastic supporter of Helmholtz's idea that it is as high-minded and commendable an enterprise or calling for a theoretical physicist to engage in the study of music and music theory as to study physics and theoretical physics. Planck's training in piano playing and choral music in Munich, alongside his active participation in both a cappella and accompanied choral groups in the *Gesangverein*, afforded him with

---

[31] Ernst Wölfflin, "Begegnungen mit Max Planck," *Ciba-Symposium*, *8* (3 Aug. 1960), 117–121. This article was called to my attention by my advisor and close friend Hans-jörg Rheinberger, Director, Max-Planck-Institut für Wissenschaftsgeschichte, Berlin. Hans Jörg is Joseph Rheinberger's grand-nephew.

the ideal background to explore music from a theoretical point of view. In Berlin Planck's interests in music entered a new phase. Helmholtz, the doyen of musical acoustics, had become his colleague and in an involuntary way had set the stage for Planck to engage in music study and research by giving him free access to the Eitz-constructed harmonium that he had commissioned. The first Eitz harmonium had been put to use in experiments with Joachim at Heidelberg in the 1860s; the second had been installed in the Berlin physics department in the late 1880s.

Planck first presented himself to the professional music world in written form in a review of Leopold Zellner's *Vorträge über Akustik* (1892) – a set of 62 lectures that Zellner had presented over a number of years at the *Konservatorium der Gesellschaft der Musikfreunde* in Vienna.[32] The review of this ambitious work represents for Planck an entrée into the world of musical acoustics that his 30-years-older colleague Helmholtz already had explored in considerable depth since the 1840s. Largely critical of Zellner's form of presentation, it is evident that Planck's review is written against the background of Helmholtz's *Tonempfindungen.* The Zellner *Vorträge* comprises two volumes: the first on the physical aspects of tonal matters; the second on the analysis of hearing and the physiological and aesthetic applications of tonal materials. The almost 800-page text includes 331 illustrations, many musical examples, and 20 appendices. Planck's overall criticism is that in the case of music students, for whom the lectures were intended, Zellner's approach is structured too deductively; instead of approaching the acoustical analysis of musical phenomena from a theoretical point of view followed by musical examples to illustrate those principles, the reverse would have been more effective pedagogically speaking. Planck acknowledges that the deductive approach has great merits when it is possible to take hold of a problem and treat it from a theoretical point of view that is comprehensive, convincing, and completely free from objections. But "in the present, case where an education in music, but not in physics, is assumed on the part of the reader, it would be preferable to use the deductive way throughout and begin with established musical experiences." Zellner, in fact, "attempts theoretically to derive propositions that are not capable of a general demonstration." He does this, for example, with the fundamental law of acoustics (of Mersenne) according to which pitch depends on rate of vibration. But, "this law is no more than a physiological experience that is said to have no

[32] Max Planck, "L. A. Zellner, Vorträge über Akustik, Zwei Bände, Wien, Pest, Leipzig, 1892." *Vierteljahrsschrift für Musikwissenschaft*, Leipzig, *8*, (1892) 536–541. Leopold A. Zellner (1823–1894) was a Croatian-born Austrian music pedagogue and composer. He was the foremost master of music theory in Vienna and had succeeded Simon Sechter (1788–1867) (teacher of Schubert and Bruckner) as professor at the Conservatory in Vienna. Zellner was general secretary of the *Gesellschaft der Musikfreunde* in Vienna, founded in 1812. Its *Konservatorium* became Vienna's chief music school. Gustav Mahler and Hugo Wolf taught there toward the end of the century. Its *Singverein*, one of Vienna's principal choirs, included among its conductors Brahms, Hans Richter, and Wilhelm Furtwängler.

exceptions, and it would be entirely hopeless to want to devise a proof based on purely physical considerations."[33] There are numerous references to Helmholtz's *Tonempfindungen* in Planck's review of Zellner's *Vorträge*. Planck intimates that in place of consulting Zellner's volumes as a source of information on matters such as tone quality, interference, beats, combination tones, the functioning ear, and intonation, music students and musicians should be advised to rely on Helmholtz's classic text. All such topics, as Planck remarks, stand in need of being examined from a point of view that seeks to understand the subtle way in which physics, physiology, and aesthetics interrelate.

The final comments of Planck's review pertain to an examination of Zellner's presentation of the notion of temperament. It is important to spell them out in some detail because the temperament problem is directly connected with the principal issue that Planck was to explore more incisively in two subsequent papers on musical acoustics. Planck wrote:

> In regard to questions concerning the advantages of the various temperaments that are now much debated, the author (Zellner) warmly supports the 12-tone tempered system in contrast to pythagorean and so-called natural intonation. He attacks the latter with all the logic and wit at his command. On closer examination, however, one finds that most of the stated reasons for defending the tempered system.... sidestep the crucial question. They deal not with the more or less complicated construction of the scale, not with the more or less restricted ability of modulation, and also not with the more or less difficult execution, but only with the decision regarding the intonation in which the music sounds best. Every theory must orient itself in relation to this consideration, no matter how it turns out, provided it does not set itself above it. As long as it is not shown, always and everywhere, that tempered intonation will produce better or at least as good music as natural intonation, it will not be possible to subscribe for all times to come, as the author does, to a verdict of the annihilation of natural intonation.... A moderately well trained listener can readily sense that a gently sustained natural triad [*Dreiklang*], that is produced independent of any theory, sounds better than a tempered tone, and sounds as beautiful as anything one actually can hear; and not only for the conscious critic but also for the uninhibited enjoyment-devoted listener. Thoughtful choir directors know well how to appreciate this situation and promote the occurrence of such effects; which, incidentally, can be accomplished more reliably and in more diverse ways by enlarging ones knowledge and the commensurate practice of singers.[34]

Planck conjectured that Zellner was well aware of the enormous difficulties associated with any attempt to put a system of natural intonation into practice. In exaggerating this difficulty in his *Vorträge* he nevertheless had overshot the mark and rejected natural intonation outright. Planck wrote:

> [T]he author must be given credit for realizing that even if one totally looks beyond the technical difficulties, it would be misguided to want to execute a piece of music in natural intonation that had been conceived and composed in tempered intonation. In many places one would encounter insoluble difficulties since the demand that natural intonation makes on polyphony in order to reconcile the harmonies are far more formidable than what is required for compromise in tempered intonation.[35]

[33] Planck, "Zellner Vorträge," (1892), 537.

[34] Planck, "Zellner Vorträge," (1892), 540.

[35] Planck, "Zellner Vorträge," (1892), 541.

It is evident on reading Planck's comments about Zellner's preference for the 12-tone tempered system that he construes Zellner's position as one that is based on comparing the relative merits of the differences, for the listener, of music that is performed in the natural system of intonation with the 12-tone equal temperament system. Planck takes the former to be the norm among music theorists while Zellner apparently takes the latter as the system of preference. Planck's argument revolves around the idea that the 12-tone equal temperament system is the one that most listeners have adopted or have been forced to adopt because of formidable difficulties encountered in performing music in any context that involves keyboard instruments. It is the same basic problem that Planck will struggle with in his subsequent papers that deal, first, with experimental investigations on an harmonium constructed to approximate natural intonation, and, second, in a paper on music theory in vocal music.

Within the year of his review of the two Zellner volumes, Planck became totally caught up in exploring what might be learned about temperament with the help of an harmonium that had been constructed to approximate just intonation. His findings led him to undertake a closely argued analysis of the role that intonation plays in the examination of vocal music. What he learned from experiments on the harmonium is recorded as follows: *"Hr. M. Planck demonstrated a new harmonium in just intonation based on the system of C. Eitz.* The instrument was designed by Carl Eitz in Eisleben, built in Stuttgart by the Schiedmayer piano factory in Stuttgart on commission of the Prussian state government, and placed in the Institute for Theoretical Physics in Berlin, where, with the kind permission of Mr. A. Kundt, it is installed in the physical institute of the university."[36]

In the two-page published account that features Planck's demonstration of the Eitz harmonium it is recorded that he was able, by means of specific examples, to illustrate a number of musically significant features of the instrument. Additional and more significant information about the circumstances surrounding Planck's experiments with the Eitz harmonium come from Planck's *Scientific Autobiography* published after his death. It is there that he mentions more explicitly what the experience with the Eitz harmonium had meant to him personally in an academic environment in which, as a theoretician surrounded by experimentalists, he had been accepted with a degree of reservation. One of the colleagues "invariably treated me in a very friendly manner but I nevertheless always had the feeling that he really considered me to be rather useless [*überflüssig*]. At that time I was, by far and wide, the only theoretician, a physicist *sui generis* so to speak; and that made my entry into the job not entirely easy. I also believed to sense that the assistants at the physical institute met me with a

[36] Max Planck, "Ein neues Harmonium in natürlicher Stimmung nach dem System C. Eitz," *Verhandlungen der Physikalischen Gesellschaft*, Berlin, *12* (1894) 8–9. The article also appeared in *Annalen der Physik und Chemie*, *49* (1893) 8–9, and is reproduced in Max Planck, *Physikalische Abhandlungen und Vorträge*, Braunschweig, vol. 1 (1958), 435–436. The construction and functioning of the Eitz harmonium that Helmholtz had commissioned before the publication in 1863 of *Tonempfindungen*, as well as the one that was installed in the Institute in Berlin in the 1890s, are discussed in connection with Chapter 8 in Helmholtz's *Tonempfindungen* on pp. 105–106 and 125–127.

certain pronounced reticence." But all was not lost, for as Planck relates, it was music that rescued him from thinking of himself as having been abandoned.

> As chance would have it, right at the beginning of my Berlin appointment, my complete attention was given over for some time to work in an area that lies far removed from my special physical interests. In fact it was just at this time that the Institute for Theoretical Physics had come in possession of a harmonium in just intonation as an article of inventory. It was designed by the genial elementary school teacher Carl Eitz in Eisleben. It had been commissioned by the Ministry, and built by the piano company Schiedmayer in Stuttgart. I had the task of carrying out investigations on just intonation with this instrument. I did so with great interest especially in connection with the question concerning the role played by just intonation in modern vocal music free from instrumental accompaniment. *While doing so I reached what for me was the somewhat unexpected result that our ear under all conditions prefers tempered intonation over just intonation* [emphasis added]. Even in a harmonic major chord the natural third sounds dull and expressionless in contrast to the tempered third. Without doubt this fact can be attributed in the last analysis to years and generations of habituation. For before Joh. Seb. Bach, tempered intonation, in general, was unknown.[37]

In anticipation of the argument that I wish to present in connection with Planck's position on the problem of intonation, it is necessary to unpack what Planck had to say in the sentence that is set out in italics in the above. In this way I will be able to put in proper context what Planck has said about music theory in his earlier writings. Planck's *Scientific Autobiography* (*Wissenschaftliche Selbstbiographie*) was written shortly before he died. The point I wish to make, in brief, is that the sentence set in italics in the above seems offhand to fly in the face of everything that Planck wrote about music theory elsewhere. There is first, Planck's review of the 1892 Zellner *Vorträge* in which, as we have shown, Zellner is roundly criticized for offering nothing but the flimsiest of arguments for preferring tempered intonation over natural intonation.

The following year Planck demonstrated the functioning and capabilities of a new keyboard – an harmonium, that had been designed by Carl Eitz with the sole intention of exemplifying and illustrating the production of intervals and chords in just intonation. Planck's experiments with the Eitz harmonium were first published in Wiedemann's *Annalen* and reproduced in Planck's *Physikalische Abhandlungen und Vorträge*. As it turned out this was the only one of Planck's three writings on music that was published in his collected *Physikalische Abhandlungen* and taken into account in the reflective writings of those of Planck's colleagues and associates who were intimately affiliated with his scientific work. The key idea behind Planck's experiments with this unique instrument was to demonstrate, as Helmholtz had done so circumspectly in his *Tonempfindungen*, that just intonation is uniformly preferred by the human ear, provided it has not, by habituation, been conditioned or been blunted (*abgestumpft*), as Helmholtz refers to this manner of conditioning the sense of sound.

[37] Max Planck, *Wissenschaftliche Selbstbiographie*, Leipzig, 1948. Reprinted in *Physikalische Abhandlungen und Vorträge*, Braunschweig, 1948, Band III, 374–401. Quotations on 383–384. *Dabei kam ich zu dem mir einigermaßen unvermuteten Ergebnis, daß unser Ohr die temperierte Stimmung unter allen Umständen der natürlichen Stimmung vorzieht.*

Planck's most important writing on music was a 22-page paper entitled "Die natürliche Stimmung in der modernenVokalmusik" (Natural Intonation in Vocal Music). Although the article was published in the foremost music journal of the time, the *Vierteljahrsschrift für Musikwissenschaft*, it did not find its way into Planck's collected works, the *Physikalische Abhandlungen*. In this closely reasoned essay on natural intonation in modern vocal music, as in the two other articles mentioned in the above, Planck argues the merits of natural intonation from a music-theoretic point of view.

The Planck study on vocal music was published in the same musical quarterly that 3 years earlier had seen the publication of the Helmholtz-inspired Tanaka study on pure intonation and received wide distribution in a pamphlet distributed by the distinguished music publisher Breitkopf & Härtel in Leipzig.[38] In the paper on intonation and vocal music Planck confronted, head-on, problems connected with temperament that from the standpoint of the physiology and the psychology of musical perception had been at the core of Helmholtz's *Tonempfindungen*. Planck's decision to tackle the temperament problem – one of the most difficult and controversial problems in music theory – represents a unique case in which a person, in this case the physicist Planck, was convinced that he would be able to provide new insights to a problem in a discipline that was located outside the bounds of his own physics discipline but not – so he assumed – outside the bounds of what he as a physicist would be able to bring to bear on the subject by way of intellectual perspective and scientific expertise. As already mentioned, a surprising number of scientists and mathematicians, as the history of music discloses, have assumed that the science of music and its theoretical foundations are best amenable to investigation by means of the intellectual and technical tools that belong to the armamentariums of scientists and mathematicians.

In his paper Planck approached the subject of intonation on the presupposition that within most music circles of his day questions connected with natural tuning (*natürliche Stimmung*) were being given widely conflicting coverage. He felt that the majority of musicians were holding fast to the opinion that just intonation is mainly a matter of theoretical interest; that for the practical pursuit of music there probably was no sufficiently resolute motive for adopting a tuning system other

[38] Max Planck, "Die Natürliche Stimmung in der modernen Vokalmusik." *Vierteljahrsschrift für Musikwissenschaft*, Leipzig, 9 (1893) 418–440. The article is not included in Planck's 3-volume *Physikalische Abhandlungen und Vorträge* (Braunschweig, 1958). In view of Planck's deep-seated scientific and humanistic interests in music (a topic which is touched upon by his colleague Max von Laue in the preface to Planck's *Physikalische Abhandlungen*), this omission would appear to be unwarranted, especially since volume 3 of the collection includes 33 of Planck's non-technical writings on subjects as varied in nature as his inaugural and celebratory public addresses, his historical and philosophical reflections, and his obituaries and personal reminiscences. I found no mention of this 1893 paper in any of the books, papers, or chronological lists on Planck with the single exception of *Planck-Bibliographie. Zum Gedenken an seinem 50. Todestag am 14. Oktober 1997*, Petra Hauke (ed.) (Archiv zur Geschichte der Max-Planck-Gesellschaft, Berlin-Dahlem) and published in *Berichte und Mitteilungen*, Heft 4/97 of the Max-Planck-Gesellschaft, Munich, 1997, 18.

than the tempered tuning (*temperirte Stimmung*) that for more than 200 years had experienced brilliant successes. But Planck also recognized that there were exceptions to this one-sided view. There were violinists who know from experience "where a double stop sounds better"; and there were "choir directors who know that in a cappella singing the third is low compared to the same interval on the piano." In string quartets and choral groups, natural tuning normally was taken for granted. Some theoreticians, he remarked, even go so far as to question the justification of tempered tuning because to do otherwise amounts to "giving the ear the lie [*dem Ohr etwas vorlügen*]."[39]

Planck proposed to formulate the temperament problem by examining two different but related issues: first, the listener's priorities and opinions in regard to mode of tuning, and second, the optimal system of tuning for concert hall and theater performances. He recognized that personal priority and opinion in these matters although of prime importance were too complex humanly speaking, too dependent upon changing times, and too difficult to settle without a proper exposé of the more practical issues of time and place of performance. He believed that too much attention had been devoted in the musical literature "to what shall be rather than to what actually is." He felt that the outcome of the deliberations about temperament among musicians had not led to noteworthy positive results. Accordingly, Planck decided to isolate a single factor associated with natural tuning by focusing on the human voice as the instrument which, even more than the strings, is able to temper musical intervals and couple the temperaments with direct assessment by the human ear. "Above all," Planck maintained, "the ear must learn to know intervals other than the tempered intervals of the piano; that is, intervals such as the well-defined differences offered by natural or Pythagorean tuning." To this end Planck recognized that the large Eitz harmonium was superbly suited [*vorzüglich geeignet*] for experimentation. As he suggested, "On this instrument one can achieve what no similarly constructed instrument has been able to achieve: eight natural thirds... as well as the other natural and Pythagorean intervals. In addition, any tempered or other desired interval can be demonstrated on this instrument in a firm tone since the intervals which are present in so large a number can be put into suitable combinations whose deviations from the desired interval is imperceptible for the finest ear."[40]

One of the problems in musical acoustics that Planck saw fit to explore with the help of the Eitz harmonium – a problem that by the end of the century had received considerable attention on the part of psychologists – was the extent to which the human ear is able to accommodate pitch and pitch variation. He first turned his attention to what one is able to learn about the listener's innate ability to mentally accommodate tones that are slightly off pitch. This, as Planck remarked, is a discernment or accommodation phenomenon that varies among musicians as well as for the same listener under different listening circumstances. For children, whose

[39] Planck, "Die Natürliche Stimmung" (1893), 418.
[40] Planck, "Die Natürliche Stimmung" (1893), 420–421.

ears have not been blunted by exposure to music in equal temperament, there is, as Planck recalled from experiences in his own youth, a sharper power of pitch discrimination. The main point of emphasis for Planck was that with an exertion of effort the listener generally is able, even if necessary by the cultivation of fantasy, to accommodate for what is referred to as "deception of the ear [*Täuschung des Gehörs*]" – a charitable thing for the listener – "otherwise no practical music could exist." Whatever else there was to be said on the subject, Planck insisted on maintaining the premise that accommodation is no excuse for badly tempered music. He noted that accommodation to pitch more readily becomes a matter of scrutiny for music critic than for music listener, and normally is more conspicuous for the music listener than for the music performer. Temperament, as he remarked, also makes itself sensed more directly in slow than in fast music passages and is sensed more readily when the opportunity for comparison with notes in natural tuning is possible. With the Eitz harmonium it was possible to check all such claims by gradually altering the pitch of any note by about a comma – that is, about 1/9th of a full tone – using beats to monitor the pitches. In describing his own reactions to tuning phenomena Planck seems to have been carried away with exuberance when he stated: "Of special interest is the moment when the natural interval emerges and the beats are totally lost. For me this point of transition, plainly, is a source of artistic pleasure; it is as if a certain last earthly residue had vanished while a veil is raised that opens up to the fantasy of an insight reaching a world of unending remoteness."[41]

The phenomenon of psychic after-effect (*psychische Nachwirkung*) was also put to test in connection with tuning. Pitches were adjusted up and down by a comma, while the ear accommodates to what is heard with the aid of memory and the power of imagination. The ear anticipates the sequence of tones that follow. The study lends itself to investigating the effect of tonal sequence on the listener and is illustrated by demonstrating that in playing a tempered minor third, followed by a natural minor third, the natural minor third sounds too high and bright, while in the reverse sequence the natural minor third sounds too low and dull. Commenting on his own experience, Planck claimed that the sensitivity of the ear diminishes with time and age: "Nowadays, after I have heard many kinds of music in many modes of tuning [*Stimmungen*], my sensitivity to tone [*Tongefühl*] no longer is so secure." As Planck was quick to point out, all observations connected with the ear's listening capabilities reveal not only that the ear continuously accommodates to intervals and sequences of intervals but becomes accustomed to the accommodation. "It is habituation [*Gewohnheit*] that here, as in all spheres of life, so frequently plays the powerful but inconspicuous role in the arts.... As in the case of other unusual stimuli one either becomes used to them and then misses them when they fail to appear, or they act disagreeably and intensify to the point of intolerance."[42]

[41] Planck, "Die Natürliche Stimmung" (1893), 424–425.

[42] Planck, "Die Natürliche Stimmung" (1893), 428–430.

It is known from biographical accounts that Planck participated in choral groups both before and after he entered academic life; and it comes as no surprise to see that discussions about vocal music and the choral ensemble take pride of place in his essay on vocal music and just intonation. Planck most likely was motivated to express his views on the subject of vocal music since he felt, as had Helmholtz, Tanaka, and others in the Berlin environment, that equal temperament had conditioned vocalists who perform solely in collaboration with the piano. They had gained the upper hand in music circles and were threatening to intimidate the theory-oriented just intonationists. "It is the power of habit, manifest notably in the frequent use of the piano, that has caused tempered tuning to get into the flesh and blood of the singer. . . . and it has been theory and not practice that has led to the revival of natural tuning."[43] For Planck it was evident that the just intonation problem was not an issue that belongs solely to music theorists and scientists dealing with music theory. The problem as he saw it was that "tempered tuning is inconsistent [*unkonsequent*], as everyone knows, and is in contradiction with itself, because it demands pure octaves, tempered fifths (that are very pure), and tempered thirds; and these three demands cannot be fulfilled at the same time since every tempered third results in overtones and combination tones that are equally real, although weaker than both the fundamental tone and the markedly impure fifths and octaves."

It was Helmholtz, Planck emphasized, "who had demonstrated the physiological, and therefore the artistic, weakness of tempered tuning." In view of a situation that had developed over centuries and in which the piano with its fixed-tone keyboard had come to play so dominant a role in music, Planck sensed that it was important for singers and choral ensembles to strike a balance between tempered singing and the "sometimes achievable" ideal of natural temperament. He recognized the inherent difficulty connected with maintaining natural temperament since sequential singing in natural temperament gradually leads to an increase of pitch – a phenomenon that usually is explained under the heading "circle of fifths [*Quintenzirkel*]."[44]

Much of what Planck had to report in his studies on natural intonation for singers and choral groups relates to focused self-listening, ensemble singing that enables vocalists to listen to one another [*Aufeinanderhören*], and the alertness and sensitivity of a choir conductor. To test some of his ideas Planck was privileged, as he reports, to engage the five-voice a cappella choirs of the Royal Academy of Music in Berlin when singing with and without piano accompaniment. He concluded that ultimately "Art locates its foundation in itself and no matter how logically founded and consistently implemented, no theoretical system is in a position once and for all to define all the demands of the human spirit and its everlastingly changing art. . . . For after all, the educational establishment must conform to art and not the reverse."[45]

[43] Planck, "Die Natürliche Stimmung" (1893), 431.

[44] Planck, "Die Natürliche Stimmung" (1893), 432.

[45] Planck, "Die Natürliche Stimmung" (1893), 438–439.

Planck's final assessment is neither optimistic nor constructive: "One thing is certain – that this will come about only when a genius arises who will know more about what to say in the language of natural intonation than anyone else. Surely, no principal misgiving will be able to stand firm against him."[46]

Experimental investigations with the Eitz harmonium and the publication of an article on just intonation in a highly respected musicology journal constitute but a small fraction of Planck's intellectual output during the first decade of his Berlin career. His escapades into music theory may be looked upon as peripheral, scientifically speaking; but for a man of Planck's scientific stature they were, humanistically and aesthetically speaking, singular and provocative. In the annals of science it was Planck's quantum of action that took the stage front and center. What Planck accomplished just prior to his enunciation of the quantum of action theory, during his fleeting escape from theoretical physics to music theory, has been pushed aside in the history of science literature in order to highlight the contributions to a black-body radiation theory that virtually made Planck world-famous almost overnight.

It was just at this juncture in the new life in Berlin, as Planck has related, that an interest in music theory provided him with the opportunity to come to terms with himself as a theoretician in both music and in theoretical physics in an environment that was predominantly experiment directed. When he first arrived in Berlin he was preoccupied with tying together various ideas on thermodynamics and the second law's irreversibility component. Thermodynamics had been the focus of Planck's concerns as student in Munich, and it had continued to occupy his attention during the 4-year associate professorship in theoretical physics at the University of Kiel. Planck's thermodynamic deliberations, in conjunction with results realized in experimental investigations on black-body radiation conducted at the PTR, where Planck had close connections with the experimental physicists, led him to totally unanticipated conclusions. He came to recognize that by combining the second law of thermodynamics with Kirchhoff's theorem (that light and heat radiation in thermal equilibrium are independent of the nature of the radiating substance), he could formulate a new radiation law that embodied what he referred to as the "quantum of action." The historical reconstruction of what followed that 1900 quantum adventure has been the subject of countless other scientific advances, controversial philosophical world views, audacious and mathematically productive cosmological speculations, and a veritable crush of technological life-changing inventions.

In 1918 *Die Naturwissenschaften* devoted one of its issues to the celebration of Planck's 60th birthday. Coming at the end of a war, when scientists were beginning to return to their teaching and research positions, the publication provided an occasion for scientists to reflect on developments in physics that bear on quantum theory such as the Nernst heat theorem, spectral theory, and probability. The authors of the papers, notably, were scientists who had been engaged in drawing out the

[46] Planck, "Die Natürliche Stimmung" (1893), 440.

consequences and implications of the quantum theory but some of them were sensitive as well to what was going on in music circles and more specifically in Planck's musical life in Berlin. Having worked in close collaboration with him, they were familiar with his intellectual outlook on life and had experienced at close hand his passion for music. These scientists were mostly colleagues who were aware of his command of music theory and were apprised of his music-making potential in home and concert environment. A number of them wrote about making music with Planck and about participating in music sessions that he had organized and conducted.[47]

Several of the authors pointed to the long road that lay ahead in working out the details, refinements, and implications of the Planck radiation theory. Wilhelm Wien (Würzburg) referred to the unprecedented demands that were being laid on experimental and theoretical physicists engaged in putting the newly won experimental discoveries and supportive theoretical lines of thought into an acceptable and internally consistent quantum physics.[48] Paul Epstein (Munich) referred to the many gaps in the existing quantum theory. He suggested that the theory might still be in its initial phase of development but believed that the gaps would vanish with the growth of information concerning matters such as atomic spectra, the structure of matter and the nature of the ether – all issues which just a few years ago seemed to be unreachable.[49] Querulous about the essential relationship of quantum theory to nature, Fritz Reiche (Berlin) remarked:

> Anyone who studies the quantum theory will encounter a certain disappointment. For we must answer to the fact that in spite of all its far-reaching rules we scarcely have taken one step closer to the core of the matter. It seems certain that there are well-defined quantum-like states for mechanical and electrical systems; but where lie hidden the deeper reasons for the cause of these erratic quantum jumps in nature?... We still do not know the nature of the radiation. Does it spread out in the way that is required by the undulatory theory or does it also have a quantum character?... All of these problems await solution, but wherever the ultimate solutions may come to rest they will have to be based in any case on an enormous amount of empirical and theoretical information that lies at hand.... We lack the guiding thought that holds all this scattered information together.[50]

Arnold Sommerfeld's lead article in the issue of *Die Naturwissenschaften* was given over more pointedly to celebrating Planck's 60th birthday and is a veritable classic in the history of quantum physics. More than other scientists who were involved in work on quantum physics during the first two decades of the century

---

[47] Max Planck. Zur Feier seines 60. Geburtstages. *Die Naturwissenschaften*, 6, (1918) 194–263. The issue includes articles by Arnold Sommerfeld (Munich), Emil Warburg (Berlin-Charlottenburg), Wilhelm Wien (Würzburg), Walther Nernst (Berlin), Max vs. Laue (Frankfurt/Main), Fritz Reiche (Berlin, Paul Epstein (Munich), and Marian Smoluchowski (Krakow).

[48] W. Wien, "Die Entwicklung von Max Plancks Strahlungstheorie," *Die Naturwissenschaften* 207.

[49] P. Epstein, "Anwendungen der Quantenlehre in der Theorie der Serienspektren," *Die Naturwissenschaften* 252–253.

[50] Fritz Reiche, "Die Quantentheorie. Ihr Ursprung und ihre Entwickelung," *Die Naturwissenschaften* 225–230.

from specific points of view, Sommerfeld was in the thick of all its developments. With his contributions, first, to the study of atomic spectra and elliptical orbits for the Bohr atom, and, second, for his proposal of an azimuthal quantum number to explain the Zeeman magneto-optic effect, he recognized that the mathematical formulation for a quantum-theoretic model had reached an essential turning point. In addition to his seminal contributions to mathematical physics, as well as to theoretical physics and chemistry, Sommerfeld was an inspiring teacher and research director. During his 25 years in Munich he had taught more advanced students and turned out more doctorates in theoretical physics than any other physicist. Offers to become Hasenöhrl's successor in theoretical physics in Vienna in 1917, and Planck's successor in theoretical physics in Berlin in 1927, were both turned down.[51] Sommerfeld never received the Nobel Prize. Few theoretical physicists did. The theoretical physicists Planck and Einstein were the exceptions and presumably were granted the prize by stretching, to some extent, the guidelines that its benefactor Alfred Nobel had stipulated in his will.[52] Planck received the physics prize in 1918 not for the formulation of the quantum theory, but for "his discovery of energy quanta." Einstein received the physics prize in 1921 not for relativity theory but for "his discovery of the law of the photoelectric effect."

Arnold Sommerfeld was the ideal theoretical physicist to undertake a comprehensive evaluation of the status of quantum theory at the end of the first world war. He chose to draw up a concise synopsis of the line of reasoning that Planck had taken in going from thermodynamics to formulation of the quantum of action in 1900 and showed how, subsequent to the work of Willy Wien, Planck had set himself the task of exploring the intersection of radiation theory and electrodynamics by way of the principles of thermodynamics. "This program," as Sommerfeld remarked, "was carried out with singular determination beginning in 1896. . . . It is a beautiful example of scientific concentration to see how over a period of several years Planck kept his eye on the goal without blinking right or left. The measurements carried out at the *Physikalisch-technische-Reichsanstalt* provided him with the occasion for testing the theoretical results with experience."[53] Sommerfeld's 1918 synopsis turned out to be prologue to a much more comprehensive exposition of quantum atomistics published the following year as *Atombau und Spektrallinien* – a work that almost immediately became the "Bible" of atomic physics. In successive editions, published during the early 1920s, *Atombau*, more than any other single work, chronicled the entire gamut of progress in the field of quantum physics

---

[51] Paul Forman and Armin Hermann, "Arnold J. Sommerfeld," *DSB*, *12* (1975) 525–532.

[52] Alfred Nobel (1833–1896), the Swedish inventor and industrialist, who made a fortune in the production and marketing of explosive devices at the end of the nineteenth century, stated in his will of 1895 that his entire fortune was left for the awarding of prizes "to those who, during the preceding year, shall have conferred the greatest benefit on mankind." Torsten Althin, "Alfred Nobel," *DSB*, *10* (1974) 132–133.

[53] Arnold Sommerfeld, "Max Planck zum sechzigsten Geburtstag," *Die Naturwissenschaften*, *6* (1918), 196.

prior to the final developments that led to Heisenberg's uncertainty principle, the correspondence principle, and quantum mechanics proper as it came to be interpreted in Bohr's Copenhagen school.

Sommerfeld wrote: "We see clearly today that the most subtle questions in atomistics and the most general properties of matter are rooted in the Planck quantum concept. No well-informed person can doubt that Planck's quantum of action [h] belongs, alongside the velocity of light [c], the gravitational constant [g], and the charge and mass of the electron [e], to the most important constants of nature in today's science." Sommerfeld's essay was an attempt to spell out in detail how what was going on in the Berlin world surrounding quantum theory overlapped with what was going on in the Munich world of quantum atomistics.

Immersed in all that was transpiring in Berlin at the level of the quantum, Sommerfeld apparently sensed that it was pertinent to say something about Planck's other *Lieblingsthema* (that is, other than thermodynamics), namely music. Acknowledging that as a student Planck had vacillated between taking up physics and music, Sommerfeld wrote: "Today we are able to congratulate ourselves that he finally chose physics. In any case music never was lost to him; and even after hours of productive work music remained for him the source of refreshment and rejuvenation. We have a literary expression [*literarischer Niederschlag*] for his musical interests in his paper on the [Eitz] harmonium in natural intonation. Planck not only plays the piano as a master but he is in complete control of the multiple-keyed harmonium that was commissioned by Helmholtz and placed in the physics institute of the University of Berlin."[54]

When Planck died in 1947 there was a great outpouring of essays of a reflective nature in which colleagues, scientific collaborators, and persons who had been Planck's students in Berlin made an effort to appraise not only the nature and extent of Planck's contributions to theoretical physics and their larger implications for the newly formulated quantum and relativistic world pictures but to reflect on his extrascientific activities and notably about his deep interest in music. These statements were written by authors who had known Planck, and who in some capacity had worked closely with him in academia or in community circles. Their reflections invariably touch on the way in which Planck's scientific accomplishments and his own interpretation of those contributions (which on occasion were at odds with the interpretations of his collaborators and colleagues) meshed with his life style, his academic and social environment, and his philosophical and extra-scientific interests.

In what follows we shall have much to report about the last 3 years of Planck's life and the decade immediately thereafter because it simply was not possible until the war was over for the wider circle of his colleagues and acquaintances to voice their opinions about the person who had been so much at the center of their own lives. Here it is surprising to discover how many reflective comments on Planck's life and career dealt with his interest and involvement in music.

[54] Sommerfeld, *Die Naturwissenschaften* (1918) 195.

When Planck's house in Grunewald was destroyed in the bombing raid on February 1944 he temporarily made his home in the village of Rogätz on the Elbe river north of Magdeburg. After the resistance collapsed and Germany had surrendered unofficially in May 1945, the Dutch-American astronomer Gerard Kuiper from the Yerkes Observatory in Chicago, who at the time was deputy director of the *Alsos-mission*, drove Planck and his wife Marga by jeep from Rogätz to Göttingen in the British zone. With the *Kaiser Wilhelm Gesellschaft* (*KWG*) about to fold for lack of leadership, Planck was enticed to become its interregnum president. He died in Göttingen on 4 October 1947. Two days later *Die Neue Zeitung* in Munich published a one-page obituary by Arnold Sommerfeld who was referred to by the editor of the paper as "another internationally renowned scholar and friend who during the national socialist regime lost his teaching post at the University of Munich for having supported the Planck-Einstein theory."[55] In the obituary Sommerfeld wrote: "a major figure in the world of physics. . . ., the discoverer of the quantum theory, has died at the high age of almost ninety years." Commenting about Planck's early life as a student in Munich Sommerfeld noted that at the time of completion of his doctorate and his habilitation the 21-year-old Planck still wavered between wanting to continue in physics or take up music as a career. "Into a high age Planck continued to be an accomplished pianist who was able directly to transpose musical notation from one key to another." Sommerfeld, who had been the major theoretical physicist in the development of quantum atomistics, noted that the years following Planck's 1900 quantum theory had been devoted largely to its experimental verification and theoretical elaboration and had given rise to the field of atomic physics. He recalled: "In the preface to my book *Atomic Structure and Spectra* [*Atombau und Spektrallinien*] of 1919 I wrote: In the last analysis all integral laws of the spectral lines and of atomistics emanate from the quantum theory. It is the mysterious Organon on which nature plays spectral music and according to this rhythm the structure of the atoms and nuclei are regulated."[56]

The Times of London referred to the quantum theory of Planck as "in many ways. . . . more revolutionary and far-reaching even than that of the originator of the theory of relativity." With the discovery of the quantum of action came the upset of the time-worn notion of causality.

[55] In 1937 Sommerfeld and Heisenberg had been labeled "white Jews of science" in the magazine of the SS. In 1940 Wilhelm Müller, one of the stalwarts of the Nazi movement, inherited Sommerfeld's professorship in theoretical physics at the University.

[56] Arnold Sommerfeld, Max Planck zum Gedächtnis, *Die Neue Zeitung*, (6.10.1947); facsimile offprint, p. 333 in Archiv der Max-Planck-Gesellschaft, hrsg. Lorenz Friedrich Beck, *Max Planck und die Max-Planck-Gesellschaft*, Berlin, 2008. Sommerfeld's *Atombau und Spektrallinien* first appeared in 1919 and immediately became the bible of atomic physics. It continued to appear in successive editions almost annually through the early 1920s and chronicled the progress in this field until the introduction of quantum mechanics in 1926.

> Since the introduction of the infinitesimal calculus of Leibniz and Newton, physics has been built on the assumption of the continuity of all causal chain of events. The discovery of a universal quantum of action has upset this basis; and for the discovery Planck bears the undivided glory. With Planck as the doyen of international science, his house in the Grunewald colony was a Mecca for physicists the world over. Planck was an artist no less than a man of science, and it is worth recalling that at one time a musical career was considered for him. In his work there also was something approaching a religious spirit. He decried skepticism, and insisted on the need for faith in the man of science. Professor Einstein once likened him to the devotee or lover, whose inspiration rose from a hunger of the soul[57]

At the memorial service for Planck in the Albani church in Göttingen on the 7th of October 1947 the sermon was delivered by Friedrich Gogarten (1887–1967), professor of theology at the university in Göttingen. He was known internationally for his writings on the demythologization of the New Testament and the church. Tributes were offered by Otto Hahn, president of the MPG, and by Max von Laue, Planck's closest student and longtime collaborator. In his address Laue spoke about how his almost 90-year-old colleague had lived an unusually rich and productive life in spite of tragedies in the family, two world wars, and the pursuit of the unpopular field of theoretical physics that nevertheless had given Planck immortal fame. Laue then mentioned the universities and institutions that had honored Planck by sending wreaths. In conclusion he remarked: "And here lies a simple wreath without a ribbon. I have placed it here as a perishable symbol of the imperishable love and thankfulness for all his students to which I also belong."[58]

In February 1948 a number of leading physicists met in Göttingen to plan the establishment of a society that would bring together in one organization the most important non-governmental German research institutes. To honor Planck who had been President of the *Kaiser-Wilhelm-Gesellschaft* the society was renamed *Max-Planck-Gesellschaft zur Förderung der Wissenschaften.* It was a renaming event that had been initiated by the British occupying forces to avoid reference to the Kaiser but without wanting to enervate the historically famous *Gesellschaft* as such. The physical chemist Otto Hahn (1879–1968), who had been released from British internment in Farm-Hall, became the first president of the MPG in 1948.[59]

On the day of Planck's 90th birthday, 23 April 1948, a memorial service was held in Göttingen. The city, one of the few German university strongholds that had been largely untouched by the war, had become the headquarters of the MPG. Organized jointly by the MPG, the University of Göttingen, the Göttingen Academy of Sciences, and the German Physical Society, the sessions were attended

[57] Professor Max Planck. The Quantum Theory. *The Times*, London, 6.10.1947; facsimile offprint in Lorenz Friedrich Beck, *Max Planck und die Max-Planck-Gesellschaft*, Berlin, 2008, 334.

[58] Max von Laue. Trauersprache gehalten am 7. Oktober 1947 in der Albanikirche zu Göttingen. Lorenz Beck (ed.), *Max Planck und die Max-Planck Gesellschaft* (2008).

[59] E. Brücke (ed.), "Max Planck Gesellschaft," *Physikalische Blätter*, *4* (1948), 132.

by dignitaries from German and foreign universities and scientific societies and institutions. After a wreath-laying ceremony at the graveside the gathered group proceeded to the great hall, the *Aula* of the University, for the main ceremony. Brief addresses were delivered by Otto Hahn representing the MPG, by Prof. Hans Stille, vice-president of the German Academy of Sciences in Berlin with which Planck had been associated prominently for many years, by Sir Charles Darwin, representing the Royal Society of which Planck was a foreign member, and by Hermann Deisch, rector of the University of Berlin. A tribute from Einstein was read by Otto Hahn. The main memorial addresses were delivered by Max von Laue, who stressed the early history of Planck's work, notably in thermodynamics, by Werner Heisenberg, who spoke about Planck's contributions to the quantum theory, and Richard Becker who stressed the philosophical implications of Planck's scientific work and his human qualities. Music from a number of Planck's favorite composers was supplied by Karl Klingler, violinist, with Raimar Dahlgrün at the piano. The ceremony began with J. S. Bach's Adagio from the D-major Sonata. Between the brief addresses (*Ansprachen*) and the commemorative addresses (*Gedenken*) the two men played the Brahms Adagio from the F-minor Sonata. The formal session concluded with music from the last movement, the Presto, from Beethoven's Kreutzer Sonata in A major.[60]

The occasion also gave rise to a series of reflective essays by scholars and close acquaintances who had experienced Planck's friendship and hospitality in his home in the *Villenkolonie* or elsewhere at the university or the academy. The Swiss historian of religion Alfred Bertholet (1868–1952), author of a cultural history of Israel, recalled that he first met Planck in 1928 at an evening session that had been arranged in Planck's home to welcome the Sanskritist and Indologist Heinrich Lüders (1869–1943) and his wife on return from an extended visit to India. After a glowing account of Planck's natural openness, cordiality, and unique intellectual interests, Bertholet remarked:

> There was something else that from the start brought me to appreciate him. I heard him at the piano wonderfully accompanying songs in a way that revealed his whole musical soul. At other times we combined our musical efforts. With special gratitude I remember the time I was privileged to have him accompany me playing the violin. On one occasion I could not refrain from asking him whether he never had come to the idea of dedicating himself entirely to music. In fact, he answered, I once was saddled with that thought and I even asked a prominent musician whether I should become a musician. And you know how he responded? If you have to ask that question you will never become one.[61]

Agnes von Zahn-Harnack has described the intellectual ambience that was at large in the *Professorenecke* (professor's corner) of the *Villenkolonie* (residential

[60] Lorenz Friedrich Beck, *Max Planck und die Max-Planck Gesellschaft* (2008) 215–217; E. N. da C. Andrade, Max Planck Memorial Ceremony in Göttingen, *Nature*, No. 4098 (May 15, 1948) 751–752.

[61] Alfred Bertholet, "Erinnerungen an Max Planck," *Physikalische Blätter*, *4* (1948) 161–162.

district) in Berlin/Grunewald.[62] The Harnack and Planck families, the closest of friends in the *Villenkolonie*, lived in the proximity of professionals and colleagues who represented disciplines as diverse as mathematics, biology, history, theology, and physics.[63] Agnes Harnack wrote about the spirit of camaraderie that prevailed in the intellectually rich *Villenkolonie* where, in spite of great professional diversity and differences in living habits, "deep community and solidarity rested above their houses." In all of them, she wrote, it was understood "that the *spiritus creator* (the spirit of creation) was the actual life-supporting force." And for me, she added, "the confidence in this spirit was almost embodied in its purest form in the figure of Max Planck." In the Planck home, as Agnes Harnack noted, no one was surprised to find the master of the house playing the piano or accompanying either Joachim on the violin or Marie Scherer singing Schubert, Schumann, and Brahms *Lieder*.

> I personally did not have the good fortune to hear Planck play, but something else came my way. For an entire winter I was able to participate in a choir and sing the Brahms *Liebeswälzer* conducted by Planck. Who else sang? In retrospect one could say: Nobel prizewinners and such who wanted to become prizewinners. There were the physicists (Otto) Hahn, (Gustav) Hertz, (Wilhelm) Westphal, and (Eduard) Grüneisen; and among the listeners and at every rehearsal there was Lise Meitner. Among women's voices the best were Frau Grüneisen and Planck's twin daughters. Whoever attended such a choir evening expecting something on the order of a simple social event soon got into trouble with the conductor. *Verum gaudium res severa* – practice was thorough, demanding, and strict but always corrected with an invariable charm by the conductor.[64]

The Berlin experimental physicist Wilhelm Westphal (1882–1978) referred to Planck's home as a place that over the years came to represent a circle of young like-minded academics whose gatherings often were dominated by music-making.[65] Westphal remarked: "It is known, that Planck was an extraordinarily musical person

[62] Agnes von Zahn-Harnack (1884–1950), whose interests and activities in the emancipation of women began at age 30, was cofounder of the German society of academic women [*Deutsche Akademikerinnenbundes*] in Berlin in 1926. Except for the self-imposed exile during the Nazi era Harnack was politically active for the rest of her life in the support of conscientious objection to war and the promotion of women's rights in church pulpits and in academic teaching positions. During the *Third Reich* she was a member of the private circle of dissidents that included Helmut Gollwitzer, the politically outspoken Protestant pastor and Professor of systematic theology at the university of Berlin, and the theologian Martin Niemöller, a World War I submarine commander who became the leader of the *German Emergency League* and *The Confessing Church* after the Hitler regime came to power in 1933. He was imprisoned in 1937 and liberated by the Allies in 1945. Agnes Harnack wrote: *Die Frauenbewegung, Geschichte, Probleme, Ziele*, Berlin, 1928, and *Frauenfrage in Deutschland 1790–1930*, Berlin 1934.

[63] Adolf von Harnack (1851–1930) was one of Planck's most important science benefactors and a crucial link in the establishment and growth of German science and industry. He provided the initiative for the founding of the Kaiser-Wilhelm Gesellschaft and from 1911 until his death served as its first president. By profession he was a Lutheran theologian. He authored a 3-volume *Lehrbuch der Dogmengeschichte* (1886–1890) and a 3-volume *Geschichte der altchristlichen Literatur* (1893–1904).

[64] Agnes von Zahn-Harnack, "Erinnerungen von Max Planck," *Physikalische Blätter*, *4* (1948) 165–167.

[65] The group included the theologian Adolf von Harnack, the historian Hans Delbrück, the acoustician Carl Stumpf, the physicists Eduard Grüneisen, Gustav Hertz, and Robert Pohl, and the chemists Lise Meitner, Otto Hahn, and Otto von Baeyer.

and a first-rate pianist who often and gladly made music with prominent artists such as (the violinist) Joseph Joachim, Gabriele Wietrowetz, and Hermann Diener. Planck's two daughters had inherited this music talent and played the piano and the violin well." After the death of Planck's first wife in 1909 the choir group assembled in his home to study and perform works such as Haydn's *The Seasons*, *The Creation*, and the Brahms *Gypsy Songs* and *Lovesong Waltzes*.[66] Westphal referred elsewhere to Planck's home in the *Villenkolonie* in Grunewald as being a place that "above all else stood under the sign of music." He reported that on many evenings Planck would gather a circle of young persons fond of singing, and, while directing the group and sitting at the piano, would practice with them the choral works of composers such as Haydn and Brahms.[67]

Scientists from beyond Germany who had come in contact with Planck in Berlin in connection with study and research remembered and spoke about his resolute commitment to music as a subject that demanded scientific study. The British chemist and historian of chemistry, John R. Partington (1886–1965) was engaged in research on the specific heat of gases with Walther Nernst in Berlin. He attended some of the physics lectures that Planck delivered from year to year on a cyclical basis. Partington reported that during the semester in which Planck lectured on acoustics, his discussions on music theory were enhanced with demonstrations on an harmonium that had been constructed to sound in natural temperament. His objective was to show how the introduction of fixed-tone musical instruments had fostered music in the tempered scale but at the expense of music in natural intonation.[68]

Persons who had left their imprint on the *Wilhelmian Kaiserreich* and continued to do so during the *Weimar Republic* lived predominantly in the residential village of Grunewald in West Berlin referred to as *Villenkolonie Grunewald.* Alongside a group of wealthy merchants and financiers lived university professors and other scholars and artists who would have been considered to belong to Berlin's intellectual bourgeoisie, the *Bildungsbürgertum*. *Villenkolonie Grunewald*, an intellectual haven and center of culture, was located conveniently to the University and to the center of Berlin by public transportation.[69] The area was not extensively settled until 1889 but then during the Empire and the Republic developed rapidly into a cultural communications center for persons in commerce, politics, and science. The cultural

---

[66] Wilhelm Westphal, "Erinnerungen an Max Planck," *Physikalische Blätter*, *4* (1948) 167–179.

[67] Wilhelm Westphal, Max Planck als Mensch, *Die Naturwissenschaften*, *45* (1958) 234–236. Persons belonging to the choral group included Otto Hahn and Frau Grüneisen as soloists. In the singing circle were the experimental physicist Eduard Grüneisen, director of the division of electricity and magnetism at the PTR, Otto von Baeyer, spectroscopist at the PTR, and the children of Karl Stumpf, philosopher, psychologist, and musical acoustician, as well as the daughters, nephews, and nieces of Adolf von Harnack the church historian and president of the Kaiser-Wilhelm-Gesellschaft and Hans Delbrück, church historian. The non-singing guests included Lise Meitner (faithfully), the radiation physicist Robert Pohl, and Gustav Hertz (Nobel in physics, 1925).

[68] J.R. Partington, "Erinnerungen an Max Planck," *Physikalische Blätter*, *4* (1948) 172.

[69] Kristina Behnke, Hella Dungen-Löper, and others (eds.), *100 Jahre Villenkolonie Grunewald* 1889–1989, Berlin, 1988. See in particular: Helga Gläser, Die Villenkolonie als Kulturelles Zentrum, pp. 63–93.

creativity that emerged from this center before the early 1930s is attributable not only to the acclaimed status of its individuals but can be traced as well and above all else to the growth of a lively *esprit de corps* that reached far beyond any professional or religious alignments among its citizens. Encounters between talented and resourceful persons coming from various branches of the financial, industrial, literary, artistic, and musical worlds contributed to the cultural flowering notably at the turn of the century and until Nazi times. The high level of intellectual productivity at the tail end of the Wilhelminian period seems all the more conspicuous in light of what followed under national socialism when the cultural diversity that in large part was due to the prominent role of its Jewish citizens – a sizable fraction of the *Kolonie* population that was forced first to leave Germany in 1933 and then was irreversibly shattered as a community when the *Grunewald Synagogue* was torched and destroyed on *Kristallnacht* 9 November 1938.

Planck lived in the *Villenkolonie* of *Grunewald* during the entire time of his career at the University of Berlin. From 1905 until his home was destroyed in an *Allied* bombing raid in 1944, the Planck family took up residence in a house on *Wangenheimstrasse* 21. His position, as a theoretician who lectured and conducted seminars but was not involved in research at the university, made it possible for him to spend most of his working hours at home. From Grunewald Planck was readily able to reach the university and the center of Berlin by public transportation. As we have shown, Planck's home over the years became the preferred locale in the *Kolonie* for engaging in *Hausmusik* and for organizing musical events that involved Planck, members of his family, musicians and scientists living in the *Kolonie*, musicians and scientists who came into the area from elsewhere, and fellow scientists who regularly attended the Planck *soirées* simply to enjoy the music and the company. Persons in the *Kolonie* generally knew each other on a personal level and were sensitive to the exclusivity of their relatively small environment (just a few square kilometers) and their close-living neighbors. On the other hand they valued living "outside" and yet close to the "pulse of the city." The peace and quiet of the suburb *Kolonie* offered respite from the hectic everyday life of the metropolis. A standout account of activities in the *Kolonie* in the golden years before 1933 is given in the *Grunewald=Echo* – the central organ and voice of the citizens of Grunewald. In the 1929 issue of this publication, the 30th anniversary of the official founding of the *Kolonie*, we have an on-hand account not only of the makeup and activities of its academic scholars, musicians, artists, theater persons, and literary personnel, but of its publishers, financiers, and wealthy businessmen. It was a kind of symbiosis between money and culture in which persons of both liberal and conservative persuasion worked and at times were referred to pejoratively as living in the "professors' quarter," the "village of luxury," the "Nobel quarter," or the "slum village for millionaires."[70]

The *Kolonie* was home for the Mendelssohn family, a highly cultivated group of descendants of the "Jewish philosopher of the enlightenment" Moses Mendelssohn

[70] *Grunewald=Echo*. Organ für den Amts- und Gemeindebezirk Grunewald. Jubiläums-Festschrift zum 30 jährigen Bestehen, Dec. 1899–Dec. 1929.

(1729–1786), who lived in Berlin after 1741 and became the leader of the movement for the cultural assimilation of Jews and non-Jews. The Berlin Mendelssohns, associated with the world of finance, lived in the "*Mendelssohn Palais*" in the *Kolonie* and were generous patrons of music and the arts. Unlike the performers and music-loving clientele who frequented the *Hausmusik* evenings in the Planck house and were mostly scientists and university professors, the music sessions in the Mendelssohn Palais were mostly sponsored in behalf of artists and writers. The composer Felix Mendelssohn (1809–1847) was a grandson of Moses Mendelssohn. The expressionist Erich Mendelsohn (1887–1953) was the architect of the "Einstein tower" in Potsdam built in 1920–1921.[71]

[71] The various address books of Grunewald for the years from 1900 into the 1930s constitute a virtual cornucopia of *fin de siècle* persona whose activities and published writings have become known worldwide. Among such as are given space in *100 Jähre Grunewald* we may briefly identify, beyond those already mentioned, the following: Walther Rathenau (1867–1922), social theorist, statesman, author of politico-economic writings, minister of reconciliation who was assassinated in 1922 by nationalists and anti-semitic fanatics; Lilli Lehmann (1848–1929), coloratura soprano who wrote a timely textbook on the art of singing and was known for her skillful renditions of Schubert *Lieder* and Wagnerian opera music; Fritz Mauthner (1849–1923), philosopher of language, co-founder of *Freie Bühne* and author of *Atheism and its History in the Occident*; Engelbert Humperdinck (1854–1921), teacher in the *Hochschule für Musik* (located in his home), composer of stage and opera music the most famous of which was Hänsel und Gretel (premiered in Weimar under Richard Strauss); Gerhardt Hauptmann (1862–1946), recipient in 1912 of the Nobel Prize for Literature; Samuel Fischer 1859–1934), founder of *Fischer Verlag Berlin* (the most important German publisher from 1900 to 1933), editor of *Neue Deutsche Rundschau* and publisher of modern literary works by H. Ibsen, É. Zola, A. Schnitzer, T. Mann. H. Hesse, G. Hauptmann, G. B. Shaw, and O. Wilde; Fritz Kreisler (1875–1962), composer of violin music and operettas and most popular violinist in Germany and America; Dietrich Bonhoeffer (1906–1945), Protestant theologian, assistant professor University of Berlin who advocated a non-religious interpretation of biblical concepts and the worldliness of Christianity. He was executed in 1943 for his participation in the resistance movement; Walter Benjamin (1892–1940), Marxist critic of the spiritual and moral corruption of the middle class who fled to Spain and committed suicide there in 1940; Hans Delbrück (1848–1929), liberal historian and professor at the University of Berlin whose 4-volume *History of the Art of Warfare* promoted going beyond technical problems and linking warfare to politics and economics (his son Max Delbrück (1906–1981), the California biophysicist, was co-recipient of the Nobel Prize for Medicine in 1969); Adolph von Harnack (1851–1930), Protestant theologian and church historian at the University of Berlin, author of 3-volume textbook on the history of dogma (1886–1890), influential member of the Prussian Academy of Science, director of the Prussian state library, close friend to Planck, author of the famous 1909 memorandum that led to the establishment of the *Kaiser Wilhelm Society* of which he was the first president until his death in 1951 (Harnack's daughter Agnes von Zahn-Harnack (1884–1950) was the presiding representative for German academic women and after 1945 active in the revival of the women's emancipation movement); Max Reinhardt (1875–1943; pseudonym M. Goldmann), actor and stage manager whose productions and directions of gigantic plays at the *Deutsches Theater* in Berlin, beginning in 1905, were full of experimentation (revolving stage, and use of the entire auditorium), pageantry, and rousing mob scenes. He was forced to leave Germany in 1933 and emigrated to the United States in 1938; Werner Sombart (1863–1941), professor at the University of Berlin in political economy and sociology 1906–1941 who investigated the development of business economics from capitalism to socialism, wrote a 3-volume work on *Modern Capitalism* (1902–1928) in which he was critical of Marxism. During the Weimar republic he advocated national socialism and became an exponent of the authoritarian state, accepted the Nazi ideology, but eventually sought to distance himself from the *Third German Reich*.

To all that has been reported here concerning the life and activities of the celebrated scientists, artists, musicians, and writers who lived in the *Villenkolonie* during the first third of the twentieth century there is, lamentably, a less salutary flip side – one that many *Grunewalders* did not come to grips with forcefully until the end of the war and the defeat of Nazi Germany. With all its celebrated and distinctively intellectual personalities, with all its crackerjack persons in finance and commerce, with all its liberality and apparent openness to the other less-fortunate and less-educated layers of Berlin society beyond the *Kolonie*, there were very few personal contacts between the *Grunewalders* and inner-city Berliners; and even when that seemed possible, it was largely lacking. In regard to habits of daily living the two sectors were worlds apart. Worst of all, the *Grunewalders* tended to be almost oblivious to the consequences of the political crises that were slowly but deliberately nibbling away at all that they themselves stood for. The *Grunewalders* simply were too isolated in their *Kolonie* and too involved in their own activities to engage in discussions and events occurring beyond their own self-imposed geographical and intellectual periphery. That the political conflicts were coming to a head as a result of these deteriorating conditions was recognized in its acuteness by few. For most of them national socialism was no more than the expression of a lowbrow culture; Hitler was taken to be nothing but a lousy play-actor. But then, many a *Grunewalder*, it was said in retrospect, was first yanked out of a dream world in the midst of reading books and making music while the political ground had long given way in the neighboring Nazi world. It was predominantly the middle-class Berlin Jews in the city who were alert enough to perceive the depth of the dangers that lay ahead.

The historical novelist and dramatist, social critic, socialist, and pacifist Lion Feuchtwanger (1884–1952) has vividly profiled events in the *Kolonie* in his post-1933 novels. He had made his home in the *Kolonie* in 1925 and was forced in 1933 as Jew and social agitator to flee to France. Arrested there, he nevertheless managed dramatically to escape to the United States in 1941. In his first novel in exile, *The Oppermann Siblings*, the central character Gustav Oppermann dedicates himself with devotion to Lessing studies and his little club in the *Kolonie*. Immersed deeply in the world of German culture – a world that had endowed him with all that belongs to the world of inner high-mindedness and reason – he fails to take national socialism seriously. When the German *Reichstag* is set on fire in 1933, and his friends urge him to take immediate refuge, he is so totally flabbergasted that he literally falls out of the clouds. "Together with Frischlin, Gustav chooses which of his books he wants to take along.... He fancies that he is making a fool of himself. He takes nothing – not even his manuscript; but how can he possibly work without a library? He will stay away for 14 days but no longer. In order not to take an oath against the evil forces that will make his short absence into a long one he refuses to take along what he cherishes."[72] Oppermann's dream world turns out to be one of deceit. His absence from the *Kolonie* was not short-lived but he did return.

[72] Lion Feuchtwanger, *Die Geschwister Oppermann*, Frankfurt, 1984, p. 211.

Most upper-class Jews, like the fictional Mr. Oppermann in the *Kolonie* who refrained from openly exposing himself and his reactions to what was taking place, remained in Germany for a stretch after 1933. Some of these Jews had adequate financial reserves, as well as the self-confidence offered by way of connections outside Germany, to temporarily brave the steadily deteriorating conditions. The more threatening and hopeless the situation became the more they decided to emigrate. The larger Jewish firms were arianized and Jews were forced to liquidate their properties for ridiculously small sums. On this score the Grunewalder Jews were privileged as well-to-do citizens, since they at least had the foreign contacts and the wherewithal to emigrate. Still, beginning in 1941 Jews from the *Kolonie* began to be deported to extermination camps. In the midst of the forced emigrations, property confiscations, and executions that were in process in the 1940s there were, nonetheless, isolated pockets of organized resistance among politically and religiously motivated individuals in the *Kolonie*. What ensued in the *Villenkolonie* as a cultural center is depicted as follows:

> In the *Wangenheimstrasse* lived the families of Bonhoeffer and Max Planck, [and in the neighboring street] lived the historian Hans Delbrück. Bonhoeffer had lost his two sons, Klaus and Dietrich, as well as his two sons-in-law, Rüdiger Schleicher and Hans von Dohnanyi, to the National Socialists. They all, like Erwin Planck, were executed as members of the resistance. Justus Delbrück together with Dietrich Bonhoeffer was placed under detention in Tegel and outlived the end of the war by only a few months. The Jews who were represented in the *Kolonie* by a fairly large number of prominent families, were either forced to leave or expelled. The Grunewald synagogue that had been dedicated on *Rosh Hashanah* in 1923 was torched and vandalized in 1938 and torn down in 1941. By that time Jews began to be deported from the railroad station in Grunewald to Lodz and Auschwitz. What happened to their individual properties is known only in part. The Mendelssohn palace, for example, was remodelled into a villa of guests of the *Reich*. . . . and Heinrich Himmler became a resident in the Kolonie.[73]

It goes without saying, finally, that the here-depicted image of the Grunewalder Oppermann, who lived under conditions of physical and intellectual isolationism, self-sufficiency, elitism, and political inattentiveness to the growing threat of National Socialism, fits Max Planck to a nicety and is something that anyone who deals with Planck inevitably and in reality must come to grips with. Like others, Planck adopted as his resolution: "Stay the course and keep on working [*Durchhalten und weiterarbeiten*]."

To this day the inhabitants of *Grunewald* hang on to the memories of the *Villenkolonie's* by-gone halcyon years before 1933. In 1989, 40 years after Planck's death, the citizens of Grunewald published a 4-page flier (*100 Jahre Villenkolonie Grunewald*) announcing a 2-months series of organized events that on the one hand would bring to mind the worst of what had happened during the *Third Reich*, but also to celebrate the high-level intellectual give-and-take that had been the hallmark of the *Villenkolonie* prior to 1933. Included in the events were lectures, slide and

[73] Helga Gläser, "Die Villenkolonie als Kulturelles Zentrum" in *100 Jahre Villenkolonie Grunewald 1880–1989*, Berlin, 1988, 63–94, but especially 88–93.

film shows, literary readings, panel discussions, the unveiling of commemorative tablets, museum visits, a German-British football tournament (Grunewald in 1989 was in the post-war British zone), lectures on Jewish private schools during the *Third Reich*, discussions on whether and how the German-educated middle class had collapsed as a group, the *Villenkolonie* as an ivory tower, and Grunewald as center of Jewish sports 1933–1938. The roster of events also included historic tours to places in Grunewald that in one way or another honored former heroes: Helene Lange, pioneer and spiritual leader in the early feminist movement, Friedrich Murnau, actor, stage manager, and film director in the 1920s, Walther Rathenau, Jewish industrialist, cultural philosopher, and statesman for foreign politics, Moses Mendelssohn, German-Jewish philosopher and leader in the movement for assimilation of Jewish culture, Max Reinhardt, dramatic artist and stage manager for theater and cabaret, Dietrich Bonhoeffer, Protestant theologian who was hanged in 1945 for his role in the plot to overthrow Hitler, and notably Max Planck, whose home in the Wangenheimstrasse next to the home of Bonhoeffer was destroyed in the Allied bombing of 1944.[74]

The profile of the *Grunewald Villenkolonie* that has been sketched here goes a long way toward providing credible information about the positive as well as the less salutary forces at work in the physical, intellectual, and cultural environments in which the political and otherwise inherently conservative Planck lived and worked as a theoretical physicist from 1905 to 1944. It also sets the stage for exploring the context in which Planck formulated his conception of the external world of nature – a realistic conception that he believed was within reach by means of scientific investigation. To Planck's dismay, even the quantum physics that he had set in motion stood in sharp contrast to his own realistic world view; he never was able to come to terms with it and its implications for physics. It has been suggested that Planck's conservatism in politics and general make-up was in line with and spilled over into his conviction that the mechanistic, deterministic, classical physics encapsulated in Newtonian gravitational theory was true for all times and could and would be extended to include thermal phenomena, electrodynamics, and the fine structure of matter. Planck maintained that classical deterministic physics was a secure and permanent landmark in need of refinement but that it was unlikely to be uprooted by developments in quantum physics. Planck certainly never thought of himself as a revolutionary. On the contrary, his focus was on preserving, improving, and completing the hard-won theoretical principles of the past. Indeed he unobtrusively entertained the hope that by some trick of theoretical understanding he would be able to demonstrate the truth of those principles. What Planck had to say about these matters in his scientific autobiography is clear and uninhibited:

> My vain attempts in some way to incorporate the quantum of action into classical theory stretched over many years and cost me a lot of effort. Many colleagues have seen this as a

[74] Grunewald = Echo, "Veranstaltungsprogramm 100 Jahre Villenkolonie Grunewald," Sept/Oct, 1989, 1–4.

> kind of tragedy. I have a different opinion, since for me the gain was that I came away (from those efforts) with a much more thorough clarification. Now I knew precisely that the quantum of action in physics plays a much more significant role than I at first had been inclined to suppose; and thereby I won a full understanding for the necessity of introducing totally new methods for examining and measuring atomic problems. The development of those methods, in which I personally no longer could contribute, were accomplished above all in the works of Niels Bohr and Erwin Schrödinger. With his model of the atom and the correspondence principle the former laid the foundation for meaningful connection between the quantum of action and classical theory. With his differential equations the latter created wave mechanics and the dualism of wave and particle.[75]

Reluctant to fully accept the revolutionary implications of the quantum of action even after almost a half century of its enormously impressive successes, Planck nevertheless was able in his own way of thinking to rationalize the overall developments in quantum physics that had occurred since 1900. Much of the later developments in quantum physics had been carried out in research centers beyond Berlin and in countries other than Germany. It had left the Berlin physicists, including Planck, more on the periphery than in its original center. Actually Planck's discovery was intimately connected with measurements at the PTR that were undertaken to improve on photometric methods called for by the need to improve on the science of incandescent filaments in competition with gas- petroleum- and arc lights. In any case Planck initially had attached little importance to his own discovery of the quantum of action and even downplayed its value in physics. He later remarked that he hadn't thought much about the idea at the time (*Ich dachte mir nicht viel dabei*).

During the some 40 years Planck was engaged in teaching theoretical physics at the university he was making contacts with experimentalists at the PTR and was engaged in foreign trips and lectures in America and elsewhere. He nevertheless spent most of his time in the ambience of the *Villenkolonie* whose activities and proclivities we have treated in the above. We learned something there about the image and caliber of persons with whom he came in contact: professors (in good supply), writers, artists, musicians, wealthy businessmen, and also theologians and church people who were struggling with him in problems on how to view the relationship between religion and modern science. It was in the *Kolonie* that Planck wrote his scientific papers, offered comments on the work of other scientists, and was confronted with the duty to respond in some way to threats his colleagues were receiving from the National Socialists. It is where he served for many years as a churchwarden and sought the advice of his fellow Grunewalders. It was where he hammered out his views on science and religion, on causality and free will, on determinism and indeterminism, on the relative and the absolute, on positivism and realism, and on the meaning and limits of science. It also was where Planck shaped his own beliefs and philosophy of science, commented retrospectively on his own views and writings, and composed his scientific autobiography. Of enormous importance for Planck was the *Kolonie* as the ideal locus to express his passion

---

[75] Max Planck, Wissenschaftliche Selbstbiographie, (1948) *Physikalische Abhandlungen und Vorträge*, III (1958) 397.

for music either in daily practice at the piano or organ, or with his son Erwin on the cello and daughter Emma on the violin and, as often as time would permit, with other musicians who were available for *Hausmusik* in the Wangenheim home or elsewhere in a larger setting.

In an earlier section we examined Planck's pre-Berlin exposure to and formal training in music and his writings on music theory that followed from his experimentation with an enharmonic harmonium. We also took note of what Planck's colleagues and fellow musicians had to say about his passion for music during the years he was in contact with them. In order to round out our image of Planck's public musical persona it is necessary, finally, to consider how his colleagues and acquaintances appraised, in retrospect, the musical component of his life.

Shortly after Planck died there was a great outpouring of obituaries and statements of personal remembrance on the part of those who had been Planck's colleagues and close friends, some of whom had made music with him in the *Villenkolonie*. By the 1940s Planck had become world famous for having brought about a scientific revolution that changed man's conception of the world of nature that is as fundamental as the changes brought on by the Newtonian theory of gravitation. The idiosyncratic position that Planck himself had taken on quantum theory made it all the more imperative for close friends and associates to put on record something about their own experiences with the great man. They often were experiences that had little or nothing to do with the contributions to physics that already had been given more coverage among scientists than Planck would have wanted – a theory that Planck felt could be no more than a good but temporary fix to a larger problem.

On the 100th anniversary of Planck's birth in 1957, Otto Hahn, the first President of the MPG, recalled in a memorial lecture that his contacts with Planck began in 1906 when he was a *Privatdozent* in Emil Fischer's chemical institute at the University of Berlin. Later on in the 1930s it was Hahn's collaborative research on radioactivity with Planck's assistant Lise Meitner (1878–1968) that gave unique prominence to the Planck-Hahn-Meitner relationships – a relationship that included common interests in music. According to Hahn:

> Planck was fond of (fostering) social life in the home [*häusliche Geselligkeit*]. In the years before the first world war a number of music-loving younger women and men, in a group to which I belonged, met in his house in Berlin-Grunewald every 14 days. We formed a four-part chorus under Planck's direction, and sang the music of Brahms and others. In one of Brahms's *Zigeunerlieder* I even was allowed to sing a little solo part. Planck encouraged me at the time to take proper singing lessons, suggesting that my breathing technique was bad and that my tenor voice would improve. So in July 1914 I began singing lessons; in August came the war. My singing was over. After the war the little singing circle under Planck did not reconvene, but the personal invitations resumed.[76]

[76] Otto Hahn, "Einige persönliche Erinnerungen an Max Planck," *Mitteilungen aus der Max-Planck-Gesellschaft*, Göttingen, Heft 5, 1957, 241–246.

Shortly after the publication of Hahn's reminiscences about the social and musical soirées in Planck's pre-war Grunewald home, the Physical Society of the German Democratic Republic engaged Lise Meitner from Stockholm to deliver a lecture in East Berlin on Planck as a person.[77] Meitner, who had been a student of Boltzmann in Vienna and had decided in 1907 to continue her physics studies with Planck in Berlin, remained in Germany the rest of her life except for a decade in Stockholm and retirement in Cambridge, England. In Berlin she became Planck's assistant for a number of years while collaborating with Hahn on problems connected with artificial radioactivity.[78]

Meitner remarked in her lecture that she first met Planck in the fall semester of 1907 when she sought him out at the University of Berlin to register for his lectures. Shortly thereafter she was invited for an evening to his house to meet his wife and twin daughters. "Already with this first visit I was much impressed by the elegant simplicity of the house and the entire family." At Plank's lectures Meitner at first had to struggle with a feeling of disappointment because in Vienna, where she had been Boltzmann's student, she had experienced a teacher who "was overflowing with enthusiasm about the wonders of the laws of nature and their comprehensibility by the human mind." With Boltzmann there had been no shyness about expressing himself in a personal manner that swept younger listeners along with great enthusiasm. Against this background and in spite of extraordinary clarity, "Planck's lectures came across as impersonal, and dull." Meitner soon learned that these encounters did not represent the real Planck; she merely wanted to point out that he had this same problem with other persons who did not really come to know him well. "He possessed a rare purity of noblemindedness and an inner straightness that matched his external simplicity and unpretentiousness."

As most of the persons who knew Planck well have written, Meitner was profoundly impressed by his enduring commitment to music and the extent to which his home in Grunewald became the main center in making music for persons such as Joachim and others who benefitted from Planck's "love for high-spirited and unconstrained conviviality." As Thomas Mann had written about Schiller's noble-minded naïvete and had referred to Schiller's *Glocke* as a song of religious orderliness, so too, writes Meitner, one might characterize many of Planck's decisive traits. "He also was noble-mindedly naïve and full of reverence for the marvelous lawlikeness of natural phenomena which he experienced as a religious

---

[77] Lise Meitner, "Max Planck als Mensch," *Die Naturwissenschaften, 45* (1958) 406–408.

[78] In 1944 the 1943 Nobel Prize for Chemistry was awarded to Otto Hahn "for his discovery of the fission of heavy nuclei" – a discovery, it must be added, whose theoretical interpretation as a fission phenomenon was given not by Otto Hahn and Friedrich Strassmann who carried out the experiments and published the announcements, but by Lise Meitner and her nephew Otto Frisch (1904–1979). They suggested, on the basis of Niels Bohr's drop model of the nucleus that the bombardment of uranium with neutrons had caused the uranium to undergo fission. Having been forced to leave Berlin in 1938 because of German racial laws, Meitner was working at the time as director of a research group in Manne Siegbahn's Institute in Stockholm and was able to stay in contact with Hahn in Berlin and Frisch in England by mail.

orderliness [*fromme Ordnung*]. Thus, e.g., in his lecture on science and religion, Planck remarked: "It is noteworthy that in all processes in nature a *universal* lawlikeness prevails that in a sense is discernible; nothing prevents us from identifying the world order of natural science with the God of religion."[79] Planck's religion, Meitner noted:

> was not supported by a particular religious form. He often emphasized that one should never forget that even the most sacred symbols are human in origin and that the deeply religious person is not glued to the symbols but understands that there are persons who are just as religious for whom other symbols are sacred.... Planck was religious in the same sense that Goethe was religious.

Meitner intimated that Planck's reverence for the "religious orderliness" in nature had its counterpart in a remarkably systematic daily routine. He played the piano daily at the same time. "As everyone knew, he was a passionate and distinguished pianist, and the many fine music evenings in his house were unforgettable for all who were privileged to enjoy them."[80]

According to Meitner, Planck had the intellectual capacity for understanding persons who were disposed to other ways of thinking than his own; he was able to reach beyond all outward appearances and take hold of essentials – this in spite of a penchant for moderation in all things and a strong bonding with the traditional. "That was the basis for his great friendship with Einstein." In the 1920s he bravely and staunchly stood on the side of Einstein when he was severely criticized by some of his own physics colleagues in connection with relativity theory.[81] As Planck's physics assistant Lise Meitner frequently attended the evening music sessions that took place in the Planck home in Grunewald; and although she did not personally contribute to them musically it was reported that she seldom missed participating as a spirited listener and sometime critic. She remembered an evening when Planck, Einstein, and a professional cellist played the Beethoven piano trio in B-flat major. "Listening to this was marvelously enjoyable, despite a couple of unimportant slips from Einstein.... Einstein was visibly filled with the joy of the music and smiled in a light-hearted way that he was ashamed of his dreadful technique. Planck stood quietly by with a blissful happy face and, hand on heart, remarked: 'that wonderful second movement.'"[82]

The theoretical physicist Max von Laue (1879–1960), who completed his doctorate under Planck and stayed on in Berlin as his assistant and became his colleague there, is mentioned in the history of science literature as Planck's leading pupil and honored lifelong friend. In 1914 Laue was awarded the Nobel Prize in Physics for his discovery of the diffraction of X-rays by crystals. In 1919 he arranged for an exchange of teaching positions with Max Born (1882–1970) who had been in Berlin since 1914 and had moved to the University of Frankfurt. Laue who had been in Frankfurt since 1914 moved to Berlin and joined Planck as

[79] Meitner, *Max Planck als Mensch* (1958) 406.
[80] Meitner, *Max Planck als Mensch* (1958) 408.
[81] Meitner, *Max Planck als Mensch* (1958) 406.
[82] Brian Foster, CERN, the violin and the music of the spheres, *CERN Courier*, Feb. 1, 2005.

Vice-Director of the Institute for Theoretical Physics. In 1932 Laue received the Max Planck Medal from the *Gesellschaft Deutscher Physiker*. For a period of almost 30 years Laue worked in close collaboration with Planck, first on his doctorate (1902–1903), then as an assistant (1906–1909), and finally as a colleague (1919–1943). It is no surprise to discover that among Planck's students and colleagues no other person than Laue has written so insightfully and unreservedly not only about his physics but about his innate conservatism and tenacious passion for music. These two earmarks of Planck's constitutional disposition – his conservative temperament and his fixation on music – are given full play in Laue's writings on Planck and will be discussed in some detail in what follows.

In a lecture to honor Planck's 100th birthday at the Berlin Academy of Science Laue attributed Planck's rapid ascent in the academic world to his outstanding teaching talents and to the patronage of Helmholtz. He not only had studied in Berlin and been Helmholtz's colleague, but he had worked in an environment that fostered mainline research in theoretical physics and sideline interests in music. Laue sought to show how Planck's thinking concerning the formal structure of thermodynamics, irreversible processes, statistical notions, probability, second law considerations, and queries about the reality of atoms, had led him to the quantum of action idea. These were notions. As he remarked, "that had given a new direction not only to physics but to the physical world view of all mankind."[83]

Laue portrayed his *Doktorvater* as "a scientist whose entire posture was nation-affirming [*die ganze Haltung war staatsbejahend*]." "As far as I know," he remarked, "Planck never expressed his opinion on the Hegelian philosophy, but the conception of the state as the materialization of ethics nevertheless was the very foundation of his posture as a citizen. When the Hitler regime took this conception to its thorough-going absurdity Planck sometimes stood hopelessly lost."[84] In Laue's opinion conservatism manifested itself in Planck's outlook on physics in a unique way. Long after the inception of the quantum notion Planck continued to voice an essential uneasiness about the quantum's epistemological implications and continued to take a position that led his colleagues to regard him as a "revolutionary agent against his own will [*Revolutionär wider Willen*]." Planck was reluctant to abandon the classical position and focused rather on the gap that separates quantum theory and classical physics. "For decades [he] wrestled with the idea of their closure." As Laue was quick to point out, Planck's efforts to find this closure paid off indirectly in the form of advances connected with Planck's ideas of zero-point energy, phase space in elements of magnitude h, and a revised form of Nernst's heat theorem. Planck's essential goal of restoring physics to a unitary world view proved to be even more out of reach. "For that reason," as Laue remarked, "he eventually abandoned his efforts, but not without emphasizing that they would have been necessary to safeguard, beyond a doubt, their unattainability."[85]

[83] Max von Laue, "Zu Planck's 100. Geburtstag," *Die Naturwissenschaften*, *45* (1958) 221–226.
[84] Laue, "Zu Planck's 100. Geburtstag" (1958) 221.
[85] Laue, "Zu Planck's 100. Geburtstag" (1958) 225–226.

In summarizing Planck's position on quantum physics as seen against the background of its general status among physicists in 1958 Laue remarked:

> The sharpest argument against the quantum theory, or at least the interpretation that it customarily is given today, consists in pointing out its unsatisfactory relationship with the principle of causality or to the unequivocal determinism [*eindeutige Bestimmbarkeit*] of physical processes as physicists prefer to state it. . . .
>
> With this (indeterminacy) Planck never reconciled himself. In the many lectures he delivered in various parts of Germany after he retired he repeatedly declared himself in support of unequivocal determinism. In so doing he got into trouble with many of the younger generation, although Albert Einstein, for example, was his confederate [*Bundesgenosse*] in this conflict. Planck, who apparently had taken the logical closure [*logische Abgeschlossenheit*] of the older physics to be erroneous, was unable to accept the failure of causality. . . . For this reason a tragic shadow is cast on a life's work that was ruled totally by a striving toward a unitary world view. He overcame this inwardly with the frequently expressed sentiment that it is not success that constitutes the value of the scientific investigator's life; that it depends much more on remaining true to science with conscientiousness and fidelity [*Gewissenhaftlichkeit und Treue*] – principles that he also sought to assign first place in life. May future generations of scholars realize that it was the consistent enforcement of these principles that led Planck to all of his achievements; and even to one whose outcome and significance can not as yet be fully estimated today.[86]

We turn finally to a consideration of what Laue had to say about Planck's lifelong interest in music. Laue intimated that Planck's compulsion about music as listener and as performer, and as a scientist attentive to its theoretical aspects, were traits indicative of a personality whose field of vision and frames of reference reached far beyond the discipline of physics and the life of a physicist. According to Laue:

> It was not an easy matter for him to choose a career. He vacillated between music, classical philology, and physics. As he once told me, he gave up music as a profession when he noticed that a number of his own attempts at composition reminded him too much of existing musical literature; and he came to recognize that a mediocre artist signifies much less than an average scientist who always can find a place in the scientific enterprise that needs to be filled. Confronted with a dilemma, Planck sought advice from Philipp von Jolly, the professor of physics at the University of Munich. Characteristic of a widely disseminated view at the time, Jolly dissuaded him from taking up the study of physics because the discovery of the energy conservation principle had left physicists with nothing much to anticipate.[87]

When Planck arrived in Berlin in 1889 it was music, as Laue recalled, that inexplicably first engaged his attention:

> It was a kind of Intermezzo during which he was occupied with music, and namely as a theoretician of musicology. At that time Helmholtz had an harmonium built in just intonation, and Planck taught himself to play on its complicated keyboard. In later years when he taught acoustics he always played it for his students. And it is characteristic that there was a difference of opinion between him and Helmholtz in regard to the appraisal of just intonation – the only one we know about. Even in the last edition of *Lehre von den*

[86] Laue, "Zu Planck's 100. Geburtstag" (1958) 226.

[87] Laue,"Zu Planck's 100. Geburtstag" (1958) 221.

> *Tonempfindungen* Helmholtz regretted that pure intonation more and more was being displaced by tempered intonation. Planck by contrast described a pure triad as less valuable aesthetically than the tempered one. Of course, by that time Planck had acquired so sensitive an ear, as he once told me, that concerts no longer were giving him full satisfaction; he constantly heard the apparently unavoidable little errors (in temperament). Later, to his pleasure, this hypersensitivity was lost.[88]

What Laue had to report in connection with Planck's experiences with the harmonium in the physics department at the University of Berlin is consistent with other reports that already have been cited, but the comments concerning the Helmholtz/Planck differences of opinion in regard to just intonation need to be unpacked. It is evident from Helmholtz's discussions on the relative merits of just and tempered intonation in *Tonempfindungen*, and from what is known about his position on the matter in later life, that he definitely preferred just intonation from an aesthetic point of view. He also maintained that just intonation should be fostered in vocal music and in instrumental music if and when musical instruments would become available that are able to sound in pure intonation, and when performers are able to master the manual skills exacted by the newly constructed instruments. However, Helmholtz by no means took a dogmatic position on intonation. Just intonation was looked upon as an aid to the enhancement of music listening. The scientific investigation of problems connected with intonation were of concern to him for various reasons. The upcoming generation of loosened-up listeners and composers had developed a new sensitivity to intonation and was searching for greater freedom in tonal expression. In addition, by mid-century music theory had become a branch of the discipline in which musicians were taking advantage of advances in the sciences to improve musical instruments while alluding to scientific knowledge to justify what they were doing. The arsenal of conceptual ideas and research tools that had been designed for the disciplines of physics, physiology, psychology, and the neurosciences had become accessible to music. The occasions on which to promote just intonation and settle for a given degree of tempered intonation, therefore, were seen by Helmholtz and others to hinge not only on factors such as musical taste, training, and experience, but on technological feasibility. If this meant that Helmholtz wanted to have it both ways – just intonation and tempered intonation – then it is safe to conclude that he would have accepted such a verdict. For Helmholtz the choices and occasions on which to opt for just or tempered intonation depended on the musical composition being performed, the types of musical instruments available for the performance, and the musical taste and level of musical knowledge and experience of the listener and listening audience.

Consider where Planck stood on these issues. On the basis of what he wrote and from what Laue and others have inferred from his writings and personal contacts, Planck's position on intonation seems to be ambiguous. On the one hand his interest in exploring what was to be learned from experiments with the Eitz harmonium, and from the study of vocal music and the voice as an experimental musical

[88] Laue, "Zu Planck's 100. Geburtstag" (1958) 223.

instrument, indicate an acknowledgment of the central significance of intonation in musical acoustics and on music theory in general. Planck's explicit remarks on the vocal issue in fact represent something of an extension of Helmholtz's more implicit remarks in *Tonempfindungen*. Planck evidently was convinced that it is important for musicians to come to terms with intonation simply as a matter of practice apart from theory. The writings of Planck that have been examined exhibit an unmistakable persuasion that accommodation to tempered intonation was an issue that musicians, unfortunately, would have to tolerate and live with in spite of the fact that they might have preferred just intonation from the standpoint of music theory and the aesthetic standards of the sensitive listener.

In Planck's *Selbstbiographie* of 1958 there is one isolated comment that can be interpreted to show that he personally preferred, or at least came to prefer, tempered intonation over just intonation.[89] It is this particular comment and not Planck's various published papers, as we have shown, that is cited repeatedly in the secondary literature as evidence for Planck's preference for tempered intonation. The comment is made in connection with Planck's experiments on the Eitz harmonium designed to explore the role of just intonation in instrumentally unaccompanied vocal music. He wrote: In these studies "I came to the somewhat for me unexpected result [*einigermassen unvermuteten Ergebnis*] that our ear prefers tempered intonation [*temperierte Stimmung*] over natural intonation [*natürliche Stimmung*] under all circumstances. Even in an harmonic major triad natural intonation sounds dull and expressionless [*matt und ausdruckslos*] compared to tempered intonation. In the last analysis this fact is to be attributed to years and generations of habit, because before Joh. Seb. Bach tempered intonation was not generally known."[90]

The implication of this discovery on the part of Planck was that before Bach's time natural or just intonation was in use simply because there was no need to temper the musical scale. Vocal music and music for instruments in which the pitch could readily be adjusted manually (as in instruments with one or more struck strings built on the model of the monochord) were performed in natural or close to natural intonation. With the introduction of fixed-tone (non-adjustable-tone) instruments such as the clavichord, it became impossible to play in natural intonation for reasons connected to the difficulties of construction and playing of instruments provided with enough tones to achieve natural intonation. After Bach's time, and with the introduction of keyboard instruments, the human ear was forced to accept and adjust to music written in tempered intonation. Acceptance and adjustment to tempered intonation led to development of the habit (accommodation) of preferring tempered over natural intonation. After Bach, it commonly came to be said that the human ear had become sufficiently blunted to prefer tempered over natural intonation.

---

[89] Planck's scientific autobiography was written in Göttingen the year during which he participated as the only German representative at the Royal Society of London's celebration of Isaac Newton's 300th birthday.

[90] Max Planck, "Wissenschaftliche Selbstbiographie," *Physikalische Abhandlungen und Vorträge*, *3* (1958) 374–401. Quotation 383–384.

But why would Laue and other physicists have called attention to this detail and refrained from mentioning that in his experimentally-based writings on music theory Planck unambiguously preferred natural intonation as supplied by naturally untempered and unaccompanied vocal music; that Laue and others apparently had taken the effort to examine Planck's *Wissenschaftliche Selbstbiographie* but that as physicist his writings on music were ignored was deemed to be of little or no importance when compared with his monumental contributions to theoretical physics? Planck's writings on music theory are not mentioned in the list of his publications in Planck's three-volume *Physikalische Abhandlungen und Vorträge*. In fact, Planck's writings on music theory are seldom included in any of the chronological lists of his publishers. The outstanding example to the contrary is the *Chronoligisches Werkverzeichnis* included in the *Max-Planck-Gesellschaft Berichte und Mitteilungen* published in Munich in 1997 on the occasion of the 50th anniversary of his death.

The difference of opinion between Helmholtz and Planck concerning the natural and the tempered triad, and about just intonation and tempered intonation in a more general sense, has been the subject of mention in several works by physicists and historians of physics – all written after Planck's *Selbstbiographie* had been published.[91] This assertion sometimes is followed by the comment that this difference of opinion was the only one that Planck ever had with the 37-years-older Helmholtz – a colleague whose authority, we recall, was so great that he came to be referred to as the *Reichskanzler* (Chancellor) of German physics. Such evaluations on the whole are too simplistic and are based solely on a comment, or rather on an interpretation of a comment, that Planck made in his *Selbstbiographie* shortly before he died at age 89. They do not support the position on intonation that Planck expressed in any of his other writings connected with music. Rather he was claiming, I suggest, that tempered intonation had become so widespread and dominant a phenomenon in the twentieth century that everyone's ears had become so blunted, dulled, or deadened (in his paper on vocal music Helmholtz uses the term "*abgestumpft*") – a claim that Planck had shown to be true by putting it to test on the ears of others and himself – that the question concerning temperament preference no longer was open for argument from an acoustical point of view; that as far as he was concerned, music listener and music performer might just as well become accommodated to tempered intonation, learn to live with it, and learn to prefer it as he himself already had done. This merely amounts to assertion that by 1900 the ears of most listeners had become so sensitivity-blunted, so *abgestumpft*, to intonation as to prefer tempered intonation.

Virtually all of Planck's biographers and commentators have had something to say about his interest and involvement in music. Those who have undertaken to lay out the history of quantum mechanics and appraise his revolutionary

[91] See, e.g., Max Born, "Max Karl Ernst Ludwig Planck," *Obituary Notices of Fellows of the Royal Society*, *6* (1948) 165; Max von Laue, "Zu Max Planck's 100. Geburtstag," *Die Naturwissenschaften*, *45* (1958), 223; Armin Hermann, *Max Planck*, Reinbeck bei Hamburg, 1973, 25.

contribution to quantum theory invariably have mentioned that Planck by nature was a conservative man and that this conservatism manifested itself in his approach not only to physics but to philosophy, politics, and life in general. Planck was a deeply and implicitly religious person but by no means so in a sectarian sense. He descended from a family of pastors, scholars, and professors of law. He was born into a rising Prussia that was on its way to unification and to bringing the Empire to the pinnacle of its scientific excellence and industrial power. Spokesman of science that he became under the Wilhelmian Empire and the German Reich, Planck was a German patriot and a man of devotion to duty who, as Heilbron demonstrates, supported frontier research in theoretical physics while quietly but tenaciously identifying with the unified and strong imperial Germany that had evolved from an authoritarian to a democratic state and then, as if to throw all gains aside, became a totalitarian state that Planck abhorred.[92]

Beyond the moral dilemmas in politics and state that Planck was confronted with during his lifetime, he faced a dilemma in his own scientific thinking at the beginning of his Berlin career that remained with him for the rest of his life. The dilemma was provoked by developments in radiation physics in which he had been the prime mover and for which he has been honored ever since as main play-actor. In his early career Planck had achieved something akin to a reconciliation between thermodynamics and classical mechanics, but then as a result of his own path-breaking interpretation of black-body radiation the mechanical physics that he had preserved in that reconciliation was shaken to its roots. Planck was reluctant to come to terms with the discovery that the laws of mechanics and the principles of classical determinism do not adequately account for phenomena occurring at the level of the micro world. Holding fast to the absolute primacy of the law of causality, he was jolted by the Heisenberg uncertainty principle and the Copenhagen interpretation of quantum mechanics. He had derived the equation that accounts for the distribution of energy by a heated black body – a derivation that required, as was shown, the formulation of a notion of energy quanta. The energy quanta became the impetus for the energy quantum's resolution, namely quantum mechanics. It was an outcome that Planck simply could not fit into his realistic philosophy of science; paradoxically so, because it was the quantum revolution, combined with his unique and venerable personal stature and temperament, that made him a leading spokesman for German science and an outstanding administrator of its institutions.

Planck's conservative outlook on life as well as on physics has been the subject of endless characterizations in which he has been referred to as a philosophical realist, an idealist devoted to church and state, a brilliant thinker for whom there was no gap between scientific and philosophical or religious convictions, and above

[92] In *Dilemmas of an Upright Man. Max Planck as Spokesman for German Science*, Berkeley, 1986, the historian of physics John Heilbron has presented a sensitive and meticulously researched reconstruction of Planck the father of quantum theory; he has examined in a balanced and critical way the moral dilemmas Planck faced as spokesman for science under the Wilhelmian Empire, the Weimar Republic, and the Third German Reich.

all one who held a strong belief in the harmony of the laws of nature and the laws of the mind. Planck began his *Selbstbiographie* with the following sentiment:

> What led me to my science, and from youth on filled me with enthusiasm, is the not at all self-evident fact that the laws of our thinking [*Denksgesetze*] conform with the lawfulness [*Gesetzmässigkeit*] that exists in the wider impressions we receive from the outer world; so that it is possible for man to gain information about that lawfulness by pure thinking [*reines Denken*]. In that case it is of essential significance [to recognize] that the outer world represents something with which we are confronted and that is independent of us and is absolute; and the search for the laws that constitute the absolute appeared to me as the most splendid scientific life-task [*die schönste wissenschaftliche Lebensaufgabe*].[93]

Early in his research on the irreversibility of radiation phenomena, and even before he became involved with quantum ideas, Planck referred to the importance of identifying the natural, universal, invariant, and everlasting singularities of matter that are independent of material bodies and substances and represent deeper-lying existent realities than the atoms of matter. Planck was uncomfortable with the notion of atoms and preferred to speak about oscillations in order to avoid dealing with atoms; although he was not certain about the real existence of oscillations they did represent, he believed, properties one could work with. In 1899, in connection with a problem in radiation theory, he spoke about the existence of universal and absolute constants in nature that retain "their meaning for all times and for all cultures, including extra-terrestrial and nonhuman ones."[94] The search for the absolute runs like an Ariadne thread through Planck's scientific writing, and it is essential to explore what he may have meant with that term. His scattered reflections on the subject provide evidence to suggest that he had in mind something close to what Plato referred to as the "idea." In the *Timaeus* Plato stressed the importance of ideas and their general form as a basis for true reality – permanent and secure behind the appearance of experienced things. According to Plato real knowledge about nature is attained by passing over, getting beyond earthly things, and latching on to mathematical ideas that reveal the real underlying structure of the world of nature. Recognizing that seventeenth-century neo-Platonists such as Galileo, Kepler, and Newton had achieved an impressive comprehension of the natural world based on mathematical reasoning, Planck may well have chosen to devote his life to theoretical physics by acknowledging on the one hand the revolutionary outcome of seventeenth-century creative mathematical thinking and realizing on the other that eighteenth- and nineteenth-century physics was becoming increasingly dominated by empiricism. Planck, the inveterate idealist, believed it opportune for physicists to return to mathematical ways of thinking. As has been shown in an earlier section of this chapter, Heisenberg characterized Planck's approach to physics as one that was permeated by the Platonic mathematical outlook. In Planck's theoretical analysis of the black-body radiation phenomenon, his

---

[93] Max Planck, "Wissenschaftliche Selbstbiographie," *Physikalische Abhandlungen und Vorträge, 3* (1958) 374.

[94] Max Planck, "Ueber irreversible Strahlungsvorgänge" (1899), *Abhandlungen und Vorträge 1* (1958) 599–600.

thinking in terms of absolutes led to an outcome that he was hard-pressed to accept. He was unable to effect reconciliation with his own offspring, the quantum theory.

In the lectures at Columbia University in 1909 Planck remarked:

> The goal for physicists everywhere, for all times, all people, all cultures, is none other than the integrity [*Einheitlichkeit*], the solidarity [*Geschlossenheit*] of the system of theoretical physics, and the unity of the particulars of the system. Yes, the system of theoretical physics lays claim to validity not only for the inhabitants of this earth but also for the inhabitants of other celestial bodies. Whether the inhabitants of Mars, if such exist, have eyes and ears like ours we do not know; it probably is rather unlikely. But for most physicists it is self-evident that the inhabitants of Mars will know the laws of gravitation and the principle of energy provided they possess the necessary intelligence; and the one to whom that is not obvious had better not count himself among physicists because then it basically would remain for him an incomprehensible riddle to realize that in the United States one creates the same physics as in Germany.
>
> We may therefore say in summary that the characteristic trait [*Merkmal*] in the actual development of the system of theoretical physics is its progressive emancipation from anthropomorphic elements with a goal toward as complete as possible a separation of the system of physics from the individual personality of the physicist.[95]

As Max von Laue suggested, this thinking in terms of absolutes cast a shadow over Planck's philosophy and his methodology. On the other hand Max Born has offered a somewhat more open-ended and conciliatory interpretation of Planck's position. He has lauded Planck for his confidence in the efficacy and trustworthiness of unflinching logical thinking; and he has shown how, in the results of experimental investigations, one can rely on providing the occasion for putting logical thinking about those experiments to test. Born saw that in the black-body problem there was something absolute that caught the attention of Planck.

> Planck's philosophical mind was always directed towards the search for the 'absolute.' From 1896 on the explanation of the [radiation] law was the aim which he pursued with amazing persistency, always in contact with the experimental investigations at the Reichsanstalt. The series of papers (1897–1901) that dealt exclusively with this problem and ended with complete success are a testimony not only to Planck's skill and ingenuity, but also to his character, his unbending will and untiring industry, and his cautious patience combined with the greatest audacity. His was, by nature, a conservative mind; he had nothing of the revolutionary, and was thoroughly sceptical about speculations. Yet his belief in the compelling force of logical reasoning from facts was so strong that he did not flinch from announcing the most revolutionary idea which ever has shaken physics.[96]

In the *Selbstbiographie* Planck claimed that one is able on the basis of "pure thinking" to gain information about the outer world; that one's thinking conforms with the lawfulness that exists in the sequence of impressions received from the outer world and represents something that is independent of man and is "absolute." Proceeding from a position of innate conservatism, logical thinking, and the outspoken realist assumption that scientific theories are true descriptions of external

---

[95] Max Planck, *Acht Vorlesungen über theoretische Physik*, Leipzig, 1910, 6–7. *Erste Vorlesung 23. April 1909. Einleitung, Reversibilität und Irreversibilität.*

[96] Max Born, "Max Planck," *Obituary Notices of the Fellows of the Royal Society*, *6* (1948), 167.

reality, it is tempting, if speculative, to ponder and explore whether similar guidelines were at work when Planck turned his attention to examining music from a scientific point of view. Serious and cultivated music-making in family and social settings was an activity that Planck nurtured long before he took on the position in theoretical physics at the University of Berlin. During the early years of his academic career he considered music to be as though-provoking from an intellectual point of view and as enticing as the discipline of physics that he decided to pursue. When Planck arrived in Berlin as a 40-year-old theoretical physicist he discovered that the almost 40-years-older theoretical physicist Helmholtz, now his colleague, was agreeable to becoming his associate on the scientific investigation of music.

The point in question that naturally comes to mind in this context is whether Planck's approach to music reveals the same commitment to realism that is so evident and forthright in his approach to physics. More specifically: how did Planck deal with the problem of intonation, that on the one hand had been examined and analyzed by Helmholtz in great detail, and then with attention to its checkered history in *Tonempfindungen* had been investigated in his own papers on the Eitz harmonium and in relation to vocal music? Did Planck in his own investigations come down on preference for tempered intonation as the passing comments in his *Selbstbiographie* of 1948 seem to suggest? That is the interpretation that Laue and others after Laue have called attention to in spite of the fact that nothing of that sort is mentioned by Planck in his articles on the Eitz harmonium, on vocal music, and in the Zellner book review written some 50 years earlier.

Laue's interpretation of Planck's position on intonation unfortunately flies in the face of Planck's deep-rooted commitment to a position that asserts man's ability, by means of pure thinking, to comprehend the lawlikeness of the world – in this case in the world of sound that defines the nature of music. Planck's papers on music, much in the tradition of Helmholtz, support the idea that the degree of harmoniousness of musical intervals is given by the simple whole-number ratios of idealized vibrating strings or frictionless columns of air. An interpretation that comes down on the side of preference for tempered intonation in place of just intonation would seem to fall back on the argument that in spite of music-theoretic imperfections that are not easily eliminated from tempered music instrumentally speaking it nevertheless is advisable to cast one's lot on the side of the enormously rich heritage of nineteenth- and twentieth-century music that is structured on tempered intonation. The argument is made that this monumental musical heritage is rooted historically in the tempered intonation that was thrust upon musicians when fixed-tone musical instruments such as the clavichord, the harpsichord, and the pianoforte came to dominate music-making, music-composing, and music-listening by the end of the nineteenth century. The latter interpretation on the side of just intonation (frequently referred to as pure or natural intonation) rests on the historically prior and theoretically sound notion of just intonation developed in Helmholtz's *Tonempfindungen*. Although it may be seen to lie beyond the scope of this analysis to definitively and unambiguously locate Planck's position of preference on intonation, it has been thought to be worthwhile to raise the issues here. The problem of intonation continued to be a major concern for music for music theoreticians and for music-oriented scientists long after Planck's time. It probably is too far-fetched to entertain the idea that Planck's characteristically conservative personality traits and overlapping Prussianism spilled over into his musical tastes and writings on music theory but it nevertheless has been tempting to offer some reflections on that topic.

References

Andrade, E.N. da C. 1948. Max Planck Memorial Ceremony in Göttingen. *Nature* 4098: 751–752.

Beck, Lorenz Friedrich. 2008. *Max Planck und die Max-Planck Gesellschaft*. Berlin: Archiv der Max-Planck-Ges.

Bertholet, Alfred. 1948. Erinnerungen an Max Planck. *Physikalische Blätter* 4: 161–162.

Born, Max. 1948a. Max Karl Ernst Ludwig Planck. *Obituary Notices of Fellows of the Royal Society* 6: 165.

Born, Max. 1948b. Max Planck. *Obituary Notices of the Fellows of the Royal Society* 6: 167.

Botstein, Leon. 2008. Einstein and music. In *Einstein for the 21st century: His legacy in science, art, and modern culture*, ed. Peter Galison, Gerald Holton, and Silvan S. Schweber, 161–175 and 331. Princeton: Princeton University Press.

Brücke, E. (ed.). 1948. Max Planck Gesellschaft. *Physikalische Blätter* 4: 132.

Dinkler, Erich. 1959. Max Planck und die Religion. *Zeitscrift für Theologie und Kirche, Göttingen* 56: 201–223.

Feuchtwanger, Lion. 1984. *Die Geschwister Oppermann*, 211. Frankfurt am Main: Fischer Taschenbuch Verlag.

Feyerabend, Paul, and Max Planck. 1967. *The encyclopedia of philosophy*, vol. 6, 313.

Forman, Paul, and Armin Hermann. 1975. Arnold J. Sommerfeld. *DSB* 12: 525–532.

Gläser, Helga. 1988. Die Villenkolonie als Kulturelles Zentrum. In *100 Jahre Villenkolonie Grunewald 1880–1989*. Berlin: Bezirksamt Wilmersdorf.

Hahn, Otto. 1957. Einige persönliche Erinnerungen an Max Planck. In *Mitteilungen aus der Max-Planck-Gesellschaft*, Heft 5, 241–246. Göttingen: Das Gesellschaft.

Heilbron, John L. 1986. *The dilemmas of an upright man. Max Planck as spokesman for German science*. Berkeley: University of California Press.

Hentschel, Klaus, and Renate Tobies. 1999. *Brieftagebuch zwischen Max Planck, Carl Runge, Bernhard Karsten und Adolf Leopold*. Berlin: ERS Verlag.

Hermann, Armin. 1997. Max Planck. In *Die Grossen Physiker*, vol. 2, ed. Karl von Meÿenn, 143–156. Munich: Beck.

Hoffmann, Dieter, and Max Planck. 2008. *Die Enstehung der modernen Physik*. Munich: Beck.

Jánossy, L. 1958. Plancks philosophistische Ansichten in der Physik. In *Max-Planck-Festschrift*, 389–407. Berlin: V.E.B.

Meitner, Lise. 1958. Max Planck als Mensch. *Die Naturwissenschaften* 45: 406–408.

Partington, J.R. 1948. Erinnerungen an Max Planck. *Physikalische Blätter* 4: 172.

Planck, Max. 1893. Die Natürliche Stimmung in der modernen Vokalmusik. In *Vierteljahrsschrift für Musikwissenschaft*, vol. 9, 418–440. Leipzig: Breitkopf und Härtel.

Planck, Max. 1894. Ein neues Harmonium in natürlicher Stimmung nach dem System C. Eitz. In *Verhandlungen der Physikalischen Gesellschaft*, vol. 12, 8–9. Berlin: Georg Reimer.

Planck, Max. 1910. *Acht Vorlesungen über theoretische Physik*, 6–7. Leipzig: S. Hirzel.

Planck, Max. 1918. Zur Feier seines 60. Geburtstages. *Die Naturwissenschaften* 6: 194–263.

Planck, Max. 1947. The quantum theory. *The Times*. London, 6 Oct 1947.

Planck, Max. 1948a. *Wissenschaftliche Selbstbiographie*. Leipzig: J.A. Barth.

Planck, Max. 1948b. Wissenschaftliche Selbstbiographie. *Physikalische Abhandlungen und Vorträge* III(1958): 397.

Planck, M. 1958a. Wissenschaftlische Selbstbiographie. *Physikalische Abhandlungen und Reden* 3: 374–375.

Planck, M. 1958b. Vom Relativen zum Absoluten. *Physikalische Abhandlungen und Vorträge* 3: 145.

Planck, Max. 1958c. *Physikalische Abhandlungen und Vorträge*, vol. 1, 435–436. Braunschweig: F. Vieweg.

Planck, Max. 1958d. Wissenschaftliche Selbstbiographie. *Physikalische Abhandlungen und Vorträge* 3: 374–401.

Planck, Max. 1958e. Wissenschaftliche Selbstbiographie. *Physikalische Abhandlungen und Vorträge* 3: 374.

Planck, Max. 1958f. "Ueber irreversible Strahlungsvorgänge" (1899). *Abhandlungen und Vorträge* 1: 599–600.

Sommerfeld, Arnold. 1918. Max Planck zum sechzigsten Geburtstag. *Die Naturwissenschaften* 6: 196.

von Laue, Max. 1958a. Zu Planck's 100. Geburtstag. *Die Naturwissenschaften* 45: 221–226.

von Laue, M. 1958b. Zu Max Planck's 100. Geburtstag. *Die Naturwissenschaften* 45: 223.

von Zahn-Harnack, Agnes. 1948. Erinnerungen von Max Planck. *Physikalische Blätter* 4: 165–167.

Westphal, Wilhelm. 1948. Erinnerungen an Max Planck. *Physikalische Blätter* 4: 167–179.

Westphal, Wilhelm. 1958. Max Planck als Mensch. *Die Naturwissenschaften* 45: 234–236.

Part IV
Adriaan Fokker. Theoretical Physics and Just Intonation Keyboards

Chapter 19
Fokker: Theoretical Physicist

Our examination of the way in which Helmholtz, Tanaka, and Planck formulated the problem of just intonation and approached the construction of just intonation keyboard instruments represents but a fraction of the imaginative and innovative attempts by others who approached the just intonation keyboard problem from different perspectives. In the work and thought of the theoretical physicist Adriaan Fokker and his musical colleagues in the Dutch school of 31-tone per octave composers we have a singularly imposing instance of the fascination with which just-tempered keyboard construction and 31-tone per octave musical compositions were explored well into the twentieth century by scientists, instrument builders, technologists, and composers. As in our account of the life and contributions of Helmholtz and Planck, an investigation of Fokker's devotion to scholarship and his intellectual legacy rests heavily on a life devoted in large part to theoretical physics. As will become evident in what follows, it is pertinent to explore Fokker's contributions to theoretical physics in order to establish his identity and credibility as an intellectual who saw fit at mid-career to throw himself as wholeheartedly into the study of music theory as he had thrown himself into the study of theoretical physics.

Adriaan Daniel Fokker (1887–1972) was born of Dutch parents in Java in the Netherlands Indies. He first studied mining engineering at the Polytechnical School in Delft (1904–1905), entered the University of Leiden as a physics student in 1906, and completed a doctoral dissertation on Brownian movement under Henrik Antoon Lorentz (1853–1928) in 1913. Fokker belonged to the younger generation of upcoming Dutch theoretical physicists such as Paul Ehrenfest, H. A. Kramers, and Pieter Zeeman, who had learned their electrodynamics from Lorentz and who became involved in exploring the hidden implications of relativistic and quantum ways of thinking that were being debated during the first decade of the century. After defending a dissertation on Brownian motion in Leiden, Fokker moved to Zürich where he first met Albert Einstein and became caught up in extended calculations of the mean energy of a rotating dipole in equilibrium with a radiation field governed by Planck's law. Einstein evidently was impressed with the promise

E. Hiebert, *The Helmholtz Legacy in Physiological Acoustics*, Archimedes 39,
DOI 10.1007/978-3-319-06602-8_19, © Springer International Publishing Switzerland 2014

of the 26-year-old Dutchman and wrote to his friend Paul Ehrenfest (1880–1933) in Leiden that he was much looking forward to actually working with Fokker.[1] On another occasion Einstein wrote Ehrenfest : "Fokker pleases me very much.... One notes in particular his excellent schooling with Lorentz." He also mentions that he and Fokker already had found something "interesting and curious" to work on – namely the application of mechanics and electrodynamics to rotating dipoles.[2] In a letter to his teacher Lorentz, written a month after arrival in Zürich, Fokker remembered: "Einstein received me with the kindest cordiality. Before long we were engrossed in long discussions as to whether a case could be made for some kind of independent quanta in empty space, or whether the solution to the difficulties should be expected from filling a gap in our knowledge about the mutual interaction of charged particles in matter.... The impossibility of reaching absolute zero was another hobby horse we rode.... We have not spoken much yet about the theory of gravitation, but I think that will gradually have its turn. Einstein mentioned his plan to travel to Haarlem in order to hear your opinion and criticism of his theory, which would be very important for him."[3]

The Fokker-Einstein discussions in Zürich led to the publication in 1914 of a joint paper on gravitational theory.[4] In this paper the two men developed "a generally covariant formalism from which Nordström's (gravitation) theory follows if the assumption is made that it is possible to choose preferred systems of reference in such a way that the velocity of light is constant. The point of Einstein and Fokker's paper was to demonstrate how the powerful machinery of the absolute differential calculus allows one to reduce to a minimum the physical assumptions required for deriving Nordström's thesis."[5] The joint paper had demonstrated that by adopting the constancy of the velocity of light as a formal principle the Nordström gravitational theory could be formulated without invoking any other physical hypothesis; and this gave the theory an advantage over other gravitational theories based on the same formal principle. The theory also supported the law of equivalence of inertial and gravitational mass.[6] The collaborative work on gravitational theory that had brought Einstein and Fokker to work together in Zürich spilled over into their mutual interests in quantum theory. With this stimulus from Einstein, Fokker was enticed to pursue the problem of determining the

---

[1] Letter Einstein to Ehrenhaft, before 7 Nov. 1913, *Einstein Papers*, *5*, Doc. 481, 564.

[2] Letter Einstein to Ehrenfest, second half of Nov. 1913, *Einstein Papers, 5*, Doc. 484, 568.

[3] Letter Fokker to Lorentz, 4 Dec. 1913, *Einstein Papers*, *5*, Doc. 490, 578–579.

[4] A. Einstein and A. D. Fokker, "Die Nordströmische Gravitationstheorie vom Standpunkt des absoluten Differrentialkalküls,"*Annalen der Physik*, *44* (1914) 321–328. Reprinted in *The Collected Papers of Albert Einstein*, Princeton, *4* (1995) doc. 28, 589–597.

[5] *Einstein Papers*, *4*, 299–300.

[6] Gunnar Nordström (1881–1923) was a Finnish physicist at the University of Helsingfors. In 1914 he suggested a formal unification of the electromagnetic field and his own gravitational theory by conceiving of the four-dimensional Minkowski space-time idea as a hypersurface of the five-dimensional antisymmetric field tensor whose 15 components were interpreted as the gravitational force.

average energy on rotating dipoles in a radiation field at low temperatures. Fokker discovered in the theoretical calculations that the method that had been applied so successfully to the derivation of the specific heats for oscillating solids by Arnold Eucken failed in the case of rotating hydrogen molecules. He remarked that he was completely at a loss to explain why this method had been successful in the Eucken case but of no help to the latter. The negative results reported by Fokker and reported by Einstein vividly demonstrate something of the drive and spirit with which theoretical arguments in quantum theory were being pursued in Holland with the encouragement of persons such as Einstein and Planck.[7]

In his theory of Brownian motion of 1906 Einstein had developed a law for the stationary state of a large number of equally constituted systems subjected to small, fast, external fluctuations. Building on what Fokker had learned in his doctoral dissertation – a work that treated Brownian motion in a radiation field and probability considerations in radiation theory – Fokker in 1914 generalized Einstein's law for the case in which the action of the external fluctuation depends essentially on the particular condition of the system in question. In 1917, after additional thought had been given to the dipole problem, Max Planck, in his examination of Fokker's generalization of the Einstein fluctuation theory, provided a proof for its basic equation.[8] Fokker's contribution to this law is what he mainly came to be known for in the age of the new physics. He had formulated his equation within the context of Einsteinian reflections on fluctuations. Planck had demonstrated its validity. The equation alternately came to be referred to as the Fokker-Planck or the Fokker-Einstein equation. It played an important role in the formulation of quantum mechanics, and is listed as such in most handbooks and dictionaries of physics.

In 1913 while Fokker was working with Einstein at the Polytechnic in Zürich, Planck and Nernst were engaged in drawing up plans to establish a professorship for Einstein at the University of Berlin – a position he accepted before the end of the year. While still in Zürich, Einstein wrote to his friend Heinrich Zangger mentioning that he would be travelling to Berlin; that he was looking forward to having an assistant "which for me is a special blessing because of my predilection for cooperation."[9] As it turned out, Fokker was one of the candidates being considered as Einstein's assistant in Berlin. Einstein wrote Ehrenfest, "I am sharing my trip

[7] A. D. Fokker, "Die mittlere Energie rotierender elektrischer Dipole im Strahlungsfeld," *Annalen der Physik. Vierte Reihe 43* (1914) 810–820.

[8] Max Planck, "Ueber einen Satz der statistischen Dynamik und seine Erweiterung in der Quantentheorie." (1917) *Physikalische Abhandlungen und Vorträge*, Braunschweig, 1958, vol. II, 435–452.

[9] Letter to Zangger, 10 Oct. 1914, *Einstein Papers, 5*, Doc. 513, p. 602. Heinrich Zangger (1874–1957), close friend to Einstein in Zürich, was a medicolegal expert who became known as the founder of emergency medicine. According to Carl Seelig (*Albert Einstein, Leben und Werk eines Genies unserer Zeit*, Zürich, 1960, 184–185), Zangger for some years played the role of Einstein's father.

(to Berlin) with Fokker (an outstanding fellow)."[10] In the spring of 1914 Einstein moved to Berlin, was elected to membership in the Prussian Academy of Science, and became director of the *Kaiser Wilhelm Institute* for Physics. He was free to lecture at the University of Berlin or not according to choice. In 1913, following the collaborative work with Einstein at the Polytechnic in Zürich, Fokker went to England where he spent the year in experimental research with Ernest Rutherford in Manchester and W. H. Bragg in Leeds. Fokker returned to Leiden in 1914 as Privatdozent. Except for 3 years of military service in the Dutch army during World War I, he briefly taught physics at the Gymnasium and then held professorships for the rest of his life, first at the University of Delft and then at the University of Leiden. By 1918 the theory of general relativity was receiving wide European support. Among the first to contribute to the development of the theory in the early 1920s was the group in Leiden (frequently visited by Einstein) that had come under the influence of Lorentz: Willem de Sitter, Gunnar Nordström, Johannes Droste, and the young Adriaan Fokker.[11]

For several years Fokker corresponded with Einstein in Berlin on relativity, quantum physics and other scientific topics; but the letters also touched on political matters that were then at large. After the war, when Fokker was recuperating from tuberculosis at the sanitarium in Arosa, Switzerland, he wrote Einstein that he had seen his name mentioned in connection with an international appeal calling for reconciliation among intellectuals of all countries. He remarked: "It is frightfully difficult to do something for one's feelings of reconciliation. There are mostly Germans here in the sanatorium. . . .(and) it is surprising how much we differ from each other. The other persons always have heard other facts. . . .and as much as I loath the Parisian intrigues, at the same time I could feel no sympathy for the others since with their screaming and protesting and their cliché that this dictated peace *must* bring on a new war, they start again their stupid warmongering." Following this anti-war invective, which certainly fell on receptive ears, Fokker brought up a number of problems that he wished he could have "clarified in conversation": the relativity of inertia, or what Einstein refers to as Mach's principle. The main body of the letter, however, was given over to a discussion of Weyl's unified theory of gravity and electromagnetism ("Gravitation und Elektrizität," 1918) and its conceivable explanation of the apparent failure of relativity to agree with solar redshift measurements – a view held by Fokker and other physicists such as Arthur Eddington and Arnold Sommerfeld. Another problem Fokker was hoping to clarify in discussion with Einstein was the compactness of the world and the cosmological constant – a problem Fokker said he had not yet "found the inspiration to work on closely."[12] Einstein replied: "Life is hard for all of us. But it is a

[10] Letter Einstein to Paul Ehrenfest, 19 March 1914, *Einstein Papers, 5*, Doc. 515, 605.

[11] For other early contributors in Göttingen, Zürich, Cambridge, Rome, and Vienna, see *Einstein Papers, 7* (2002) 101.

[12] Letter Fokker to Einstein, 26 July 1919, *Einstein Papers, 9* (2004) Doc. 75, 110–113; and Einstein to Besso, 26 July 1920, *Einstein Papers, 10* (2006) Doc. 85, 349 fn 7.

matter of luck that, to some extent, one is able to get outside of one's own uncomfortable skin and strive toward the objective things to whose high heights the miseries of life are unable to penetrate." Einstein noted that Fokker was having an opportunity in Arosa to observe "the German political mentality," complaining that the German defeat in World War I and the Treaty of Versailles represented a "forced peace" (*Gewaltfrieden*). "I am pessimistic about the future," wrote Einstein, "but believe in a gradual success of the League of Nations. . . . It is particularly welcome that America, which is not burdened with the fatal European traditions, has the leadership." And "as for relativity, I also originally sought to achieve the relativity of inertia by letting $g_{\mu\nu}$ (the cosmological constant) go to infinity. . . .but the equations don't allow this. . . .and also violate experience." The letter went on to offer comments about Hermann Weyl's *Raum, Zeit, Materie* (1918) and Erwin Freundlich's work on the redshift in the spectra of fixed stars.[13]

Following an announcement that the British expedition had provided a proof of the relativistic deflection of light by the sun, Fokker congratulated Einstein on the "magnificent confirmation of your expectations. . . . But how lost to the world you must have felt yourself in this moment [when you read the notice]! Are you not deeply thankful to the sun for having been so good to you?" Fokker remarked that he still was mulling over gravitational problems and relativity corrections and hoping that Einstein would stop off to visit him in Arosa on his next trip to Zürich.[14] Einstein responded saying that all his time was being taken up in mere reflexive motions and trivialities, that his political optimism had suffered a blow, and that everyone was becoming Prussian for opportunistic motives. "It seems to me in general that the (administrative) machinery is so hopeless that I begin to sense that all that is achievable and worth striving for is the political disinterest and international disposition of the intellectually more advanced earlier centuries." He added that he was sending the reprints that Fokker had requested, was dubious about what he had said there about the electron, but was excited about Fokker's new work on the theory of bound electrons in the interior of the atom. "It is so fantastically noteworthy that electrodynamics holds its ground exactly in certain trains of thought whereas in others it totally comes to nothing."[15] Some months later Ehrenfest wrote to Einstein: "Fokker sent me a very elegant work about the bound electrons; he is a fine analytician."[16] When Einstein was elected to the Royal Academy of Sciences in Amsterdam Fokker wrote to congratulate him and expressed the wish that he would see him more often. "As an enthusiastic advocate of relativity theory I probably would still have much to ask about the formulation of the relativity of inertia; it still is a puzzle for me, especially as you conceive it. When will you call on an outstanding experimentalist to demonstrate that moving clocks run slower than clocks at rest; resp. that moving atoms are redder than

[13] Letter Einstein to Fokker, 30 VII [1919], *Einstein Papers, 9* (2004) Doc. 78 117–119.

[14] Letter Fokker to Einstein, 8/11/19, *Einstein Papers*, *9* (2004) Doc. 168, 236–239.

[15] Letter Einstein to Fokker [after 1 Dec. 1919], *Einstein Papers*, *9* (2004) Doc. 187, 264–265.

[16] Ehrenfest to Einstein 6 June 1920, *Einstein Papers*, *10* (2006) Doc. 46, 297–299.

atoms at rest? It surely is scarcely bearable that the rel. theory still has to limp on one leg."[17]

While Einstein in Berlin and Fokker in Arosa and then in Delft were discussing problems in theoretical physics by mail, Fokker's colleague Paul Ehrenfest (1880–1933) at the University of Leiden was passionately involved in trying to entice Einstein to accept a professorship in Leiden. He wrote to Einstein in Zürich where he erroneously thought Einstein was at the time:

> In connection with other circumstances we now suddenly all are in accord that we need to pin you down *([festlegen*] here in Leiden. The matter is extremely simple. In case you just say yes it will be possible – at least according to all human foresight – to arrange everything according to your wishes in the shortest possible time.... Here you immediately will have close to you or be in good contact with: Lorentz, De Sitter, me, Kuenen, Droste, De Haas, his wife, Fokker, Burgers, Julius, Zeeman; and you often will see intelligent and humanly agreeable, noble young fellows; in addition to guests such as Nordström, Bohr, etc.... . In case you move outside [the city] you will be able to isolate yourself as much as you like.... Dear, dear Einstein, don't say no but come to us for 14 days or at least for a week as soon as you can – in any case, answer me directly and tell me that it is not so stupid what you propose.... . Einstein, remember that you come here among a group of persons who very personally love you, and not only the brain juice [*Gehirnschmaltz*] that oozes out of you! No one here expects great results from you; we simply want you to be close by.... (So) dear dear Einstein don't ruin all my dreams and hopes – Help me in my efforts by immediately giving me a favorably enough answer so that I can put all else in motion.[18]

In another letter to Einstein 6 days later, this time to Berlin, Ehrenfest spelled out more of the details that were being offered to Einstein if he came to Leiden: income, no teaching obligations, and complete freedom to give lectures elsewhere. "All that one would wish is that we would have you in Leiden in general. And that one would say: Leiden is Einstein. Einstein is in Leiden!" As for the Dutch language, Ehrenfest assured Einstein that nothing else but German would be required of him "since you are no Prussian."[19]

Einstein replied, "Your offer is so fabulous and you wrote so cordially and friendly that you scarcely can conceive of the turbulence your letter has evoked in my head." As Einstein remarked that it would not be a simple matter for him to turn his back on Berlin and Planck, he was convinced, however, that there would be occasions to visit Leiden. "You have no conception of what kind of cordiality surrounds me here; there are not merely those here who catch the juice that sweats from my brain."[20] As is well known, Einstein did not accept the Leiden offer.

---

[17] Letter Fokker to Einstein, 2 June 1920, *Einstein Papers, 10* (2006), Doc. 40, 287–289. In another letter to Einstein 5 months later (by which time Fokker had been released from Arosa and was in Delft), Fokker wrote that a certain problem was "going through his head," namely the orientation and precession of the center of gravity of a moving gyroscope in Euclidean-Minkowskian space-time resulting from Lorentz-contraction. "Has this ever plagued you?" Fokker to Einstein, 2 November 1920, *Einstein Papers, 10* (2006), Doc. 189, 476–477.

[18] Letter Ehrenfest to Einstein, 2 Sept. 1919, *Einstein Papers*, *9* (2004) Doc. 98, 145–146.

[19] Letter Ehrenfest to Einstein, 8. *IX*. 19, *Einstein Papers*, *9* (2004) Doc. 101, 150–151.

[20] Letter Einstein to Ehrenfest, 12. IX 19, *Einstein Papers*, *9* (2004) Doc. 103, 154–155.

He remained in Berlin until 1933 and thereafter spent the rest of his life at the Institute for Advanced Study in Princeton while holding an honorary professorship in Leiden from 1920 to 1945.

The 3 years spent in convalescence in Switzerland following 4 years of military service did not, as we have shown, prevent Fokker from active participation in the interchange of ideas on relativity and quantum theory that were taking place among persons such as Lorentz, Planck, Einstein, and Ehrenfest. While in Frankfurt Born wrote Einstein: "I am writing in haste to you in Holland because I should like to have Fokker's address. He has sent me a fine paper. . . .(and) I would very much like to thank him; Ehrenfest will know where he lives." On return to the Netherlands from Arosa in 1921, Fokker taught at the Gymnasium in Delft. Einstein in Berlin had high hopes for Fokker as a theoretical physicist and wrote to Max Born in Frankfurt in 1922: "It now is extremely difficult to create a position for a theoretician. Holland suffers from overproduction. . . .(where) there are excellent theoreticians (e.g. Fokker) in unpretentious Gymnasium teaching posts."[21]

Apart from the collaborative efforts of the 30-year-old Fokker with Einstein and Planck that led to the so-called Fokker-Einstein equation, Fokker's most significant overall contributions to physics, before he became *Curator* of the Teyler Museum, can be examined under three headings: statistical mechanics, gravitational and relativity theory, and the mathematical extension of Lorentz's treatment of Maxwellian electromagnetic theory. An insightful and interpretive account of Fokker's contributions to physics as it relates to the structure of theoretical physics in the 1930s is given by Fokker's younger Leiden colleague, the theoretical physicist H. A. Kramers (1894–1952). At the time of his writing Kramers was considered the Dutch master of the quantum theory of the atom. His appraisal of Fokker's work reveals the high esteem Fokker had acquired by the late 1930s.[22] Fokker's interest in statistical mechanics had first been explored in a preliminary way in his doctoral dissertation on Brownian motion.[23] In his appraisal of Fokker's work, Kramers showed how Fokker, while building on a dissertation devoted to the investigation of the statistical principles of light quanta and working in intellectual collaboration with Einstein and Planck, was guided in searching out the relativistic and quantum implications of the classical conceptions that had characterized Lorentz's thinking and that had led to Fokker's work on Einstein fluctuation theory and the

[21] *Albert Einstein, Hedwig und Max Born Briefwechsel, kommentiert von Max Born.* Munich, 1969, Letter no. 17, 70/1, and letter no. 42, 103–104.

[22] H. A. Kramers, "De dissertatie, het wetenschappelijk oeuvre en de publicaties van A. D. Fokker," *Nederlandsch tijdschrift voor natuurkunde*, 's Gravenhage, *5* (1938) 204–215. Included in Kramers's essay is a list of 77 of Fokker's publications prior to 1938. Several of them impinge on the field of musical acoustics. They will be mentioned later but are largely peripheral to the aspects of musical acoustics that are our main concern. As professor of physics at the Technical University of Delft from 1923 to 1927, Fokker continued to work on problems in gravitation, inertia, and relativity, but he also published articles on subjects as varied as optical reflection and refraction, the conductivity of electricity in noble gases, and the movement of fish and snakes.

[23] A. D. Fokker, *Over Brown'sche bewegingen in het stralingsveld en waarschijnlijkheidsbeschouwingen in de stralingstheorie*, Leiden, 1913. A French résumé appeared in the *Archives néerlandaises* of 1917.

Planck-Fokker equation. The equation expresses the connection between the mobility of particles and their diffusion constant. Their magnitudes are proportional and the constant that relates them is the product of the Boltzmann constant and the temperature. Discussions on gravitational theory with Einstein led in 1929 to Fokker's book on relativity theory. Another treatise in 1960 dealt more conclusively with time-space and inertia-gravity problems (*Tijd en Ruimte, Traagheid en Zwaarte*). In 1920 Fokker supplied one of the first extensions of Lorentz's celebrated derivation of the 1902 Maxwellian macroscopic comparisons with the microscopic electron theory – a theory in which a medium susceptible to electric or magnetic fields is regarded as made up of electrons embedded in an ether that is the seat of an electromagnetic field obeying Maxwell's equations. In 1927 Fokker provided key ideas to the solution of the difficult problem of defining the center of gravity of an ensemble of particles in relativity theory.

Adriaan Fokker belongs to the relatively small but influential company of Dutch physicists who inherited, or one might say "came to be saddled with," the classical world view of physics that had been shaped by physicists such as Maxwell, Kirchhoff, and Helmholtz. He also rubbed elbows with theoretical physicists such as Lorentz, Planck, and Einstein, and made a number of important theoretical contributions to statistical mechanics, relativity, and quantum physics. The figure of Fokker that emerges from the appraisal that Kramers made of his colleague in 1938 (recall that this was prior to the time that Fokker's interest shifted from theoretical physics to musical acoustics) is one in which Fokker is situated at the center of a Lorentz-Einstein-Planck triangle witnessing and reacting in his own unique way to the making and breaking of links connecting classical physics, gravitational relativity, and quantum physics. As a doctoral student with Lorentz, and then alongside Lorentz as colleague at the University of Leiden and at the Teyler Museum for a total of some 15 years, Fokker was in about as good a position as any of his Dutch colleagues to take advantage of Lorentz's scientific thought and output. As an insider, he also was able to undertake an even-handed appraisal of the scientific spirit and exemplary manner of Lorentz's communications with students and colleagues. In all of Fokker's writings it is evident not only that Lorentz was for him a father figure but that he was at the intellectual center of the revolutions in gravitational and quantum theory as Lorentz envisioned them. In his seminal article on Lorentz, Russell McCormmach has captured the essence of Lorentz's contributions to modern physics: "Lorentz brought classical physics to a state of development that made the need for reform in its fundamental principles evident and urgent to Einstein and other followers; this is the essential historical significance of Lorentz's work."[24]

[24] Russell M. McCormmach, "Hendrik Antoon Lorentz," *DSB*, *8* (1973) 487–500 provides an excellent account of Lorentz's life, scientific contributions, and influence on the development of modern physics. The work that most directly provides proximate insights into Lorentz's personal life and work as seen from the perspective of his own family and countrymen is G. L. De Haas-Lorentz (ed.), *L. A. Lorentz. Impressions of his Life and Work*, Amsterdam, 1957. Included are essays by Einstein, W. J. De Haas (assistant and father-in-law to Lorentz), Fokker, Ehrenfest, and Casimir. The editor G. L. De Haas is Lorentz's daughter, and in her "Reminscences" she provides pertinent information about her father's life.

During the first three decades of the twenty-first century, Lorentz, more than any other physicist of his generation, was at the center of developments that led to an understanding of light, matter, and relativity. He especially was admired by students and colleagues for the consummate skill with which he conveyed the sense of excitement and reward associated with working on open, unsettled, and problematic frontiers of physics. The fermentation of ideas that was in the air in the 1920s and that Lorentz communicated at the time is revealed in the lectures at the California Institute of Technology toward the end of his career. Delivered in semi-popular and clear language, they liberally demonstrate the enormous range of Lorentz's comprehensive outlook on the problems *and the difficulties* encountered in post-1900 physics.[25] Central to the Lorentz lectures were efforts to reconcile classical Maxwellian electromagnetic theory with Bohr's formulation of quantum theory. Although he was unable to resolve the difficulties inherent in that effort – difficulties that he was able to spell out with great clarity and sharp boldness – he nevertheless succeeded in presenting a lucid and concise formulation of Bohr's correspondence principle. In its original form it asserted that the behavior of atomic systems approximates to what is predicted by classical physics in cases where specific circumstances are unimportant.

When Lorentz retired from his teaching duties at the University at the age of 70 on 18 July in 1923, Fokker, who had just taken up his professorship at the Technical University in Delft, wrote an essay featuring his retired professor as a teacher. It was to be the first of three essays in which Fokker, speaking as a deeply grateful student, presented a sensitive characterization of the teaching methods and research orientation that had captured the minds and research perspectives of an entire generation of upcoming Dutch physicists. The central idea that most vividly identified Lorentz's way of thinking, according to Fokker, was his focus, historical in essence, on problems in physics in need of theoretical clarification and resolution. Lorentz was a troubleshooter who identified and opened up so many dark and unexplained corners in physics that his students came to believe that the main advances in the physics of relativity and quantum mechanics would depend not so much on new discoveries as on clarification of the problems encountered at the borderline where classical physics and the new physics intersect.[26] A second essay at the time of Lorentz's death in 1928 is given over largely to the line of reasoning that Lorentz took to enunciate his electron theory of 1892. He accomplished this by exposing the defects of Maxwell's dispersion theory and its inability to predict Fresnel's experimentally confirmed formula for the convection of light in moving media.[27] The last of the three essays is a lecture delivered in 1931 on the occasion of the unveilung of the Lorentz monument in Arnheim.[28]

---

[25] H. A. Lorentz, *Problems of Modern Physics*, Boston, 1927.

[26] A. D. Fokker, "Hendrik Antoon Lorentz, 1859–18 July 1923," *Physica. Nederlandsch Tijdschrift voor Natuurkunde*, *3* (1923) 201–205.

[27] A. D. Fokker, "H. A. Lorentz, 18 July 1853–4 Feb. 1928," *Physica*, *8* (1928) 1–13.

[28] A. D. Fokker, "Rede bij het Lorentzmonument," *Physica*, *11* (1931) 241–247.

During the last year of his teaching in the physics department at the Technical University of Delft, Fokker was appointed by Lorentz to the position of *Curator* at the Teyler Museum in Haarlem. When Lorentz died in 1928, Fokker succeeded Lorentz as *Conservator* at the Museum. It was a position that included the Teyler professorship at the University of Leiden and the obligation to deliver Saturday noontime demonstration-lectures at the Teyler Museum designed to describe and illustrate the newest discoveries and theoretical developments in physics. The new appointment fundamentally altered Fokker's professional life. Apart from teaching duties at the University and the management of activities at the Teyler Museum, he henceforth would find only limited opportunity to reflect on the problems in statistical mechanics, relativity, and quantum physics that had engaged his interests for two decades and with which he had been involved since the time of his student days with Lorentz. He managed. however, to publish articles on various more popular topics in physics. They were based mainly on lectures delivered at the Teyler Museum.

In May 1940 the Germans invaded the Netherlands and crushed the Dutch resistance. There followed a sequence of disastrous political events that radically altered Fokker's life as it did the lives of an entire generation of individuals who lived through the second World War. During the Nazi occupation of Holland (1940–1945) and the resultant ban on university teaching, Fokker at age 53 decided, abruptly, to turn his attention to the study of music and take advantage of the research facilities at the Teyler Foundation in order to investigate problems in musical acoustics.

The death of Lorentz and what his intellectual world-picture embodied coincides with the end of an era in classical physics that not only had exposed the exigency for change but in significant ways had set the stage for the coming of the new physics. It has been said, and aptly so, that Lorentz was the most highly respected theoretical physicist of his time. Although he had lived his entire life within the environs of Leiden he was the most cosmopolitan of physicists. The extent to which classical physics was in need of reform and the crucial points at which it had become evident that the traditional notions of space, time, matter, light, and energy had become inadequate became evident precisely because of the place to which Lorentz had been able to take physics. The *Collected Papers* of Lorentz were published by Martinus Nifjhoff in 9 volumes over a period of 5 years under the editorship of Pieter Zeeman and Fokker.[29]

[29] H. A. Lorentz, *Collected Papers*, P. Zeeman and A. D. Fokker (eds.), 9 vols., The Hague 1935–1939. Lorentz and Pieter Zeeman (1865–1943) shared the second Nobel Prize in physics "in recognition of the extraordinary service they rendered by their researches into the influence of magnetism upon radiation phenomena."

References

De Haas-Lorentz, G.L. (ed.). 1957. *L. A. Lorentz. Impressions of his life and work*. Amsterdam: North-Holland.

Einstein, A., and A.D. Fokker. 1914. Die Nordströmische Gravitationstheorie vom Standpunkt des absoluten Differrentialkalküls. *Annalen der Physik* 44: 321–328.

Fokker, A.D. 1913. *Over Brown'sche bewegingen in het stralingsveld en waarschijnlijkheids-beschouwingen in de stralingstheorie*. Leiden.

Fokker, A.D. 1914. Die mittlere Energie rotierender elektrischer Dipole im Strahlungsfeld. *Annalen der Physik. Vierte Reihe* 43: 810–820.

Fokker, A.D. 1931. Rede bij het Lorentzmonument. *Physica* 11: 241–247.

Lorentz, H.A. 1927. *Problems of modern physics*. Boston: Ginn.

Lorentz, H.A. 1935–1939. *Collected papers*, 9 vols., ed. P. Zeeman and A.D. Fokker. The Hague: Martinus Nijhoff.

McCormmach, Russell M. 1973. Hendrik Antoon Lorentz. *DSB* 8: 487–500.

Seelig, Carl. 1960. *Albert Einstein, Leben und Werk eines Genies unserer Zeit*, 184–185. Zürich: Europa Verlag.

Chapter 20
Fokker and the Teyler Foundation

With the Nazi invasion of the Netherlands in 1940, and with Fokker's decision to devote his efforts to study and research in musical acoustics, the Teyler Museum and the Teyler Foundation taken together became the determinative factor in the development of Fokker's career. It is pertinent at this point in our story to provide a prospectus and an historical overview of the Teyler connection. For 30 years Fokker's life and work came to be aligned with all that transpired in Haarlem at the Teyler Museum.

The Teyler Foundation traces its origins and functioning to the vision and generosity of a Dutch entrepreneur and benefactor in Haarlem. Pieter Teyler van der Hulst (1702–1778) was a highly respected member of the Anabaptist-Mennonite (*Doopsgezinde Mennoniet*) tradition whose characteristic tenets included non-violent resistance to war (*bewetensbezwaarde*) and the practice of believer's baptism in contrast to child baptism. Profoundly influenced by the rationalist enlightenment that swept over Europe in the eighteenth century, the Mennonites, in addition to their traditional humanitarian and ecumenical focus, became actively engaged in the support of scientific trends of thought, the arts, and culture.[1]

By profession Teyler was a silk manufacturer and businessman who by inheritance and success in his own business ventures acquired great wealth. Although decidedly frugal in private life, he was a benefactor to the poor and a generous promoter of associations committed to acts of charity. In 1750 Teyler supported an orphan home; in 1756 he founded a home for women. With a deep and consummate interest in the natural sciences, his stately house in Haarlem became the location of a select collection of books, paintings, physical instruments, and valuable art works. In 1756, after the death of his two children and his wife, Teyler bequeathed his total fortune to the establishment of a Foundation, the *Teyler Fundatie*, devoted to the assistance of the poor and the promotion of the arts and science. Following his death

[1] For a comprehensive history of the Dutch Mennonites that also includes information about Teyler: S. Groenwald, J. P. Jacobszoon, and S. L. Verheuis (eds.), *Wederdopers, Menisten, Doopsgezinden in Nederland, 1530–1980*, Zutphen, 1981.

E. Hiebert, *The Helmholtz Legacy in Physiological Acoustics*, Archimedes 39,
DOI 10.1007/978-3-319-06602-8_20, © Springer International Publishing Switzerland 2014

in 1778 the Haarlem house, with its library, art collection, and valuable rarities, together with a sizeable sum of money, came under custody of the Teyler Foundation and five directors appointed to oversee: (1) the retirement home for women that Teyler had founded in 1756, (2) a museum, the *Teyler Stichting* in Haarlem, whose 1885-constructed building housed paleological, mineralogical, and physical divisions, a coin and medal assortment, and a collection on the graphic arts that included a series of Rembrandt's books and art works, (3) the publication of a periodical, *Teyler's theologische Tijdschrift* (1903–1946), and (4) two learned societies, one of which represented religious and theological interests while the other oversaw the general scientific, literary, and artistic interests of the Foundation. Both societies presented annual prizes for the authors of selected manuscripts in a special field.[2] Editor of the Dutch journal of physics *Physica* from 1921 to 1933, President of the Dutch Physical Society 1923–1928, Fokker was elected to membership in the Royal Netherlands Academy of Sciences in Amsterdam in 1949. Until his retirement in 1955, Fokker's duties at the Teyler Museum included regular noontime lectures on physical topics, physics education, and the role of physics in society.[3]

As already mentioned, in 1927 Fokker succeeded Lorentz as Curator of the laboratory at the 170-year-old Teyler Museum – a position that included the Teyler professorship at the University of Leiden. By that time, as we have attempted to show, Fokker's contributions to relativity theory, quantum theory, electromagnetism, and atomic physics placed him in the ranks of the preeminent theoretical physicists in the Netherlands of the 1920s and 1930s. Among the lectures that Fokker delivered during the 27 years of his curatorship at the Teyler Museum, a number were in the field of acoustics. They were, in part, semi-popular and general in nature and had little in common with the musical acoustics projects on intonation, music theory, and the 31-tone per octave keyboards that were first launched in the 1940s and continued into the 1960s. They nevertheless merit brief mention here to the extent that they indicate that Fokker's interests in some aspects of physical acoustics were astir in the 1920s.

Fokker first was enticed to explore problems in acoustics in the 1920s on reading the papers of William Clement Sabine (1868–1919). His collected papers on the acoustic properties of enclosures appeared in 1922. Sabine, who was a professor of physics at Harvard, had been asked by president Charles William Eliot whether something could be done about the poor acoustics in the lecture hall of Harvard's new Fogg Museum of Art. Sabine's experimental investigation of the situation led to the discovery of the relation between the reverberation time, absorbent capacity, and volume of the Fogg lecture hall. In 1900 there appeared in the *American* Architect a series of articles in which Sabine provided an analysis of the conditions

[2] The source for the above information about Pieter Teyler and the Teyler Foundation is N. Van der Zijpp, "Pieter Teyler van der Hulst" and "Teyler Stiftung," *Mennonitisches Lexikon*, Frankfurt am Main/Weierhof Pfalz, *4* (1913) 303–304.

[3] Rudolph Rasch (ed.), *Corpus Microtonale*, Adriaan Fokker, Haarlem, 1987, p. 15.

necessary to secure good hearing in an auditorium; it was one of the first such studies ever to be undertaken by scientific methods. In determining the time that a given sound reverberates in the hall, Sabine found that a single syllable of speech endures long enough to overlap and produce confusion in what follows. The acoustical quality of the room could be improved, he discovered, by covering the walls with sonically absorptive materials. In 1898 these findings were put to use in the design of Boston's Symphony Hall. By showing that the product of the reverberation time and the total absorptive power of the walls and furnishings is a constant, and that this constant is proportional to the volume of the room, he came to enunciate a law known since as Sabine's law – a law that hinges on the three basic criteria of "adequate loudness" (that the simultaneous components of a complex sound should maintain the proper relative intensities) and that the successive sounds in rapidly moving articulation, either of speech or music, should be "clean" and "distinct" (free from each other and from extraneous noises). In architectural acoustics the unit of sound-absorbing power is called the *sabin*.[4]

In a research paper on reverberation (nagalm), Fokker was able to show, on the basis of theoretical arguments, that the Sabine law connecting reverberation, volume, and absorbing power of a room is as dependent on the shape of the room as on absorption by the walls; that the number in the reverberation law is to be regarded merely as a mean value for cases usually met in practice.[5]

In a second paper on the Sabine reverberation law – a law that had been established empirically – Fokker presented a theoretical deduction that started with the assumption that sound energy is uniformly distributed through space, traveling isotropically in all directions.[6]

A 1929 lecture on comparing sound from the human voice and musical instruments in the open air and in an auditorium was directed toward exploring the optimum conditions from the transmission, reception, and reverberations of sound from the point of view of the listener. Echo sounds from the center of a circular room and from other non-central foci were treated as a special case.[7] The problem of constructing sound reflectors that respond to speech into an auditorium in such a way as to prevent "the spreading of the sound waves in upward directions where they wander astray through vaulted spaces and return as troublesome reverberation" was taken up by Fokker the following year.[8] With the help of personnel from the Teyler laboratory and the Physical Laboratory of the Philips' Gloeilampenfabriek in Eindhoven, Fokker designed parabolic and hyperbolic reflecting surfaces

---

[4] All of Sabine's articles on acoustics are contained in *Collected Papers on* Acoustics, Cambridge, Mass., 1922. See also Leo Beranik, "Wallace Clement Sabine and acoustics," *Physics Today* (Feb. 1985) 44–51, and Daniel Kevles, Sabine, *DSB 12* (1975) 54.

[5] Adriaan Fokker, "Over de nagalm," *Physica*, *4* (1924) 262–273.

[6] Adriaan Fokker, "Sabine's formule voor den nagalm," *Physica*, *7* (1927) 198–200.

[7] A. D. Fokker, "Over de akoestick van zalen, van muziekinstrumenten en van de menschelijke stem." *Archives du Musée Teyler*, Haarlem (2) *7* (1929) 1–64.

[8] A. D. Fokker, "On the Construction of Sound Reflections," *Archives du Musée Teyler* (3) *7* (1930) 73–76. Quotation on p. 73.

that could be tested in the great St. Bavo cathedral in Haarlem. The objective was to learn something about speech intelligibility as a function of reflectivity. The forward direction from the source when reflected from the walls of the cathedral was compared with the intensity of sound arriving from different reflectors. Fokker concluded that "intelligibility-measurements with common speech showed that reflectors increased the intelligibility radius by a factor of 2 or 3 compared with a common church pulpit."[9]

On the initiative of the International Electrotechnical Committee (I.C.E.), delegates from the centers of acoustical research from Germany, Great Britain, France, Italy, Belgium, Norway, Czechoslovakia and the Netherlands assembled in Paris in 1937. Fokker represented the Netherlands. His report on what had transpired at the 4-day conference seems to have been somewhat peripheral to any of his own acoustical interests at the time but put him in contact with an international constellation of acousticians who were engaged in the standardization of acoustical methods, nomenclature, pitch, and apparatus.[10] The international conference on acoustics in Paris in 1937 marks the end not only of Fokker's career in theoretical physics but in the science of acoustics.

From here on Fokker is dealt with as a scientist engaged in the field of musical acoustics, which in itself is a part of music theory more than it is a part of mechanics and physics. On the theoretical side, the study of sound considered as the physical cause of the sensation of hearing is a part of the much larger field of the mechanics of solids and fluids. Nevertheless it should be pointed out that in physical acoustics the agreement between theory and experiment usually is less exact than in other branches of physics such as optics and electromagnetism. This is due, at least in part, to the fact that in many cases it is impossible to set up experimental conditions in keeping with assumptions made in deriving the theoretical solution. In addition, the theoretical solution to a general problem may be obtained in mathematical expressions whose numerical values can be arrived at only approximately.

In musical acoustics, where physiological and neurophysiological components often become dominant, the problems of agreement between experiment and theory become more difficult by an order of magnitude. Fokker, who had lived for almost four decades in an environment of theoretical Dutch physics and had established a firm foothold in the ways of theoretical thinking about relativity and quantum

[9] A. D. Fokker and M. J. O. Strutt, "Measurements on the Acoustic Efficiency of Specially Designed Sound Reflectors," *Ibid.*, 77–87. Quotation on p. 87. Fokker's links with the Philips physics laboratory reach back to 1913 when he recommended that Kamerlingh Onnes's assistant at the Physics laboratory in Leiden, G. Holst, create a physics laboratory at Philips and find out as much as possible about the physical aspects of the incandescent lamp. For several years Philips at Eindhoven, alongside the Auergesellschaft in Berlin, was the largest manufacturer of tungsten lamps in the world. A. Heerding, *The History of N. V. Philps' Gloeilampenfabriek*, Vol. 2, Cambridge, 1983, 314–15 and 340–42.

[10] A. D. Fokker, "Verslag van de internationale conferentie voor akoestiek de Paris," *Nederlandsch Tijdschrift voor Natuurkunde*, *4* (1937) 249–253.

physics, had simultaneously, and for over a decade, become involved in Teyler Museum lectures on a number of theoretical aspects of architectural acoustics that were of public interest but only peripherally challenging from a theoretical point of view. In 1940, when Fokker learned that his famous seventeenth-century Dutch countryman Christiaan Huygens had made fundamental contributions to the theory of music, he turned his attention to the study of music theory and spent the rest of his life working and writing about intonation and the Huygenian 31-tone system.

References

Fokker, Adriaan. 1924. Over de nagalm. *Physica* 4: 262–273.

Fokker, Adriaan. 1927. Sabine's formule voor den nagalm. *Physica* 7: 198–200.

Fokker, A.D. 1929. Over de akoestick van zalen, van muziekinstrumenten en van de menschelijke stem. *Archives du Musée Teyler*, Haarlem (2) 7: 1–64.

Fokker, A.D. 1930. On the construction of sound reflections. *Archives du Musée Teyler* (3) 7: 73–76.

Fokker, A.D. 1937. Verslag van de internationale conferentie voor akoestiek de Paris. *Nederlandsch Tijdschrift voor Natuurkunde* 4: 249–253.

Groenwald, S., J.P. Jacobszoon, and S.L. Verheuis (eds.). 1981. *Wederdopers, Menisten, Doopsgezinden in Nederland, 1530–1980*. Zutphen: Walburg Pers.

Rasch, Rudolph (ed.). 1987. *Corpus microtonale*, 15. Haarlem: Adriaan Fokker.

Chapter 21
Fokker Music Theorist

Fokker's abrupt move from theoretical physics to theoretical aspects of music when he was 50 is unique in the history of music – especially when placed alongside the way in which the other three main *persona* examined in this monograph balanced their professional lives in theoretical physics with a profound and critical interest in music and music theory. The nature of Fokker's commitment to the new life in musical acoustics after 1940 became so resolute that he never returned to a revisitation of the problems in physics for which he had been prepared and in which the most illustrious theoretical physicists of his generation had given him their unreserved support. We recall that regardless of lifelong interests and firm commitments to exploring the field of musical acoustics neither Helmholtz nor Planck considered it expedient to abandon their chosen professional careers in physics while being actively being engaged in the pursuit of music. Tanaka, who had a good background in physics and went to Berlin to pursue his interest in musicology, nevertheless on his return to Japan became involved in electromagnetic studies and railway engineering while being engrossed in one musical venture after another.

It is evident that we have had rather much to report on in connection with Fokker's career as a theoretical physicist – not only because he made a name for himself in physics but because his physics gave him the mathematical and scientific outlook and the intellectual tools of investigation to pursue music theory in a direction that more or less had been displaced by the introduction of tempered intonation since the middle of the eighteenth century. The additional rationale for having provided a précis for the salient points in Fokker's career in theoretical physics, and for having shown how his work interlocked with the work of other Dutch notables who were at the frontier of post fin-de-siècle revolutionary thought in theoretical physics, is that it netted him the intellectual credentials, the professional status, and the institutional freedom – actually the leverage – to engage *ad libitum* in investigations on a topic in music that impinge on the problem of just intonation on fixed-tone keyboards. That particular topic became of interest to Fokker for reasons that will become evident in the examination of his writings on music theory, his efforts to demonstrate the phenomenon of just intonation,

E. Hiebert, *The Helmholtz Legacy in Physiological Acoustics*, Archimedes 39,
DOI 10.1007/978-3-319-06602-8_21, © Springer International Publishing Switzerland 2014

and his encouragement of the musical compositions for 31-tone per octave keyboards. As with the work of Helmholtz, Tanaka, and Planck, Fokker's efforts were motivated by the scientific and mathematical challenge inherent in a problem in temperament that is intrinsic to traditional music theory and dates back to ancient Mesopotamia, Egypt, China, and Greece.

Given the exigencies of the time – a world war and its interference with teaching and research at the university – it was entirely plausible for Fokker to anticipate that the situation at the Teyler Museum would provide him with the opportunity to engage full time in the pursuit of a topic that would be freed from external control. In that situation it was much to his advantage to believe that the support he had received from his peers in physics would continue in his new venture into music theory. It signified that for the years during which he had occupied the dual position of Curator at the Teyler Museum and the position of Professor at the University of Leiden he would be able to anticipate the encouragement of his university colleagues, the assistance of the staff at the Teyler Museum, and unlimited access to its laboratory facilities. One of the most salutary by-products of the demonstration-lectures on just intonation keyboard physics at the Teyler Museum was that Fokker, now the experimenting physicist, had at his command an inquisitive audience of perceptive and critical listeners – scientists and music lovers – who in their responses to and criticisms of the ideas and the newly constructed instruments being featured at the Teyler Museum became a valuable resource for Fokker's ongoing research.

In our discussion of the distinct approaches that Helmholtz, Tanaka, and Planck took in their studies and research ventures into just intonation and fixed-tone keyboards we noted that the performance of music as such and the composition of music for the instruments they constructed entered but peripherally, if at all, into their writings. The emphasis, rather, was on the exploitation of the keyboard as a research tool in the investigation of the problem of just intonation and on the use of these special keyboards as teaching aids in music theory. Helmholtz and Tanaka explicitly underscored that emphasis. In *Sensations of Tone* Helmholtz indicated that among musical instruments the harmonium lent itself most readily to experimental study of the problem of just intonation. It was seen to be uniquely sensitive to inaccuracies of intonation and delivered uniformly sustained sounds that are piercing in their timbre and unparalleled in their tolerably distinct combination tones. The emphasis was explicitly on the experimental study of just intonation.[1] And so it was in the case of Shohé Tanaka. Although he was associated with the publication of a number of elementary musical compositions and arranged public demonstrations on 31-tone per octave organs, he did not attempt to establish a school of just-intonation performers or composers. In Max Planck the motives for pursuing the study of just intonation on the Helmholtz-type harmonium are not entirely transparent. His experiments were directed to the study of issues such as aural accommodation but there is no evidence to support the notion that he advocated a move in the direction of musical performances in just intonation or musical compositions oriented toward just intonation. In the case of all three

[1] Helmholtz, *Sensations of Tone*, Dover ed., 1954, p. 316.

investigators the emphasis was on the construction of keyboard instruments that could be used to study the problem of just intonation.

In several ways Fokker's interests progressed along very different lines of thought. He promoted and arranged for the building of 31-tone per octave organs, organized concerts for these special instruments, and tried his hand at microtonally-conceived compositions. Only a limited number of these compositions were performed during or after his lifetime, and while his reputation as a music theorist was substantial, then as now, the compositions were known at the time mainly among a limited circle of Dutch intimates, mostly other 31-tone composers. Musicologists have referred to Fokker as one of the "Dutch five" composers who wrote for the 31-tone keyboards or other microtonal instruments. Toward the end of his career in Leiden Fokker became more deeply implicated in promoting performance on 31-tone keyboards and supported the establishment of institutional facilities for nurturing the microtonal tradition that he had brought to life in the Netherlands.

Unlike Helmholtz, Tanaka, and Planck, all of whom had the benefit of family- and community-cultivated participation in music. Fokker seems not to have had similar opportunities in music prior to his decision to study music theory from a scientific point of view. Undoubtedly the unique opportunities and research facilities placed at his disposal at the Teyler Foundation played an important part in the decision to engage in music studies. A conspicuous rationale for Fokker's focus on the problem of just intonation and multiple-tone per octave keyboards was his discovery in 1940 that the mathematician-physicist-astronomer-music theorist Christiaan Huygens – the most famous of Dutch scientists – had produced an important mathematically structured work on music theory and the 31-tone scale almost three centuries earlier. Beyond such a proximate enticement, the records suggest that Fokker's interest in musical acoustics was generated from within the mental framework of some three decades of thought at the frontiers of theoretical physics; that these encounters had engendered in him a love for science at all levels. Fokker's publications on and related to musical acoustics, some 50 in number (not counting compositions) constitute an impressive corpus of theoretical and technical papers and books mostly in Dutch, but also in French, German, and English. They stretch from a 1941 paper on Huygens's division of the scale into 31 notes and terminate with a 1975 (posthumous) discussion about new music that had been composed for performance on a 31-note per octave harmonium.[2]

[2] The most important publication on Fokker's contributions in the field of musical acoustics is Rudolf Rasch (ed.), *Corpus Microtonale. Adriaan Daniel Fokker (1887–1974). Selected Musical Compositions (1948–1972)*, Utrecht, The Diapason Press in Cooperation with the Huygens Fokker Foundation, Haarlem, 1987. Other useful appraisals of Fokker's contributions to music theory include Bouw Lemkes, "Over zuivere intonatie," *Mens en Melodie. Algemeen Maandblad voor Muziek*, *23* (1968) 37–42; Rudolf Rasch, "The Theories of Adriaan Fokker," *The Quarterly Journal of the Just Intonation Network*, *4*, no. 1 (1988) 4–7, 14 and *4*, no. 2 (1988) 6, 10–12. Rudolf Rasch, "A. D. Fokker, *NG*, *9* (2001) 59 and *MGG*, Personenteil *6* (2001) Col. 1405/6. Rudolf Rasch, b. 1945, is a Dutch musicologist at the University of Amsterdam. He completed a doctorate in psychology at the University of Groningen with a dissertation on polyphonic music and a doctorate in musicology at the University of Utrecht with a dissertation on seventeenth-century southern Netherlandish polyphonic carols. He has published theoretical works on tuning, temperament, and mictrotonal music, and was secretary of the Huygens-Fokker Foundation for Microtonal Music (1986–1995).

Fokker's initial motivation to study the problem of pitch intervals came in 1940 when volume 20 of the *Oeuvres Complètes* of Christiaan Huygens (1629–1695) was published by Martinus Nijhoff in The Hague almost 60 years after they had been commissioned by the Société Hollandaise des Sciences.[3] Among works on music by Huygens that were completed in his own time, there is little that appeared in print apart from some letters on the harmonic cycle. It became known, nevertheless, that Huygens had developed a mathematical description of the system of the then-current mean-tone tuning system that came to his attention during visits to Paris in the 1650s and 1660s.[4] Huygens's description of the Parisian mean-tone tuning system included the calculation of string lengths and tuning instructions entitled *Divisio monochordi*. It made use of the logarithmic measurements of intervals along with the analogy between mean-tone tuning and the division into equal parts.[5] These preliminary ideas were brought together in 1691 in Huygens's *Lettres touchant de cycle harmonique*, a small text that contained the description of an imaginary transposing harpsichord with 31 strings to the octave and a 12-tone alternate keyboard which allowed for the selection of various intervals from the 31-tone keyboard. In Huygens's unpublished manuscript *Nachlass* it was discovered that there were several other texts on music in addition to numerous notes, drafts, and sketches that treat music theory in a much expanded form. It is these items that became available to Fokker in 1940 when volume 20 of the *Oeuvres* appeared in print.[6] What Fokker discovered in these writings on musical theory was

---

[3] In 1882 the Netherlands Academy of Sciences at Amsterdam organized a preparatory committee for a comprehensive edition of Huygens's works. In 1885 the Society of Sciences at Haarlem took on the responsibility for the publication. Sixty years of editorial commitment to the project led to publication of *Oeuvres complètes de Christiaan Huygens, publiées par la Societé Hollandaise des Sciences*, 22 vols., The Hague, 1888–1950. H. J. M. Bos, "Christiaan Huygens," *DSB*, *6* (1972) 597–613. See fn. p 613. The prominent Dutch historian of science Henk Bos has referred to the work as "the best edition of any scientist."

[4] Mean-tone temperament, widely used in a number of different ways on keyboard instruments from the beginning of the sixteenth century to about 1830, replaced the earlier Pythagorean and just tunings while preserving some features of both systems.

[5] Huygens expressed musical intervals in terms of logarithms because equal intervals that are used in equally tempered scales can then be more simply expressed as equal distances on a geometric scale that avoids a cumbersome number system. For example, in the usual common logarithmic scale (where $\log MN = \log M + \log N$ and $\log M/N = \log M - \log N$) the various octaves are indicated by the equidistant figures of 1, 2, 4, 8, etc. In the cent system that first was proposed by Alexander J. Ellis (translation into English of Helmholtz's *Tonempfindungen*) the various octaves are indicated by the equidistant Figures 0, 1200, 2400, 3600, etc. cents with each semitone indicated by 100 cents. Huygens used the logarithmic system to represent both the 12-tone per octave and 31-tone per octave equal tempered scales. To provide an example of additional advantages that follow from the logarithmic system we may note that the ordinary measurement of the fifth g (1½, 3, 6) lies exactly midpoint in the octave, although this place actually should be occupied by the well-tempered f-sharp as it is on the logarithmic scale.

[6] Volume 20 of the *Oeuvres* bears the subtitle *Musique et Mathématique. Musique Mathématique de 1666 à 1695*. The Hague, 1940.

that Huygens had tackled the technical matters of partitioning the octave so as to create the best temperament; he had explored the links between music and mathematics in order to propose a more genuine theory of Baroque harmony.

References

Bos, H.J.M. 1972. Christiaan Huygens. *DSB* 6: 597–613.

Helmholtz. 1954. *Sensations of Tone*, 316. Dover.

Lemkes, Bouw. 1968. Over zuivere intonatie. *Mens en Melodie. Algemeen Maandblad voor Muziek* 23: 37–42.

Chapter 22
Temperament and the Circle of Fifths

In what follows an effort is made, first, to locate the essence of what Fokker understood and appropriated to himself from Huygens's views on music theory, and then, to demonstrate how Fokker implemented and reconstructed Huygens's ideas in connection with the objectives that he came to set for himself in his own research on musical acoustics. Most of what Fokker gleaned from Huygens will emerge in the writings that are being examined here, but it is necessary in advance to explain the meaning of a number of musical terms used by Huygens and by Fokker. First there is the seemingly magic number 31 that shows up prominently in discussions about keyboards designed to achieve just intonation by using more than 12 keys per octave. As Huygens explained it, there are seven basic notes to the octave, A, B, C, D, E, F, G, which with each of their flats and sharps and each of their double sharps and double flats comprises 35 notes; but since among these notes there are two double flats and two double sharps and two double sharps and two double flats that are identical, the total number of notes reduces to 35–4, or 31. With that particular number of options on a keyboard, as Huygens had proposed, and as Fokker came to assert, it is possible with the right combination of intervals to achieve something very close to just intonation in all modulations. The context in which the discussion on the 31-tone per octave scale enters is Huygens's reference to "le cycle harmonique." The harmonic cycle corresponds to an arrangement of pitches that in the history of music came to be called "the circle of fifths." The circle of fifths (*der Quintenzirkel*), first described and illustrated by Johann David Heinichen in *Der Generalbass in der Composition* (Dresden, 1728), results from the fact that a succession of 12 acoustically pure fifths produce a pitch that is about one quarter of a tone higher than the starting point.[1]

[1] Acoustically pure fifths (the fifths that belong to the harmonic series) are equal to 702 cents and are larger by 2 cents than equally tempered fifths (700 cents).

E. Hiebert, *The Helmholtz Legacy in Physiological Acoustics*, Archimedes 39,
DOI 10.1007/978-3-319-06602-8_22, © Springer International Publishing Switzerland 2014

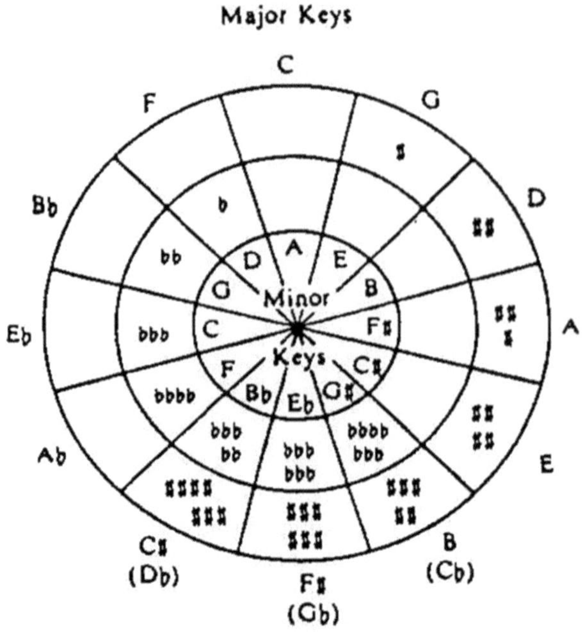

Circle of Fifths

To illustrate the principle of the circle of fifths using the above figure: if on a piano C to G is tuned as an acoustically pure fifth, and D is tuned as a pure fifth above the G, and A is tuned as a pure fifth above the D, etc., the C that is produced as a pure fifth above the preceding F on the circle would not be the same as the C on which the circle began. In order to end on the C where it began some or all the fifths would have to be made smaller than pure fifths by amounts totaling to about one-quarter of a semitone. Historically the fifths either were not reduced at all, or were reduced according to one or another of the mean-tone temperaments that adapt the Pythagorean scale so as to preserve the just major third, or they were all reduced by exactly the same amount as they are in the system of equal temperament.

In 1941, shortly after Fokker began his studies of music theory, he presented two Saturday noontime lectures at the Teyler Museum on "harmonic music." In expanded form they were published as a lengthy essay in the *Archives du Musée Teyler*.[2] The caption used to introduce Fokker's lecture on harmonic music comes from Leonhard Euler's *Du véritable caractère de la musique moderne* (Berlin, 1764). "We now have arrived at our goal which is to assign modern music its true character. No longer can one doubt that this true character consists not in employing a new kind of consonance that has been entirely unknown in the music of times past. . . . As a matter of fact rather than having introduced new types of

[2] A. D. Fokker, "Harmonische muziek,." *Archives du Musée Teyler*, Serie III, vol. IX, Fascicule 5, (1942) 449–506.

consonances (to achieve the true character of modern music) one must take the consonances to a much higher degree of perfection."[3] Embodied in this Euler quotation lurks the motivation for a programme of research in musical acoustics that Fokker came to pursue for three decades – the search for the "much higher degree of perfection (*plus haut degré de perfection*)" that characterizes just intonation.

In the Teyler lecture on *harmonische muziek*, the format of which had become the regular responsibility of the Curator of the Museum, Fokker essentially examined the basic principles of musical harmony that treat of fundamental tones: the harmonic series and the structures derived from them in the form of melodies, scales, chords and inversions, the central importance of the seventh harmonic as leading tone, and the lesser importance of the eleventh and thirteenth harmonics. It is worth noting that already in this early stage of venturing into the theoretical branch of music that treats intonation, Fokker touches on the meaning of all the basic terms that over the next several decades would come to acquire importance in his writings on just intonation, 31-tone tuning, and the use of intervals smaller than a half tone that are referred to as the diësis or microtones.[4]

On reading the introductory remarks to the comprehensive research programme in musical acoustics that Fokker was envisioning in his essay on "harmonic music" at the Teyler Museum in 1942, it becomes unmistakably apparent that he is full of enthusiasm about revisiting for himself and his Dutch and French countrymen a field of study and investigation that boasts a venerable history. His aim is to call attention to the cultural and artistic legacy and the expectations of an earlier era of towering seventeenth- and eighteenth-century Dutch and French accomplishments in science and music. On the other hand, at the time of this writing, in 1942, Holland and France were under attack from their nearest European neighbor. For Fokker, the transfigured theoretical physicist, the lecture on *harmonische muziek* furnished the occasion for an inauguration into theoretical and experimental investigations

[3] Fokker, "Harmonische muziek," p. 451. "C'est donc en effet un plus haut degré de perfection auquel on a porté la Musique, y ayant introduit cette nouvelle espèce de consonances."

[4] Historically, the *diësis* was introduced into Greek theory to represent an interval smaller than a semitone; in medieval treatises it is referred to as *semitonium minus*. In the Pythagorean scale, which derives all the tones in its scale from the interval of the pure fifth, the diësis was characterized as one of three categories of tuning. They are denoted as the diatonic, chromatic, and enharmonic genera. The diatonic scale is based on an octave divided into five tones and two semitones. The chromatic scale is based on an octave of 12 semitones in contrast to the 7-tone diatonic scale. The enharmonic scale is based on pair pitches that are not absolutely the same (such as G# and $A_\flat$) and is exploited for purposes of modulation. Enharmonic keyboards employ separate keys for some of the pairs of pitches that would otherwise be equivalent. Fokker and others who worked with keyboards having more than 12 keys per octave sometimes referred to the enharmonic keyboard as an instrument designed to produce different types of microtones. Microtones are smaller than a semitone and are distinguished from the several types of *diesis* by being some numerical fraction of a semitone rather than a tone in a particular collection of pitches that makes up a scale. Microtones have served both melodic and intonational functions in Western music since antiquity.

into the problem of just intonation and fixed-tone keyboards that would flesh out the Huygens system. Fokker introduced his subject with the following remarks:

Our subject [harmonic music] leads us back to mankind's past. Topically speaking, however, it actually is in the present and is important for the future of music. Far into the past....but I am not thinking of the incantation songs of the primitives. I also am not thinking of the school of Pythagoras several centuries before Christ. Closer to us lie the sixteenth and seventeenth centuries and a period of musical revival. In the context of our subject I come upon the scholars of those days and what they investigated and scrutinized. I think of Simon Stevin and Isaac Beeckman, of René Descartes, of Marin Mersenne and Christiaan Huygens.

On the other hand we are confronted at the present time with questions concerning the significance of harmony and with composers and performing musicians pursuing new expressions and new music. Some of them discard the inspiration to which the rules of the theory of harmony are bound. Others search for renewal of a discipline in the laws of beauty [*schoonheidswetten*] that lie above the strong trend toward the mania of sound [*klankbezedenheid*]. At the present our subject is situated at the border of past and future.

It also is situated at the border of two worlds, two mental worlds. It exists where the world lives and sings in voice and where the emotions and tensions of countless sounds of the heart change to the world that perceives and measures; and then it distances itself in numbers from the world of the senses so as to reveal a unity that no longer belongs to the world. Our subject belongs to the domain where music and the science of calculation are able to join hands [*waar de muziek en de rekenende wetenschap elkander de hand kunnen reiken*].[5]

Against the background of these reflective thoughts on the state of affairs in regard to the future of harmonic music, Fokker draws up a balance sheet of potential problems connected with linking music and science at the level of harmony. The task has an attractive side: one "acquires information about beauty and truths that are new, and one experiences the sense of discovery in a new territory." But there also is a danger: "The world is so old and so many excellent heads have thought about these matters that whatever appears to possess the glow of novelty will be thought of as old fare [*oude Kost*]." Perhaps one nevertheless can formulate a self-evident axiom that touches on a point where minds and intellects are anxiously divided. That should not keep investigators from raising questions that lead to cooperation and perhaps the "dissolution of the boundaries between music as a profession and physics as an occupation."[6]

The problems in musical acoustics that Fokker explored during the 1940s followed directly his own discovery that long unknown works on musical acoustics by Huygens had been put into print. He remarked: "In the year that has just expired

---

[5] A. D. Fokker, "Harmonische muziek," 1942, p. 451.

[6] A. D. Fokker, "Harmonische muziek," *Inleiding* p. 452.

[1940] the Dutch Society of Science in Haarlem has released the 20th volume of the collected works of Christiaan Huygens. Christiaan Huygens was a fine musician; and he has thought much and also written about the theoretical foundations of music. His studies and annotations concerning these matters are included in the 20th volume of his works. The unnamed authors Dr. A. J. Vollgraff and Dr. E. J. Dijksterhuis have foreseen that these musical writings by Huygens hold a treasure of historical information."[7] Huygens's discussion on the division of the octave turns out to be the crucial point of departure in Fokker's deliberations on the acoustics of just intonation. It is essential therefore to examine in some detail exactly what it was that Fokker learned from studying Huygens's writings on the division of the octave and how he took possession of Huygens's ideas in developing his own thought on keyboards designed to deliver just intonation. "It is remarkable," writes Fokker, "that Huygens breaks a lance for the *seventh* harmonic. He does not want to concede that consonance is exhausted [*uitgeput*] with the combinations that are possible using only the first to the sixth harmonic. He says, for example, that if one considers the subject without prejudice "one will find that the number 7 is not powerless to bring forth consonance." It shows up prominently in the course of the progression from dominant seventh to tonic – a familiar cadence in music termed the dominant seventh. "To the question about the seventh harmonic one can add the eleventh harmonic which unquestionably is used in practical music."

Huygens brings up another point in his investigations that is "of the greatest importance for the establishment of all tonal systems," namely the problem that arises when instruments are tuned to perfect fifths. Fokker illustrates it as follows: "Everyone knows that if all the strings in a string quartet are tuned to perfect fifths the e-string of the violin is higher in comparison with the fifth harmonic on the c-string of the cello. Accordingly Huygens's musical contemporaries had come to agree on taking each fifth a fourth of a comma smaller so that all major thirds in the tone scale remain pure. Since a major and a minor third together form a fifth, the small thirds should be longer by as much as the fifths are too small, which is just a quarter of a comma. With this tuning the difference between the large major seconds (9/8) and the small major seconds vanishes so that it comes to lie halfway between c and e."[8]

As Fokker explained, a solution to some of the difficulties inherent in the Pythagorean system had been offered earlier by "Simon Stevin who adapted the Aristoxenian division of the octave into 12 equal parts and thus provided the foundation for the normal semitone system, the so-called even-suspended [*gelijkswevende*], or better the proportionally-suspended [*evenredig-zwevendel*] temperament with the help of which Bach could write his famous fugues that still

[7] A. D. Fokker, "Christiaan Huygens' oktaafverdeling in 31 gelijke diëzen," *Caecilia en de Muziek, 98* (1941) 149–152.

[8] Fokker, "Christiaan Huygens oktaafverdeling," pp. 149–159. The Pythagorean comma as here used is the interval by which the sum of six whole tones, each designated by the ratio 9/8, exceeds the octave 2/1. A small interval, the comma, usually is taken to be about one-ninth of a whole tone.

are the basis for all of today's instrumental music-making."[9] From a mathematical point of view, Fokker remarked that the 12-tone chromatic scale is perfect. Musically speaking, however, this scale is a "compromise; or so at least one puts it in order to avoid using the word violence [*geweldpleging*]." In a letter to Stevin, the organist Verheyen in Mijmegen expressed his objections to speaking of this as a compromise because he used both small and large semitones; he treated the small seconds as the difference between a fourth and a major third and the augmented second as the difference between a major third and a minor third.

Huygens, who lived a half century later, examined the 12-tone scale of Stevin in another way. Although the fifths in the scale of Stevin are approximately the same, his seconds are larger than two tones by as much as a tone and a half is smaller than a minor third. The difference is 2/3rds of a comma. One recognizes that so large a defect in this harmonically important interval (the second) is much worse than the defect of a quarter comma in the fifths. Naturally Huygens was not blind to the great advantages of Stevin's scale, especially in regard to transposition and modulation. He therefore searched for a way to provide equal divisions of the octave that would give unlimited transposition and modulation possibilities, with minimum distortion of the most important musical intervals. The mathematical solution he reached embodied an idea that already had appeared in the writings of the Spanish theorist and organist Francisco de Salinas (1513–1590), the Italian composer and theorist Nicola Vicentino (1511–1576), and the fourteenth-century Italian theorist and composer Marchetto da Padova (fl. 1305–1326). In these systems an octave is made up of three major fourths plus one diësis, that is, about 31 dièses. Similar reasoning must have led Huygens to the conclusion that all desired tonal possibilities could be reached directly by dividing the octave into 31 equal diësis intervals. As Fokker remarked, this division offers many advantages. Ten dièses are almost precisely a major third. Eighteen dièses are a little smaller than a fifth; the difference is less than a quarter of a comma. Eight dièses are a little smaller than a minor third. The increased purity by which the diësis-tuning achieves these harmonically important intervals is not the only benefit of the Huygens scale. In the Huygens diësis scale it is not only the first six traditional harmonics that are improved; the 7th, 11th, and 13th harmonics also are more precisely realized than occurs with thirds in semitone tuning. The Huygens system provides complete freedom in transposition from one key to another since its circle of fifths is made up of 31 rather than 12 divisions A keyboard instrument with this mode of division was referred to in the seventeenth century as the archicembalo.[10]

[9] Aristoxenus (b. ca. 375–360, d. ? Athens) was a Greek musical theorist who challenged the Pythagorean tonal system of the musical scale based on mathematical ratios and rather placed the emphasis on aural reception.

[10] The term archicembalo was used by Vicentino in 1555 for a harpsichord or an organ with divided keys on a second manual that permitted the playing of intervals smaller than a semitone; that is, it permitted the playing of pure as against tempered intervals in a variety of keys. These instruments were conceived with the intention of making possible the performance of the diatonic, chromatic, and enharmonic genera of ancient Greek theory.

"The influence that the adoption of the twelve semitone scale had on the development of music," writes Fokker, "is difficult to measure. It worked to the good and to the bad – to the good in that it has promoted the more perfect unfolding [*de volmaaktere ontplooing*] of the art of composition; to the bad in that it blunted [*afgestompt*] our yearning for pure harmonies. There are persons – and they are not among the least competent musicians – who point out that endeavours in the direction of so-called atonal music creations [*atonale muziekscheppingen*] must be looked on as the eventual consequence of the failure to appreciate pure harmony in this scale. From these endeavors it seems to be the case that once again one has come full circle in search of a more permanent foundation."[11] Fokker suggested in conclusion that if one can believe that the endeavors in question have proceeded from a dissatisfaction with the current tonal system, then it may well be the case that the harmonies possible with the 7th harmonic, and perhaps as well with the 11th and the 13th harmonics, could be put to use in the creation of a new field for music wide open for exploration. It would open up the search for a richer, more nuanced inner life for music than is possible by mastering and developing music within the limited theoretical structure in existence. "The normal diësis-tuning of the *nouveau cycle harmonique* of Huygens offers a subtler and more powerful aid to music than the normal semitone tuning of Stevin."[12]

In the early 1940s Fokker extended his *Caecilia en de Muziek* series of studies on Huygens's 31-tone per octave tuning system in order to spell out more precisely what the system entails in terms of the selection of tones that provide musically pure intervals and combinations of intervals. Fokker wrote: "The reservoir of tones is inexhaustible [*toonschat is unuitputtelijk*]. A continuous series of tones is at our disposal. In order not to drown in this richness one chooses a number of distinct and recognizable tones and makes music with them. How does one go about making these choices?" This for Fokker is the question that confronts the musician in search of the purity (*zuiverheid*) of musical intervals. With adroit and rigorous mathematical maneuvering it is possible, according to Fokker, to make musically significant interval choices and partition the octave into units that in combination, and at many levels of the harmonic series, approximate theoretical purity, or just intonation, and satisfy the most stringent demands of a sensitive ear. How Fokker achieves this in his *Caecilia* essays cannot be outlined here since they are given over largely to a point for point analysis and comparison of the degree of musical purity or harmoniousness resulting from the use of intervals based on the equal-temperament scale of Stevin and the use of intervals based on Huygens's partitioning of the octave into microtonal units – units that correspond to the diëses already recognized in early Greek times. What is central to these essays is the argument used to buttress his position, namely that a satisfactory solution to the just intonation problem is coupled with the hard fact that no numerical set of sequential fifths can be made to coincide, pitchwise, with any number of sequential octaves.

[11] Fokker, "Christiaan Huygens oktaafverdeling," p. 151.

[12] Fokker, "Christiaan Huygens oktaafverdeling," pp. 151–152/

Although 12 pure fifths almost coincide with 5 octaves, there is a small and musically important difference. As Fokker remarked, one might well continue to draw on the unlimited reservoir of tones in the octave to search for other possibilities of tonal combination, but this does not get at the root of the intonation problem. It is possible to get rid of the small difference between 12 acoustically pure fifths and 5 octaves (one six-hundredth of an octave) by dividing each of the 12 fifths into 2 semitones in such a way that a self-closing series of tones is reached.[13]

References

Fokker, A.D. 1941. Christiaan Huygens' oktaafverdeling in 31 gelijke diëzen. *Caecilia en de Muziek* 98: 149–152.

Fokker, A.D. 1942a. Harmonische muziek. *Archives du Musée Teyler*, Serie III, vol. IX, Fascicule 5: 449–506.

Fokker, A.D. 1942b. De keuze der muziektonen. *Caecilia en de Muziek* 99: 25–29.

[13] A. D. Fokker, "De keuze der muziektonen," *Caecilia en de Muziek, 99* (1942) 25–29; "Grundtoon, gidstoon en hun volledig akkord," *Ibid., 99* (1942) 72–75; "Vierklanken als getallenkwartetten," *99* (1942) 116–118; "Intervalsnippers en de verdeeling van octaf en kwint in even groode deelen," *99* (1942) 172–176; "Tartini en de zevende harmonische," *100* (1943) 78–80.

Chapter 23
Arithmetic Reflections on Music

Fokker's *Rekenkundige Bespiegeling der Muziek* (Arithmetical reflections on music) of 1944 is an historically focused overview of what was written on the subject of numerical relationships in music by the Swiss mathematician and physicist Leonhard Euler (1707–1783) and precursors of his such as Zarlino, Mersenne, Leibniz, Rameau, and Tartini. Huygens, who flourished almost a century later than Euler, is not dealt with in any significant way in this study of 1944 because his important treatises on music only became available to scholars in 1940 and would not have been accessible to Euler and his precursors. Fokker's *Arithmetic Reflections* thus seems to give an overview of what was known on this subject a century before Huygens's time.[1]

The treatise is built on the theme of numerically reasoned theories on intonation by Euler and his precursors. It covers the full range of music topics that deal with the composition of intervals, the analysis of chords, melody, melodic progressions, tonal grouping, the harmonic series, overtones, combination tones, proportional tuning, and intonation.

The *Tentamen novae theoriae musicae* (1739) of Euler is treated by Fokker as a turning point in history of music theory.[2] In the *Tentamen* Euler expanded on a mathematical theory of consonance formulated in ancient Greece. He introduced the use of partial differential equations into musical acoustics (a practice that came

[1] A.D. Fokker, *Rekenkundige Bespiegeling der Muziek*, Gorinchem, 1944. The title page prominently features Fokker as "Curator van het Natuurkundig Laboratorium van Teyler's Stichting de Haarlem." It is apparent that Fokker produced this impressively researched monograph by drawing liberally on a work that he lists as a main reference. Wilhelm Dupont, *Geshichte der Temperatur*, Inaug. Diss. Erlangen, Nördlingen, 1935. Published during World War II, Fokker's *Rekenkundige Bespiegeling* probably had a very small distribution outside of the Netherlands. I am most grateful to personnel in the Harvard University Library Interlibrary Loan division who secured a copy for my use from the University of Nebraska Library in Lincoln.

[2] At the age of 19 Euler planned a treatise on all aspects of music theory. The only part of that plan that came to fruition was the *Tentamen*, published in St. Petersburg in 1739 after Euler had been invited there by Empress Catharina I to become head of the laboratory for theoretical and practical physics.

E. Hiebert, *The Helmholtz Legacy in Physiological Acoustics*, Archimedes 39,
DOI 10.1007/978-3-319-06602-8_23, © Springer International Publishing Switzerland 2014

into wide use in modern times), established systems of scales and modes, and provided what at the time was the most thoroughgoing theory on modulation. It is evident that much of what Euler had to contribute to music theory was built upon ideas that appear in the works of earlier authors. One of Euler's most influential precursors was the Italian theorist and composer Gioseffo Zarlino (1517–1590), whose *Le istitutioni harmoniche* of 1558 is a landmark in the history of music theory for having united speculative theories of ancient sources with the compositional practices of his time. The *Traité de l'harmonie* (1702) of the French composer and theorist Jean Phillipe Rameau (1683–1764) was an influential and theoretically controversial work based on the physical properties of sound. The Italian instrumental composer and violinist Giuseppe Tartini (1692–1770) wrote on the acoustical foundations of harmony and discovered difference tones. This is not the place to follow Fokker's detailed probings into what Euler's precursors had to offer to the problem of intonation or to elaborate on what Euler learned from them in order to establish his own mathematical scale system and theory of modulation. It may suffice to bring Fokker's study on Euler and his precursors into focus by taking note of Euler's emphasis of the central importance of the harmonic seventh in relation to interval analysis and the realization of polytonality. "What is important," writes Fokker, "is that Euler's starting point diverges from that of his contemporaries. He begins not with the fundamental but with the completed chord. Think of his plea for a major seventh chord (in those times!). . ..Precisely by setting it apart his theory had something to offer to our modern contemporaries that lies outside the traditional way of thinking on current traditional music."[3]

The octave has been recognized in all cultures even if only as the difference between what men sing and what women sing. As for the octave itself, it may be said in the abstract that it is made up of an infinite number of tones; and every culture has had somehow to decide which notes from among that infinite reservoir of notes to draw on for its music. As we have seen, it was the Huygenian model of the 31-tone octave that became for Fokker the point of departure for a research programme devoted to the theoretical exploration and practical implementation of the phenomenon of just intonation. In recognition of this commanding focus on the direction that Fokker's investigations were to take it is important to recognize that the 31-tone per octave system in itself is not an isolated scale system pursued exclusively by Huygens and then by Fokker. The subject has been widely examined and debated in literature pertaining to music theory and notably by persons who have used their musical, scientific, mathematical, and technical knowledge and skills to examine the phenomenon of just intonation. It is desirable, therefore, to offer a succinct overview of the progressive steps and arguments that led to the achievement of acoustical purity by resorting to microtonal divisions of the octave and that came to rest, more specifically, on a 31-tone per octave division.

---

[3] Fokker, *Rekenkundige Bespiegeling*. pp. 1–49.

As already implied, every culture has had its own conception of what it means to be in tune. Until the beginning of the twentieth century, Western music, theoretically and generally speaking, had come to be based on consonance or the absence of beats.[4] The notion of octave division was implicit in the theory and practice of Greek music and was pursued avidly by the sixth-century B.C.E. Pythagoras and his followers the Pythagoreans, who believed that numbers were to be used as a guide to interpreting the world as exemplified in music. In the Western world the discovery that the principal intervals of the musical scale are given by numerical ratios is attributed to this early group of mathematician-philosophers. The Greek scale was made up of whole note intervals, major tones of equal size having a frequency ratio of 9/8, an interval that measures 204 cents in modern terminology.[5] The Pythagorean diatonic scale is characterized by pure fifths (3/2), pure fourths (4/3), and whole tones defined as the difference between a fifth and a fourth ($3/2 - 4/3 = 9/8$). The system has large major seconds and thirds and small minor seconds and thirds. The system was widely used during the early Renaissance.

At different times in the history of Western music individual notes within the scale were slightly adjusted in order to accommodate instrumental and compositional needs. The process of slightly modifying an acoustically pure or just interval is referred to as temperament. In reference to the musical scale, temperament is the process of tuning in which most or all of the concords, the harmonious soundings together of two or more notes, are made slightly impure in order that none or only a few are disagreeable. Equal temperament, in which the octave is divided into 12 uniform semitones, is the standard in Western music except in the performance of early music. Temperament becomes necessary chiefly in concords of triadic music because octaves, fifths, and thirds often are incommensurate in their pure form. Three pure major thirds fall short of a pure octave, four pure minor thirds, exceed a pure octave, and the circle of pure fifths does not add up to pure unison. The practice of tempering intervals has been shared by all musical cultures. For more than two millennia the desirability of acoustical purity has been championed and almost taken to be self-evident by music theorists. Temperaments, accordingly, represent the practical compromises that are made necessary by the fact that preference for musical purity and the practice of carrying out transpositions and modulation is not compatible with any closed system of tuning or with musical instruments that are not able to adjust their interval sizes to accommodate changing harmonic or melodic contexts. Historical events that have led to the need for temperature adjustment are readily available. For example, when the organ was

[4] According to the classic definition, beats are vibrations that are heard when the same note is played out of tune of when two or more notes in the harmonic series are played together but are slightly out of tune.

[5] The cent system of measuring pitch was introduced by Helmholtz in 1863 in his *Tonempfindungen* and was adapted for practical use in musicological research by Alexander J. Ellis in 1885. A logarithmic unit, the cent, is used for expressing numerical intervals and represents one-hundredth of an equal-tempered semitone. The octave in this scheme is equal to 1,200 cents since each of its semitones is equal to 100 cents.

put to common use in churches in the ninth century, Pythagorean tuning, which is based on octave, fourth, and fifth, but avoids thirds, was adopted as the norm. With the development of instrumental music, and especially following the introduction of stringed keyboard instruments such as the harpsichord at the beginning of the fifteenth century, Pythagorean temperament was adopted on a much wider scale, and that unfortunately brought on the question of how to deal with the major third. The major third had posed no problem in ancient times simply because Greek harmony avoided it. During the Middle Ages, when harmony progressed to the stage of wanting to use the major third, Pythagorean temperament had to be abandoned in order to adopt a less dissonant third. It was replaced by a scale that makes room not only for fourths and fifths but for thirds. This was accomplished by constructing a new size for a whole step by cutting the pure major third in half by taking its average or mean; therefore, mean-tone temperament. A fast shift forward timewise brings us to the widespread adoption of the chromatic scale made up of 12 equal semitones – a system that still is dominant in Western music.

Provided it can be agreed upon that the problems of temperament are generated by the human desirability of achieving acoustical purity, defined here in terms of intervals generated within the harmonic series, then it follows logically that the temperament problems that the musician makes an effort to come to grips with are inherent in the nature of things, namely in sounds that are defined in terms of vibrational frequencies. This implies that problems in temperament are not fabricated by the musician, the music theorist, the scientist, or the mathematician. Rather they are inherent in the nature of music itself in the sense that the desired beat-free intervals can not be arranged in a self-contained intonation system. Take a fifth, from C to G, ratio 3/1; it spans 702 cents. Take an octave C to c, ratio 2/1; it spans 1,200 cents. By piling up successive fifths, C to G, G to D, E to B, and so on, one cannot return to C; only after the necessary 12 steps can one complete the circle of fifths and reach B♯ which is 24 cents higher than C. Piling up thirds is even less satisfactory in that one eventually reaches G♯ which is 42 cents short of reaching C. The thirds can be made pure, but that leads to the deterioration of the fifths. The history of temperament, perforce, is a history of compromises such as are given by Pythagorean tuning, mean-tone tuning (several types), equal-tempered tuning, or various kinds of multi-tone or microtone tunings (tones smaller than a semitone) invented and maneuvered into one scale system or another as was done by Helmholtz, Tanaka, Planck, Huygens, and his enthusiastic student Adriaan Fokker.

The music-listening, music-performing, and music-composing publics of the modern Western world and of the countries that have adopted Western culture have settled, more or less, for equal temperament and to the tuning of their musical instruments and the use of the voice to scales and intervals that closely approximate equal temperament. In this system the fifths are almost pure (702 rather than 700 cents) and the thirds are pure enough to be generally acceptable (408 rather than 400 cents). In music that moves forward at a reasonably rapid tempo, and in music performed on instruments in which the tones die off rapidly after being sounded (i.e., have rapid decay times such as the piano) departure from acoustic purity is less noticeable for the listener. On the sustained notes of the organ the thirds become sufficiently intolerable as to warrant mean-tone tuning or some alternative unequal or irregular temperament. Beginning in the sixteenth century a variety of tuning expedients were introduced to evade unacceptable intervals. On some keyboards, for example, the G♯/A♭ key and the D♯/E♭ key were divided so as to provide tunings for both notes. The flutes that the flautist Johann Joachim Quantz (1697–1773) had constructed for himself and for his royal patron Frederick the Great had two keys on the foot joint of the flute to distinguish between D♯ and E♭. In order to provide tunings that went beyond the mere splitting of single keys, the Italian composer and theorist Nicola Vicentino (1511-ca.1576) gave an account in 1555 of a reconstructed harpsichord, the *arcicembalo*, and of a reconstructed organ, the *arciorgano*, in which every key was divided, some more than once, in order to achieve (perhaps mostly in paper constructions) acoustical purity. Paul von Jankó (1856–1919) in his *Eine neue Klaviatur* (1886) portrayed a piano keyboard whose 12-tone equal temperament was remodeled with two additional interlocking manuals that together provided six tiers of short narrow keys.

Many other keyboard instruments designed to improve on equal temperament are featured in the literature. Most of them were abandoned for reasons connected with expense or because of the inordinate complexity inherent in the construction and the playing of multi-key and multi-manual keyboards (the traditional 88-key piano already was the most complex machine in existence in the nineteenth century). For these reasons among others, most pianists and organists, at least since the middle of the nineteenth century, have come to put up with equal temperament and by virtue of habit simply have come to expect no more than a well-tuned instrument in equal temperament. The same holds true for other musical instruments especially when played in accompaniment with keyboard instruments. Instrumentalists have learned to stay in tune as closely as possible by making their notes a little sharper or a little flatter as needed. String players are able readily to accommodate to pitch changes by fingerboard adjustment except on the instrument's open strings. Singers, in principle, have no problem adjusting to pitch provided they have control of their vocal apparatus and are in possession of sensitive pitch discrimination. Wind players are able to make adjustments by lipping the pitch, by hand-in-bell techniques (as on the horn), and by using alternate fingerings (as on the clarinet). In addition to all that has been reported here on temperament it is well known that accommodation to temperament can also take place on the part of the listener; that the listener is able, so to speak, to "hear" the tones and intervals as if they are in tune or are sounded as the listener would like to have them sound in the context in which they are heard. Helmholtz refers to the phenomenon by which the ear accomplishes its accommodation as one in which the ear becomes insensitive (*abgestumpft*) to acoustical purity by force of habitual listening to nothing but music in equal temperament.

In our discussion we have not been detained by dwelling on concepts such as chords, modes, harmony, modulation, register, and timbre. None of them is ambiguous. The concepts of consonance and dissonance, by contrast, give rise, and for centuries have given rise to deliberately different points of view. The words consonance and dissonance themselves convey something concerning which the ear is called upon to reach a degree of repose or resolution, and the historical record reveals that at different times and in different places the ear is either too ill-prepared or too dogmatically disposed to accept the assignment with any degree of unanimity. Jeremy Montagu, the now retired curator of the Bate Collection of Musical Instruments in the Faculty of Music at the University of Oxford has captured the essence of the temperament problem as illustrated by the scale on which Western music is structured.

With our own scale so firmly fixed to the harmonic series and our musical harmony so heavily biased toward thirds, fourths, and fifths, it is difficult for us to comprehend the musics of those other peoples who do not hear things that way. And yet, around the world, it is only in the Indian subcontinent that we hear music akin to our own. There, tuning is even more precise than ours. . . .In their theoretical basis for scales, dividing the octave into 22 *strutis*, they have a similar concept to our major and minor tones, with some whole tones larger than others.

But India is the exception. In most other cultures, the fifth and third simply do not exist. This is a permanent puzzle for musical psychologists. The sounds of the natural harmonics are all around us. . . .In Africa, all the pitches produced by the musical bow are those of the harmonic series, and yet this has had little effect on their music. Why this should be, we do not know, but this, and the preference for monody over harmony, are the main reasons that so many musics sound so strange to us, and ours, of course, to them.[6]

In *Arithmetic Reflections on Music* of Fokker recognized that there would be no way to put Huygens's 31-tone system into practice without first establishing a system of notation for each one of the 31 different notes. To accompany the system he therefore introduced a two-dimensional network depicting the placement of the notes in a way that would make it convenient for the player of the keyboard to sound the requisite musical chords of the composition. Thirty-one numbered notes were given 31 specific names corresponding to the seven notes of the diatonic scale, their flats and sharps and their double flats and double sharps. Fokker wrote: "In Huygens' 'cycle harmonique' of 1691 the octave is divided into 31 equal intervals, the normal diëses, so that one has 31 tones that need to be assigned names. The classic interval of a pure fifth comes close to 18/31 of an octave, that is 12 normal diëses. The pure fourth comes close to 13 diëses. If one proceeds upwards with steps in fifths and downwards with steps in fourths one returns, after 31 steps, to a complete set of octaves above or below the starting tone. The tones that follow should have numbers from 1 to 31." The network scheme by note number and note name is depicted below.[7]

[6] Jeremy Montagu, *Origins and Development of Musical Instruments*, Lanham, Maryland/Toronto/Plymouth, U.K., 2007. 214–217. Scales and Music.

[7] Adriaan Fokker, *Rekenkundige Bespiegeling*, 1944, 50–51.

ceses : geses : deses : ases : eses : beses : fes : ces
7 , 25 , 12 , 30, 17 , 4 ,22, 9
ges : des : as : es : bes : f : c : g : d : a : e
27, 14 , 1 , 19 , 6 , 14, 11, 29, 16, 3 , 21
b : fis : cis : gis : dis : ais : eis : bis : fisis
8, 26, 13, 31, 18, 5 , 23, 10, 28
cisis : gisis : disis : aisis : eisis : bisis
15, 2 , 20, 7 , 25 , 12

The table below gives the English equivalents of the European notation used by Fokker.[8]

2. Noten und Versetzungszeichen – Notes and inflections

ceses	c	double flat	c♭♭
ces	c	flat	c♭
c	c		c
cis	c	sharp	c#
cisis	c	double sharp	cx
deses	d	double flat	d♭♭
des	d	flat	d♭
d	d		d
dis	d	sharp	d#
disis	d	double sharp	dx
eses	e	double flat	e♭♭
es	e	flat	e♭
e	e		e
eis	e	sharp	e#
eisis	e	double sharp	ex
feses	f	double flat	f♭♭
fes	f	flat	f♭
f	f		f
fis	f	sharp	f#
fisis	f	double sharp	fx
geses	g	double flat	g♭♭
ges	g	flat	g♭
g	g		g
gis	g	sharp	g#
gisis	g	double sharp	gx
ases	a	double flat	a♭♭
as	a	flat	a♭
a	a		a
ais	a	sharp	a#
aisis	a	double sharp	ax
heses	b	double flat	b♭♭
b	b	flat	b♭
h	b		b
his	b	sharp	b#
hisis	b	double sharp	bx

[8] Horst Leuchtmann, *Wörterbuch Musik*, 5th ed., Saur, Stuttgart, 1998, p. 314.

In the depicted network scheme bises and deses, note number 12, or d_{bb}, represent the same note, and geses and eisis, note number 25, or $g_{bb,}$ represent the same note. Since two of the 33 notes are duplicates, only 31 remain as different notes – namely Huygens's 31-note system. The arrangement of the notes on the keyboard network can be altered to feature network schemes better adapted either to meet the particular desiderata of the performer or to give preference to a specific chord that earmarks the seventh or some other harmonic. The value of the 31-tone system is rooted in the premise that although an infinite number of potential notes are available for constructing a scale, only a limited number of those notes can be used to construct a scale whose intervals sound in acoustical purity. As Fokker states this premise: "There is no boundary between dissonance and consonance. There is a difference in the larger or smaller amount of tension, and one can formulate an hypothesis to measure how much larger or smaller the amount of tension is. Resolution of a dissonance can mean only a move toward smaller tension."[9]

Arithmetic Reflections on Music represents a preliminary effort on the part of Fokker to explore the subject of musical temperament by arranging keys in the Huygenian 31-tone system of intonation on a keyboard network that makes it manually possible to deliver acoustically pure intervals expeditiously. Although the work was a report on research in progress on a subject that would continue to be under investigation for another two decades, it formed the backbone to Fokker's ambitiously conceived study of acoustically pure intervals, or as Fokker put it in the caption: "Zuivere intonatie het eerste nodige." What that entails for the listener, as Fokker perceived it, is the promotion of aural skills that draw on imagination and tonal fantasies (*voorstellingsvermogen en de fantasie voor tonen*) that are rooted in the cultivated ability to identify, with mathematically conceived precision, higher harmonic intervals, such as the seventh, the eleventh, and the thirteenth harmonics. But "one cannot rely on instruments with fixed tunings because all fixed tunings inevitably end in compromise. To be sure, one can control the intervals by bringing them in tune either with the voice or with another instrument, but then the responsibility of reaching the right compromise rests on the shoulders of the player."[10]

References

Fokker, A.D. 1944a. *Rekenkundige Bespiegeling der Muziek*. Gorinchem: Noorduijn.

Fokker, Adriaan. 1944b. *Rekenkundige Bespiegeling*, 50–51.

Leuchtmann, Horst. 1998. *Wörterbuch Musik*, 5th ed, 314. Saur, Stuttgart.

Montagu, Jeremy. 2007. *Origins and development of musical instruments*, 214–217. Lanham, Maryland/Toronto/Plymouth.

[9] Fokker, *Rekenkundige Bespiegeling*, Section 6.6 Dissonanten en oplossingen, p. 165.

[10] Fokker, *Rekenkundige Bespiegeling*, p. 207.

Chapter 24
Just Intonation and the 12-tone System (1949)

Having for several years devoted himself exclusively to investigations focused on what can be achieved in regard to acoustically pure intervals using the Huygens 31-tone system, Fokker chose in 1949, by way of comparison and contrast, to redirect his attention to examining the limited extent to which just intonation is achievable for music structured on the ordinary equal temperament scale of 12 notes to the octave using instruments that produced tones with fixed pitch. While the pianoforte and the organ gain the advantage of complete freedom of modulation in this system, they sacrifice the purity and beauty of perfect concords. It is for this reason, as Fokker remarked, "that stress must be laid on a special training to acquire a just intonation of perfect intervals and chords."[1] To correct for divergences from just intonation using instruments tuned to equal temperament, Fokker's treatise provided a series of exercises and historical examples designed on the one hand to assist the ear in making judgments concerning the accuracy of intonation and on the other hand to direct the voice and the fingers of string players to pitches that more closely approximate just intonation. The suggested practices, hopefully, would become a matter of subjective reflex and were intended, for example, to feature the seventh, whose importance already had been stressed by Huygens, Euler, and Tartini. In modern times the seventh had been used by Brahms in his horn trio and by Britten in his *Serenade* for tenor voice and horn. In his article on harmony in the *Encyclopedia Britannica* Donald Tovey remarked: "Harmony has not yet found a place for so simple a natural phenomenon as the 7th note of the harmonic series." To this, Fokker added: "I am attempting to find that place," by establishing "a melodic subscale of the harmonics six to ten," and by showing "the various ways in which the combination of two or three subscales gives rise to classical and to new scales," and by setting up a "structure [that] as a whole might be strong, simple, and sufficiently sound to open the way to a new musical development."[2]

---

[1] Fokker, *Just Intonation and the Combination of Harmonic Diatonic Melodic Groups*, the Hague, 1949, pp. 1–50. Quotation p. 1.

[2] Fokker, *Just Intonation*, p. 2.

E. Hiebert, *The Helmholtz Legacy in Physiological Acoustics*, Archimedes 39, 235
DOI 10.1007/978-3-319-06602-8_24, © Springer International Publishing Switzerland 2014

Recognizing that the method of acquiring just intonation is supported by supplying an accompanying sustained supporting note, Fokker adopted a modified system of notation. It was similar to the Curwen tonic sol-fa system, but with the intent of enriching the melody, deviated from the conventional tonic sol-fa series, for example, by substituting the Pythagorean minor third for the perfect minor third.[3] In details that would take us too far afield, Fokker demonstrated that various leading notes can come into existence by combining different melodic centers. He also touched on the subject of methods of dividing the interval of the perfect fifth, the most promising division being that of three very nearly equal intervals. "This is a feature of existing musical culture of the far East, of the *gamelan slendro* and *gamelan angklum*, in the isles of Java and Bali. I see no reason why it should not become an element of musical culture in Europe and the western hemisphere."[4]

In 1953 Fokker presented a short note at the International Society for Musical Research in Utrecht in which he referred to two ways of approaching temperament based on 31 fifths of a tone per octave: the seventeenth-century developments in fixed-tone keyboard instruments that led Huygens to recognize that 31 steps per octave "would afford an all but perfect rendering of the perfect seventh," and the approach motivated by reproducing the characteristics of indigenous or autochthonic folk music. Fokker comments:

> It is well known to folklorists that autochthonic popular music in its melodic tunings uses intervals which cannot be reproduced in our twelve semitones equal temperament. This fact induced Aloys Haba to invent playing with, in his mind flexible, quartertones. Belá Bartók witnesses in a paper of 1931 that in the Hungarian peasant music he so often found the perfect seventh as an element of equal importance to the fifth and the third, that he added it to the common chord in a setting of a fundamental concordant chord of four notes. Hence arises the problem to find an equal temperament which allows a satisfactory rendering of the octave, the fifth, the third, *and* the seventh. By rather simple arithmetics the solution is found to be an equal temperament of 31, thus confirming Christiaan Huygens' anticipation. One will note that here fifths of a tone are required instead of the uncertain flexible quartertones of Haba.[5]

At the Utrecht conference Fokker proceeded to describe an actual organ with a playable keyboard whose major thirds were only one part in 2200 sharp on the frequency scale, whose fifths were one part in 320 flat, and whose sevenths were perfect. He remarked that on the whole this 31-tone instrument made it possible to perform "the old classical works" so as to reveal their "nobility, gracefulness, sweet intimacy, (and) candour." The occasion provided Fokker with the optimal opportunity to present the international music society with concrete evidence concerning

---

[3] Tonic sol-fa is a type of musical notation and is associated with a method of sight-singing developed in England in the nineteenth century by John C. Curwen (1816–1880). He published music in the sol-fa system with the aim of teaching beginners to sing accurately and to foster proficiency in sight-reading.

[4] Fokker, *Just Intonation* (1949), p. 34.

[5] Fokker, "The Qualities of the Equal Temperament by 31 Fifths of a Tone in the Octave," *International Society for Musical Research* Fifth Congress Report, Amsterdam, 1953, pp. 191–192.

what his music studies had led him to accomplish since he first learned about Huygens's seminal contributions to music. He capped his concise remarks about the 31-tone octave with an invitation. "The members of the conference are kindly invited to come and hear the organ in Teyler's Museum, on Monday July 7th, the last day of the conference."

Following the international meeting in Utrecht it did not take long for American scientists to recognize that a Dutch theoretical physicist, who was "known for his early work in quantum statistics and for the Fokker equation," had constructed a special organ and had captured the attention of the European music community. In response to this acclaim on the part of American scientists – an acclaim directed to a moderately well known theoretical physicist but focused on a scientifically-anchored subject – Fokker crafted an article for the *American Association for the Advancement of Science* that was as cogently integrated and convincing in its clear text and informative illustrations as anything he had yet put into print. The programme in music that he was to lay out was based on the open-endedness of the concept of musical temperament.

> There are a great many musics among the peoples inhabiting our planet, and they show differences as great as the languages spoken by the various races and nations. . . .
>
> It is a very difficult task to represent all this music by written symbols. This task requires centuries of analyzing and systematizing work. Too soon one is eager to believe that one's own music is the most natural, and that one's own way of putting it on paper in notes on staves is the most straightforward and efficient way of recording it, and that there is no alternative. The fact is, however, that our subdivision of the interval of the octave into 12 equal semitones represents a solution for certain special needs only. It embodies a stage in the historical development of the art, a system with its virtues and its deficiencies, and one that is likely to be modified or even relinquished during the course of historical evolution. There are vast civilizations of autochthonic musics, not very remote from the idioms of Western civilization, that must remain behind the horizon of our understanding conscience, unless the system of 12 equal semitones in the octave is disposed of, and a better approximation is substituted for it.[6]

This general statement on the many musics of many peoples, and the many alternative ways of producing music and of notating music and subdividing the octave, sets the stage for Fokker's analysis of temperament. Given that there exist an infinite number of potential notes in the octave and that musical instruments whose tones result from vibrating strings as columns of air are fixed on pitch, it necessarily follows that there will be restrictions on the extent to which musical instruments are in control of pitch. Except for its open strings, pitches on instruments of the violin family can be controlled by adjustments on their fingerboards. On brass and woodwind instruments the pitches can be controlled to a limited extent by means of embouchure control, hand-in-bell adjustment, or alternate keys. Apart from sliding instruments like the trombone, the only musical instrument whose pitches can be controlled completely, at least in principle, is the human

---

[6] Adriaan Fokker, "Equal Temperament and the Thirty-one-keyed Organ," *Scientific Monthly* (Published by the American Association for the Advancement of Science, Washington, D.C.) *83* (1955).

voice. By contrast the musical instruments on which the performer has no control are the family of keyboards such as the clavichord, harpsichord, piano, and organ.

Already in the sixteenth century music theorists such as Vincenzio Galilei (1520–1592) and mathematicians such as Simon Stevin (1548–1620) advocated equal temperament as a method of compromise to avoid many of the harmonic problems created by temperament systems that were generated in earlier times. One of the seventeenth-century mathematicians who took it upon himself to challenge the 12-tone equal temperament system and suggest an alternative scale system based on dividing the octave into 31 equal intervals so that there would be 31 fifths of a tone in an octave, was Christiaan Huygens (1629–1695). Fokker had begun the study of music theory during the time that academic activities were suspended during World War II. In 1940, when Huygens's writings on music theory became available in volume 20 of the *Oeuvres Complète*, Fokker was so inspired by the notion of Huygens's 31-tone scale as a solution to the temperament problem that he adopted it as the *terminus a quo* for his studies on musical acoustics. As a mathematician Huygens discovered a new harmonic cycle (*nouveau cycle harmonique*) when, in attempting to preserve the differences between major and minor semitones, he took this difference as an elementary unit for the division of the octave. He recognized that the same unit is the excess of an octave over three perfect fifths. It is about half the minor semitone and one-third the major semitone. Thus it is one-fifth of a whole tone, and Huygens called this unit the *diësis*. For the seven basic notes in the octave, A, B, C, D, E, F, G, their seven flats, their seven sharps, their seven double flats, and their seven double sharps, it follows that there are 35–4 *dièses* to the octave since two of the double flats are identical to two double sharps and vice versa.[7]

As Fokker observed, a total of 31 undoubtedly was "a multitude of notes ... far too much for the comprehensive faculties of Huygens' contemporaries" to entertain. His discovery was left to oblivion; and so too was his discovery "that besides a faithful rendering of the perfect third, his harmonic cycle also provided a nearly exact reproduction of the perfect seventh." Huygens claimed that it was just as valid as the other perfect intervals that are used in music. We may note that during the Middle Ages the repudiation of the perfect seventh as a concord was not unlike the disqualification of the perfect third. Although Huygens's contemporaries paid scant attention to his 31- tone scale, Fokker was keen to point out that the Czech composer Aloys Hába (1899–1973) was alert to the idea that he needed a finer grain than equal semitones to reproduce the native songs of his countrymen. He therefore introduced quarter tones into his music. What was needed, as Huygens recognized in the 1690s, was not quarter tones or simply more microtones, but fifths of a tone that when put together in specific ways produce a maximum number of intervals that belong to the harmonic series and are acoustically pure. In a similar way Hungarian composer Béla Bartók (1881–1945), recognizing the significance of Hungarian peasant music as an inspiration for modern music, was struck by how

[7] Fokker, "Equal Temperament," pp. 162/3.

often he heard the interval of the harmonic seventh; so he added the perfect seventh to the common chord of four notes specified by the harmonic numbers 4: 5: 6: 7.

The notes related to harmonics 3 and 5 readily lend themselves to representation on a two-dimensional network. On the other hand the recognition of the seventh harmonic as a legitimate concord provides good reasons for representing the elementary set of 31 notes in a 3-5-7 lattice of harmonics as shown below. "By establishing identities between various notes, the threefold endless multitude of notes in the harmonic lattice is reduced to a finite set," and this is similar to the way in which the network of atoms is depicted in a crystal lattice.[8]

Elementary set of 31 notes in the 3-5-7 lattice of harmonics.

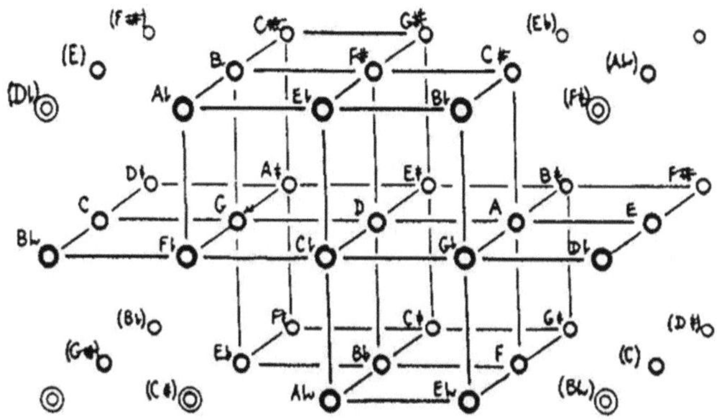

The problem of putting these historical and systematic efforts into practice reduces to being able to have all the notes available at one time. "For an organ there is no specific difficulty in providing more pipes. The valves can be opened and shut by electropneumatic devices. The problem lies in the construction of the keyboard, and in the tuning of the pipes."

Fokker's solution to the keyboard problem was to have two manuals, the second being placed behind the first and joining it at the same level, but rising at an increased inclination steeper than the first. The keys were arranged like tiles on a roof, each key midway between a lower and a higher pair of keys. The keys in a horizontal row were arranged to sound notes with distances of whole tones between them. The keys were colored to facilitate orientation. The lettered notes A, B, C, D, E, F, G were white in conformity with an ordinary keyboard. The black keys of the ordinary keyboard appeared in pairs; thus C♯ and D♭, D♯ and E♭, etc. The other notes had blue keys and their arrangement was such as to have every white key lie in a straight rising tier between two blue keys, etc. In toto there were 7 white keys, 14 blue keys, and 10 black keys, or 31 keys per octave. To facilitate playing, every note had two keys arranged so that the little finger and the thumb would reach the keys they needed in one range, and the three longer fingers could reach the keys they needed in a higher range. Pairs of keys placed in a similar

[8] Fokker, "Equal Temperament," *loc. cit.* pp. 164–165.

relative position provided the same interval, so that it was not necessary to change the fingerings when playing in different scales or melodies at different pitches. Transposition was accomplished simply by displacing the hand in uniform translation. Apart from the straight horizontal rows and the upward rising tiers, the keyboard had skewed tiers such that rising to the right provided ascending major semitones and descending to the right provided minor semitones. The tuning of the organ pipes was made by adjusting the pipes until the resulting Lissajous figures showed the required number of beats.[9]

References

Fokker, Adriaan. 1949. *Just intonation and the combination of harmonic diatonic melodic groups*, 1–50. The Hague: M. Nijhoff.

Fokker, Adriaan. 1953. The Qualities of the equal temperament by 31 fifths of a tone in the octave. *International Society for Musical Research*. Fifth Congress Report, 191–192. Amsterdam.

Fokker, Adriaan. 1955. Equal temperament and the thirty-one-keyed organ. *Scientific Monthly* 83. Washington, DC: Published by the American Association for the Advancement of Science.

[9] In 1855 the French physicist J.A. Lissajous (1822–1880) invented a way to study acoustical vibrations by reflecting a light beam from the vibrating object onto a screen thus to provide visual demonstration of the wave form and obtain precise tuning of musical instruments without using the ear. In *Tonempfindungen*, in the section on musical tones of bowed instruments, Helmholtz gave a detailed description of the use of Lissajous figures to calculate the whole motion of a string and the intensity of its upper partials.

Chapter 25
Confronting Developments in Contemporary Music

An overall review of the state of affairs in musical acoustics at the Teyler in the early 1960s leads one to conclude that the approach to music theory and instrumentation that Fokker had taken since 1940 – two decades of acoustical experimentation and publication – was very much on the periphery to what had transpired elsewhere in music theory and composition in the world of contemporary music. By that time, when he was in his mid-1970s, Fokker nevertheless had established himself within a subset of theoreticians and composers as a music theorist and preeminent spokesman for the 31-tone octave system of Huygens. Organists, in general, welcomed the opportunity to engage in experimental ventures connected with instruments especially constructed to avoid the intonational difficulties inherent to the traditional organ. Above all it was the organists who were alert to the acoustically purer temperaments that composers had begun to make available in their compositions – mostly but not exclusively by Dutch composers. Additional support for the Fokker enterprise came from composers who sought in their music to expand the essence of what they had to express by making use of the microtonal techniques to be found in non-Western music or in the works of modern composers such as Bartok who had left hints about the use of microtones in his music and in his writings that many had noted but mostly ignored. As the reading of Fokker's various writings shows, he often invoked historical precedent to support the use of microtones or non-traditional interval ratios such as the seventh, eleventh, and the thirteenth harmonics.

By 1962 it had become evident that Fokker was prepared, or so he sensed, to confront his critics and go on the offensive about the programme that he and the supporters of the 31-tone system had put in place at both the theoretical and the practical levels. The platform of discussion on which new musical compositions and new musical trends were being debated at the time, and of which Fokker availed himself in 1962, was the publication *die Reihe*, a series devoted to developments in contemporary music, edited by Herbert Eimert and Karlheinz Stockhausen, and

E. Hiebert, *The Helmholtz Legacy in Physiological Acoustics*, Archimedes 39, 241
DOI 10.1007/978-3-319-06602-8_25, © Springer International Publishing Switzerland 2014

published at irregular intervals in Vienna between 1955 and 1962.[1] The periodical's title derives from the notion of the "row" or "series" and reflects the intent of the editors to signalize "serial music" which at the time referred to the 12-tone technique of Arnold Schoenberg and his followers. Although Schoenberg was not directly connected with the particular group of avant-garde composers who looked upon *die Reihe* in Vienna as the most important platform for disseminating their views and offering critical commentary on the new music being composed, he nevertheless indirectly, as the fountainhead of the notion of "row" and the "series," often was at the focus of the discussions in *die Reihe*.[2]

In 1902, in the issue devoted to the theme "Retrospective" (*Rückblick*), Fokker published an article entitled "Wherefore and Why? Questions relating to new music." In this article he took it upon himself as a physicist "to offer a critical commentary to what had appeared on the topic of musical craftsmanship (*Musikalische Handwerke*) in the third, 1957, issue of *die Reihe*."[3]

Fokker's "Wherefore and Why" paper begins with a number of provocative assertions.

> Throughout the ages styles are constantly changing. Whenever a certain style is established by a generation of brilliant artists, and their epigones have exhausted all its possibilities, a certain dissatisfaction begins to grow. Tradition degenerates into a mere form which can no longer inspire the younger generation nor offer them the means of realising their full powers and intentions. They begin to doubt, and with their minds in a ferment of criticism,

---

[1] The individual volumes of *die Reihe* were devoted to Elektronische Musik 1955), Anton Webern (1955), Musikalische Handwerke (1957), Junge Komponisten (1958), Bericht-Analyse (1957), Sprache und Musik (1958) Form-Raum (1960), and Rückblicke (1962). An English edition of *die Reihe* was published between 1957 and 1968.

[2] The German composer Herbert Eimert (1897–1972), director of the *Studio for Electronic Music* (1951–1962) in Cologne, was a pioneer in the development of 12-tone music, for which he gave the first systematic description in 1923 (*Atonale Musiklehre*). The German composer Karlheinz Stockhausen (1928–2007) studied at the *Cologne Musikhochschule* but received his decisive stimulus for composing when, after study with Oliver Messiaen (1908–1992) at Darmstadt in 1951, he saw the possibilities of using long range serial processes. In his own teaching at Darmstadt, where he first lectured in 1953, he pursued the ramifications of serial instrumental music to denote an ordered succession of elements that are used as basic material in a musical composition. The *Internaionales Musikinstitut Darmstadt*, founded in 1946, attracted avant-garde composers such as Messiaen, Stockhausern, Boulez, Nono, Berio, Pousseur, and Cage. Proceedings of the courses were published annually as *Darmstädter Beiträge zu Neuen Musik* (1958-). The Austro-Hungarian composer (American citizen after 1941) Arnold Schoenberg (1874–1951), in his search for a personal approach to music, abandoned the traditional concept of tonality in 1907. From his position there evolved what came to be known as an "atonal" style and "twelve-tone" or "serial music." Schoenberg was an influential teacher of unusual authority and integrity who above all promoted private performance in music. As music theoretician he set the stage for a trend in the composition of music that has not yet been exhausted.

[3] The authors represented in the musical craftsmanship volume were the series editors, Eimert and Stockhausen, the American composer John Cage (b. 1912) who pioneered in the use of electronic devices, prepared piano, percussion ensembles, indeterminacy (aleatorics), and music using natural resources; and the Belgian composer and music theorist Henri Pousseur (b. 1925) who was an avantgardist in the Darmstadt and Cologne groups that cultivated stylistic heterogeneity.

> the young avant-garde elevate their doubt into a principle. All tradition is suspect and must be rejected, and at the end they are suspicious even of their own judgement. A few, however, sometimes endeavour to seek new points from which to begin afresh by appealing to supra-personal achievements in the arts, which are now being regarded almost as scientific disciplines.[4]

The "questions and observations" that Fokker's article calls to attention bear on what it means to speak of a science-motivated "musical craft – a craft that draws on scientific principles to generate music." For Fokker, like music, "physics also is an art and one that is struggling with immensely great problems (encountered) in the face of recent discoveries." Fokker approaches those great problems by directing his attention to the ways in which avant-garde composers had appropriated and misappropriated to themselves the vocabulary that had been formulated and shaped, historically, to deal with the natural sciences and mostly with physics. He singles no one out by name, but it is clear from the nature and context of his critique that he has in mind Stockhausen's article on "*wie die Zeit vergeht*" (how the time passes). Fokker, in fact, is displeased with the imprecise manner in which terms such as chance, phase, frequency, acuity of pitch, duration, and formants (a group of sound waves special to a vowel sound) are used to designate a method of composition in avant-garde music. In physics chance has a precise meaning. In classical mechanics, for example, chance is used to explain certain laws as the effect of randomness in the case of a very large number of incidents. In music it seems to have been disposed of by some kind of rules of gambling followed by transferring the outcome to the free determination of tempo and dynamics. "The composer withdraws and leaves his work to chance. It seems he has no faith in himself being able to say anything on his own account. One may believe that this randomness appears to result from the composer's distrust of himself.... Whoever elevates randomness to a first principle contradicts all human aspirations and efforts. Liberated chance breaking away is a King of Chaos. Art has its casual fortuities. Casual fortuities never create art."[5]

A discussion on the geometrical (tempered) duodecimal (12-tone) series brings us closer to problems Fokker was dealing with in his own work on musical acoustics, and it is on this score that Fokker refers critically to an author who "seems to be greatly annoyed by the dominating integer relations in the structures of durations, such as do not occur in the scale based on the root of 2 (i.e., on twelve successive semitones of equal size)." The author calls it a contradiction. "Of course it is evident that he is not responsible for this discrepancy. The sophisticated scale is to blame." Schoenberg, as Fokker points out, knew better, for in his *Harmonielehre* he referred to the geometrical tempered system as "a truce, concluded for an indeterminate time" that nevertheless will "be absorbed in a higher order."[6]

---

[4] Adriaan Fokker, "Wherefore and Why? Questions relating to music," *die Reihe*, *8*, (1962) 68–82. Quotation p. 68.

[5] Fokker, "Wherefore and Why?" pp. 68–69.

[6] Fokker, "Wherefore and Why?" (1962), p. 76.

Fokker took Schoenberg's reference to "a higher order" as support for his own efforts to move in the direction of a 31-tone system. Schoenberg had called the tempered duodecimal system "a truce [*Waffenstillstand*] concluded for an indeterminate time." It would, as Schoenberg had remarked, "be absorbed in a higher order." Schoenberg also had talked about a compromise between the natural intervals "and our incapacity to handle them" on the existent instruments.[7]

For Fokker the time had come to move on to that "higher order" that Schoenberg referred to in *Harmonielehre*. Fokker wrote:

> Today our ability to handle the natural intervals is much greater than in Schönberg's time. The time has now come to stop making do with a temporary device, and to break the constraint of the compromise. But instead we see that the propagators of "new musics" proclaim as a foundation the merely provisional means-to-an-end. One acquiesces with inadequacy, while doing away with things handed down. One brands harmonic relations as anti-stylistic, but retains a musical monstrosity, to which in practice nobody adheres, neither vocalists, nor string-players, nor wind-players. How is it possible to be satisfied by the platitudes of this characterless series, if one is in the forefront of the search for new values?[8]

In the meantime, as Fokker put it, while pianos and organs by force of historical circumstance are tempered to the powers of the twelfth root of 2, "artists with their lips shift the tone a trifle sharp or a trifle flat – often without being conscious of

---

[7] The full text of Schoenberg's remarks on temperament that Fokker drew on follows:

> The discovery of our scale was a stroke of luck in the development of our music, not only with regard to its success, but also in the sense that we could just as well have found a different scale, as did for example the Arabs, the Chinese and Japanese, or the gypsies. That their music has not evolved to such heights as ours does not necessarily follow from their imperfect scales, but can also have to do with their imperfect instruments or with some other circumstance which cannot be investigated here. Moreover, it is not to our scale alone that we owe the evolution of our music. And above all; this scale is not the last word, the ultimate goal of music, but rather a provisional stopping place. The overtone series, which led the ear to it, still contains many problems that will have to be faced. And if for the time being we still manage to escape those problems, it is due to little else than a compromise between the natural intervals and our inability to use them – that compromise which we call the tempered system, which amounts to an indefinitely extended truce. This reduction of the natural relations to manageable ones cannot permanently impede the evolution of music; and the ear will have to attach the problems, *because it is so disposed*. Then our scale will be transformed into a higher order....Whether there will then be quarter tones, eighth, third, or (as Busoni thinks) sixth tones, or whether we will move directly to a 53-tone scale. . . we cannot foretell. perhaps this new division of the octave will even be untempered and will not have much left over in common with our scale. However that may be, attempts to compose in quarter or third tones, as are being undertaken here and there, seem senseless, as long as there are too few instruments available that can play them. Probably, whenever the ear and imagination have matured enough for such music, the scale and the instruments will all at once be available. It is certain that this movement is now afoot, certain that it will lead to something.

Arnold Schoenberg. The Theory of Harmony, translated from the 3rd German ed. of 1922 by Troy E. Carter, London, 1978, pp. 25–26, pp. 22–23 in *Harmonielehre*, Vienna, 1922.

[8] Fokker, "Wherefore and Why?", p. 76.

doing so – by a conditioned feedback." Elsewhere "Turks and Arabs have an ear far more attuned than ours to the refined melodic steps in their monophony. At best we bother about semitones; they sing their melodies with commatic shades." In view of the entrenchment of fixed-tone keyboards and equal temperament in Western music, Fokker was alert to acknowledging that there is no conceivable way to move all of music in another direction. *But* is there not a glimmer of visions toward other modes of musical expression? "Can the clumsy coarseness of our system offer a foundation for further development? Who is able to believe, seriously, that the inadequacy pointed out by Schönberg must be preserved *under all circumstances* [emphasis added]?"[9]

To demonstrate the possibility of constructing functional 12-tone series that are not made up of semitones, Fokker alluded to "gamelan slendroh," a type of instrumental music with supraseconds practiced by the Javanese.[10] The particular gamelan system described by Fokker is made up of suprasecond steps corresponding to the harmonic interval 8/7. Since the fifth contains three supraseconds (in the traditional equal temperament system) one can take 11 of these steps and use the resulting 12 notes to generate a scale system that is mastered "without difficulty" with the 13-*diësis* system of Huygens.[11] According to Fokker another way to free oneself from the chains of tempered semitones is to construct chords by addition as in the Fibonacci series. In these chords the frequency of one note is the sum of the frequencies of the next lower pair of notes and the difference of frequencies of two neighboring notes is the frequency of the next lower note. "The physiological constitution of the ear is such that hearing two notes it is apt to hear a third note, with a frequency equal to the difference of two actual frequencies." The series in which each member is given by the sum of the two preceding numbers is well established, as in 0-1-1-2-3-5-8-13-, 1-3-4-7-11-18-29-, 2-5-7-12-19-31-, and is known as the Fibonacci series. The chords have a special sensation quality.[12]

It is evident from Fokker's concluding remarks that he felt indebted to the editors of *die Reihe* for giving him the opportunity to express his views on a subject that he had come to explore only after he had established himself, professionally speaking, in a career in theoretical physics.

> [My] observations have been critical but I must emphasize that I want to support these endeavours at innovation in a constructive mood. I am not too old to understand that there are great problems which have to be tackled in the face of new possibilities. If someone attempts to present a scientific account of his aims and methods, we have to

[9] Fokker, "Wherefore and Why?" (1962), p. 76.

[10] In gamelan slendroh the pitch and interval structures usually differ from octave to octave and from one gamelan to another.

[11] Fokker, "Wherefore and Why?" (1962), pp. 76–78.

[12] Fokker, "Wherefore and Why?" (1962), p. 78. The discovery of the Fibonacci series is attributed to the Italian mathematician Leonard Fibonacci (ca. 1170-after 1240) who solved the rabbit problem. It assumes that a pair of rabbits requires 1 month to mature and thereafter reproduce itself once each month. If one starts with a single pair, how many pairs will one have after *n* months? The answer leads to the famous Fibonacci series.

> consider his endeavour with sympathy and with a realisation of its promise. But we ought not to invent scientific terms with other meanings.... Pseudo-science is no better than pseudo-music. Where I have been denouncing failures these denouncements belong to the feedback (in the sense of cybernetics) which have to be activated whenever a great task is attempted.[13]

The task of responding to Fokker's "Wherefore and Why?" essay was taken on by the German composer Gottfried Michael Koenig (b. 1926). At the time he was teaching music technology at the Cologne *Musikhochschule* where he was an assistant to Stockhausen and others. In 1964, shortly after publication of his *die Reihe* commentary on Fokker, he became director of the electronic music studio at the University of Utrecht. There his compositional approach moved from serialism and electronic music to a style in program music in which the musical raw material is generated electronically before being used in more traditional compositional processes.

The context in which Koenig was enticed to write a commentary to Fokker's article apparently had been considerable confusion about the meaning and use of the serial method. Fokker's essay was taken by the serialists to represent additional confusion. Koenig as spokesman for Stockhausen decided in his commentary to provide clarification. He wrote:

> "The fact that an article [by Stockhausen] clarifying what was previously not clear meets opposition shows how few [and according to Koenig Fokker belongs to the few] understood this, although everyone composes serially." Koenig emphasized that Stockhausen's composing is "not theoretical in an analytical way, but more programmatic; yet it is analytical inasmuch as music, having once taken the first big step into the electronic terra nova, has not been able to do entirely without instrumental practices." The critics tend to confuse the two fields. If the composer is suspected of electronic interests or has realized them in an electronic studio, they accuse the composer of merely transferring electronic conceptions to instruments for which they are not appropriate. "The programme is not just wishful thinking, but is illustrated with palpable musical facts and in detail. Perhaps it is not superfluous to go over it again here – also in connection with the objections of Adriaan D. Fokker, the eminent Dutch physicist."[14]

The starting point of Stockhausen's conception is a 12-note all-interval row. Fokker had referred to the geometrical equal temperament series as "makeshift" and had argued that one is far more capable of using the natural intervals today than in Schoenberg's time. According to Koenig, "that is true [only] inasmuch as the continuum of pitches... were not unlimitedly at our disposal until electronic music." If this were the case, counters Koenig, the impression would arise all too easily that the pure intervals represent an ideal; if this were so it would not have been necessary to introduce temperament 250 years ago. The argument of Koenig is that even if the tempered system is not ideal, a musical language had been established at the time, and nothing was of more consequence than to adapt the regular divisions of the octave to this system. "Which system was better or worse

[13] Fokker, "Wherefore and Why?" (1962), p. 79.

[14] Koenig, "*Kommentär*," *die Reihe* (*Rückblick*), *8* (1962); translation in "Commentary," *die Reihe* (Retrospective), *8*, 1968, p. 80.

cannot be decided outside the music itself. When nowadays various theorists invent scales which are supposed to overcome the alleged imperfections of the present system, these scales are not any the more useful because they have been constructed outside of practice." According to Koenig, theorists also make the same claim as that imputed to the 12-tone system: that of being ideal, and Schoenberg suspected this because scales which deviate from the 12-tone temperament are usually corrected by the ear, especially steps closely resembling the octave.

In defense of Schoenberg's position on temperament from Koenig'z point of view, Koenig wrote:

> Even if Schoenberg called the tempered system a truce that had been concluded for an indefinite time, this time is now up. Most orchestral instruments are gauged to this temperament, and this will not change in the near future. On the other hand [and here Koenig shifts the playing field in an entirely new direction], every conceivable scale can be executed in electronic music. It is becoming apparent that the electronic studio has an advantage over other instruments in that any number of scales can be used instead of just one.[15]

The manner in which Koenig has expressed the temperament issue in this counterargument – and done so against the background of Schoenberg's views on temperament and Stockhausen's serialism – seems on the one hand to condone what Fokker is attempting to achieve by way of acoustically pure intervals. On the other hand he recognizes that if musicians in fact should want to move in the direction of a greater degree of purity, this can be accomplished electronically without having to resort to the 31-system complications presented by Fokker.

The frontier on which Koenig made his claims is basically that there is much more at stake in the creation of new music than temperament. It is an *idée fixe* that Fokker appears to be hung up on. According to Koenig, flexibility in compositional techniques does not demand more steps per octave and more developed mechanical instruments that are able to realize those steps, but variability of the scales. "It does not matter how one interprets the relationship of pure intervals to equal temperament; musical development has proceeded without it." Criticism about equal temperament perhaps could be justified if pure intervals were the only alternative; but then only in order to be argued from a theoretical point of view. "Now that it is possible to make music in any scale whatever (or without scales) in electronic music.... the scale becomes a historical reminiscence.... [and] one must realize that the twelve-tone temperament is perhaps not ideal but most usable." Rather, instrument builders should be asked to build instruments with variable scales. This would be of direct benefit to the music of the past, for Koenig felt that it is wrong to perform a piece of music, vocal or instrumental, that owes its construction to the possibilities of temperament. This includes practically every important musical instrument since 1700 – all built on intervals that are pure, and not tempered. This is "wrong because... performance [on such instruments] wantonly violates the elementary demands for justice's being done to the work. To a musically educated ear (educated that is to tempered music) every 'wrong' interval (for example a pure one where a tempered one was composed) is painful."[16]

---

[15] Koenig, "Commentary," p. 89.

[16] Koenig, "Commentary," p. 90.

To support his argument that musicians tend to play "wrong notes that are painful" Koenig mentioned string players who are especially "backwards in this respect." This, as Koenig notes, is a position that is strongly supported by the American violinist of Austrian birth Rudolf Kolisch, who wrote in 1958 that the instruments of string players "are the only ones still used in regular music-making that have not essentially changed since their construction nearly four hundred years ago. They have been spared the process of chromaticism which other instruments – such as keyboard and wind instruments – have experienced in order to have become serviceable for tonal music." Actually Kolisch states his case much more forcefully. He maintains that with the progression in western polyphony string players had so distanced themselves from the general state of music as to have created an essential chasm between themselves and other musicians. He states: "In a centuries-old reserve area of reactionary ideology they remain in arrears of... music practice of other instrumental groups and firmly resist all attempts to adjust their techniques to the actual state of music. They defend their anachronistic positions with the fanaticism of a religious sect and as such consider themselves to be the chosen keepers of the only true doctrine." The extreme privileged position of string players (*extreme Sonderstellung der Streicher*) therefore derives from the fact that their instruments are several centuries older than the other musical instruments in use.[17]

In response to those who comment that it "doesn't sound right" or it "doesn't sound beautiful" when piano and strings play together in equal temperament, Kolisch lashed out with invective, cynicism, and disrespect:

> Certainly this combination does not sound right if both parties use a different system of intonation and in doing so infringe in an uncompromising way on one of the most primitive conditions of concord, namely congruence in pitch. The truly initiated indulge in orgies among themselves to celebrate the false "pure" intonation and make listening to a string quartet from the Viennese classical school a painful experience. They defend their position beyond current music not only with the necessity of having to rescue the forever lost values by means of "rationalization" but then call upon nature, namely the nature of tones, the last instance of misunderstood "pure" intonation; driven into a corner by rational objection their posture becomes ever more mystical and finally flows into the harmony of the spheres.[18]

Kolisch indicated that the music of Beethoven and other composers in the Viennese school of music already required a wealth of dynamic and expressive categories for which technical correlates had to be invented. And it was during that time that

[17] Rudolf Kolisch, "Ueber die Krise der Streicher," *Darmstädter Beiträge zur neuen Musik*, Mainz, 1958, 00. 84–90. Quotation on p. 90 in Koenig, on p. 84 in Kolisch. Rudolf Kolisch (1896–1978), who studied with Schoenberg (and later became his brother-in-law), is best known as founder of the Kolisch quartet that settled in the USA in 1935. The quartet gave performances of modern works, notably music of the Second Viennese School (whose leading figures were Schoenberg, Berg, and Webern), and of Bartók. Kolisch insisted on having string groups play in equal temperament in order, as he remarked, to avoid a free-for-all that leads to chaos. This was the experience of Elfrieda Hiebert who in the mid-1960s was pianist in a string trio coached by Kolisch at the university in Madison, Wisconsin.

[18] Kolisch, "Ueber die Krise," (1958), p. 85.

traditional music pedagogy broke down with the result that the priests of the religion of pure intonation placed the ideal of the beautiful tone [*des schönen Tons*] on their altar. "In so far as no piece of art music ever was written for beautiful tones the expression itself is unmusical. Its exclusive cultivation substitutes a subjective-sensualistic, musically indifferent moment for all dynamic-expressive ones of the language of music. Under its aegis reproduction is effected beyond its actual task, that of unveiling construction, merely as cult of beauty."[19]

In summarizing the differences in perspective of Stockhausen and Fokker on the question of substituting the geometric series (Stockhausen's position) for the interval series or arithmetic series (Fokker's position), Koenig claimed that what is at stake is what has "real musical meaning" because it is not numbers that are being manipulated, but "musical events.... so that when studying theoretical articles, which the composers themselves must provide today, one must clearly distinguish between musical facts and resources of presentation frequently borrowed from mathematics." Composers of serial music, however, are not mathematicians or arithmeticians; in the case of electronic music they are "really technicians or sound engineers."[20] In conclusion: a few words about th confusion of "music" and "non-musical" assistance as viewed by Koenig:

> Anyone in need of a plumber is also happy if the latter can also get a bent lock open. But woe betide to the composer who knows how to work with logarithm tables, or who puts his nose in a book in order to find out about the physiological processes of hearing. He who has contact with music has his own ideas, however they may be about it; it is not always his fault if they are wrong.... The confusion of music with non-musical assistance keeps on penetrating the music itself.... This is confirmed by the fact that the article in question [the Fokker article] contains hardly any references to composing; how it is done is not given away technologically. The question as to musical need is misplaced with regard to the article; this need is not actually presupposed and neither does it appear at the end as something suppositious after it has already been fulfilled. The material, as Stockhausen explains it, is ambiguous. Without it, not a single tone of his music is achieved, and yet his music seems to float above it. Music could be said to be the practical superstructure of theory. But this would cede to the latter a priority which it does not possess. By penetrating the form, what is heard inwardly assumes something of its constraint. It easily degenerates to an agreeable arrangement, to applied art. It can only be otherwise by dint of self-forgetting devotion; the intention sinks into it with its entire weight. Finally it breaks through: the form is split open like the shell of a nut which has ripened into its form.[21]

Reference

Fokker, Adriaan. 1962. Wherefore and why? Questions relating to music. *die Reihe* 8: 68–82.

[19] Koenig, "Commentary," p. 90; Kolisch, "Ueber die Krise," (1958), p. 86.

[20] Koenig, "Commentary," pp. 97–98.

[21] Koenig, "Commentary," pp. 97–98.

Chapter 26
Refinement of Pitch

The problem of wanting to, or being able to, put the notion of acoustical preference or acoustical purity to rest, either by being content to use comprehensive equal temperament in all instrumental ensembles, as Kolisch and his fellow-serialists would have had it, or by adopting the Fokker-supported 31-tone per octave Huygenian system, was not destined to be settled. Theoretical discussions on pitch and its refinements have been a perennial matter for discussion since antiquity. The never-ending conundrum of the search for an answer to the problem of pitch and its refinement was brought to Fokker's attention by Stravinsky's response to a query from his friend and collaborator Robert Craft.[1] The question that Mr. Craft raised was: "Is any musical element still susceptible to radical exploitation and development?" Stravinsky's answer: "Yes, pitch."[2]

Modern composers have voiced similar sentiments about pitch. Bëla Bartók, for example, stated in 1931: "In our [Hungarian] folk melodies. . .the seventh appears as an interval of equal importance with the third and fifth. . ..[and] we often heard those intervals as of equal value in succession....We sounded the four notes together. . .the four notes were made to form a concord." As Fokker was eager to point out, although the four-note primary concord called for by Bartók is inhibited in duodecimal equal temperament it is not out of place in a uniform division of the octave that gives a true reproduction of Bartók's primary concords. On this model "there is no difficulty in singing where the range of pitch is continuous. The puzzle arises when one has to use instruments with a finite number of fixed notes." The Stravinsky – Bartók point, as Fokker saw it, had been solved three centuries earlier by Huygens when he discovered that a perfect fifth is near 18/31 of an octave, the major third is near 10/31 of an octave, and the seventh is near 25/31 of an octave. So he divided the octave into 31 units (called dièses), each dièsis, therefore, being a fifth of a tone. The scale of 31 dièses allows for the realization of

[1] The American composer and writer Robert Craft (b. 1923) has written several volumes on conversations with Stravinsky (1882–1971).

[2] Adriaan Fokker, "Refinement of Pitch," *Organ Institute Quarterly*, *10* (1963), pp. 13–14.

E. Hiebert, *The Helmholtz Legacy in Physiological Acoustics*, Archimedes 39,
DOI 10.1007/978-3-319-06602-8_26, © Springer International Publishing Switzerland 2014

Bartók's primary four-tone concord. Notations of even finer grain are made with sharps, flats, half-flats, one and a half flats, half-sharps, and one and a half sharps.

In a short note regarding the Organ Institute in Andover, Massachusetts, Fokker mentioned that the 31-keyed organ that he had commissioned was placed in the Teyler Museum in Haarlem in 1950; that around this instrument a musical repertoire was being built for organ and wind instruments by Dutch and foreign composers, and that it was time to establish international conventions of notation for microtonal music. He acknowledged that the 31-keyed system would not be the last step in the improvement of equal temperament. "Now and here is not the place for a comprehensive discussion and comparison of various divisions. Much has to be left to the future. We have to meet, and we can meet, the need of the present." This theme, the open-endedness of the division of the octave – and the dominance of certain intervals – is a theme that runs throughout Fokker's writings. History is being invoked to show that music is and always will be on the move; for example, that the musical structure that had been codified in the sixteenth century by the then-novel introduction of the major third had led to the acceptance of a 12-tone scale. To cope with the perfect harmonic seventh as a new element in musical structure, Fokker maintained, "the tricesimoprimal [31] temperament proves to be the straightforward alternate structure." Stravinsky had it right when he stated that the musical element of pitch "still [is] susceptible to radical exploitation and development."[3]

Reference

Fokker, Adriaan. 1963. Refinement of pitch. *Organ Institute Quarterly* 10: 13–14.

[3] Fokker, "Refinement of Pitch," (1963), p. 14.

Chapter 27
Neue Musik mit 31 Tönen

On commission of the Society for the Advancement of Systematic Musicology, Fokker in 1966 published a volume in the society's series on fundamental questions in music. In the foreword to the volume, entitled *Neue Musik mit 31 Tönen*, Martin Vogel, professor of musicology at Bonn University, explained why Fokker's approach to just intonation deserved to be taken into account, not by those who want to predict the direction in which the development of music may or should go (which in any case is not the concern of musicology) but by those whose task it is to explore and, if the occasion arises, to provide necessary and essential music materials.[1] As Vogel points out, the ideas that Fokker treats in his volume on new music with 31 tones arose within the complex of problems in which persons of Fokker's generation were on the lookout for an appropriate form of expressing music in the modern world. They observed that the ideas of pure thirds, pure sevenths, and the intervallic relationships of prime numbers simply arise, but how the ear relates to the higher prime numbers hardly had been explored, even though it is known that the pure thirds and pure sevenths that stand behind the prime number 5 and 7 are in fact grasped by the ear and sound "beautiful." Fokker became convinced after reading Huygens that the exploration of these options would bring a genuine enharmonic and provide the composer with new sounds, cadenzas, and methods of modulation notably in seventh-related tonality.[2]

When keyboard instruments were tuned to the third of a triad, as Vogel remarks, the thirds became the problem. With the rise of equal temperament the thirds

[1] Martin Vogel (1923-) studied philosophy and psychology. He acquired his doctorate in 1954 with a dissertation on "the number seven in speculative music theory," and completed his *Habilitation* in musicology in 1959 with a work on the Greek concept of enharmonics. He has written extensively on fundamental questions in ancient and contemporary music theory, and on musical acoustics, instrumental techniques, tonal systems, tuning, electronic keyboards, and the automatic coupling of fifths, thirds, and sevenths.

[2] Enharmonic is taken here to represent keyboard instruments that have separate keys, strings, and pipes for different enharmonic tones; that is, they are tones that actually are one and the same degree of the chromatic scale but written differently, e.g. g♯ and a♭.

E. Hiebert, *The Helmholtz Legacy in Physiological Acoustics*, Archimedes 39,
DOI 10.1007/978-3-319-06602-8_27, © Springer International Publishing Switzerland 2014

largely were relinquished and tuning in fifths was reintroduced. As for the pure seventh, its applications still were not forthcoming although it had been known for more than 300 years that nothing stands in the way of their use from the standpoint of the ear. The group of seventh intervals lies on the border between consonance and dissonance. The Fokker tonal system introduces pure thirds and pure sevenths. With its 31 steps to the octave it has twice as many tones as are at the disposal of today's keyboard instruments. The musical layman undoubtedly will assume that it is impossible to find one's way among such a large number of tones. According to Vogel:

> Our friends from the Netherlands were able to come to terms with this system and the here named composers and musicians have demonstrated that the 12-tone system can be breached insofar as one is willing to strive for a higher degree of accuracy in making music. In the Netherlands it notably became possible to establish a precision-listening foundation [*Nauluisterenheid Stiftung*] in whose name already is indicated where their goals reside.

What this trend of developments in the Netherlands reveals, as Vogel evaluated it, was that it would be possible at the time of his writing (1966) to realize without any difficulty fifths, thirds, and sevenths in choral singing and in string and brass ensembles provided the players themselves want to undertake music-making in as pure a fashion as possible.

> What has paved the way in the Netherlands essentially is the work of a single man, in fact an outsider, the physicist Adriaan Daniel Fokker. What has been achieved here we owe to his circumspection, enterprise, and his undauntedness.... For more than twenty years he has stood up for the 31-step tonal system. We have invited him to put his personal and most pertinent experiences into print, and we are delighted that he has responded to our request.[3]

Fokker's treatise on *New Music* begins with information about the Haarlem branch of the *Dutch Society of Scientists* founded during the enlightenment. One of the Society's most important achievements was to oversee publication of the *Nachlass* of Christiaan Huygens, the largest portion of which was held by the university library in Leiden. The project, begun in 1885, was completed in 1950 with the publication of its 22nd volume. Volume 20, published in 1940, contains Huygens's writings on music theory. As in several other works of Fokker, *New Music* provided the essential elements of Fokker's explication of Huygens's 31-tone per octave system. However, on this occasion he was being asked to furnish a conspectus not only of how and to what effect he had come to focus on the Huygens system and the construction of special fixed-tone keyboards but how these activities and investigations, and his writings on the subject, had been and were being supported by the Teyler Foundation and the ongoing experimental investigations at the Teyler Museum in Haarlem. Fokker therefore decided to partition his

[3] Martin Vogel, "Vorwort des Harausgebers," Adriaan Fokker, *Neue Musik mit 31 Tönen*, Düsseldorf, 1966.

treatise with an "anecdotal historical part" that draws together his own ideas and writings and those of his colleagues at the Teyler Museum, and a "systematic part" that describes the layout and functioning of the constructed keyboards and other just intonation instruments and the musical compositions expressly written for them. The anecdotal-historical part of the treatise is of main interest to us here since it provides information about Fokker's overall work on musical acoustics at the Teyler Museum that is nowhere else available. It is evident that *New Music* was published with the intent of informing the larger community of German music theorists about the experimental ventures in instrument construction and in musical compositions that had their origin in the activities organized by Fokker at the Teyler Foundation since 1940.

In his own mathematical and music-theoretical reasoning, Fokker shows how Huygens had clinched the argument that the major advantage of the 31-tone system over the equal temperament system is that it affords unlimited modulation in all tonalities without any major mistuned chords. The thirds and sevenths are almost pure, and the fifths are too flat by only 5 cents, that is, by a five-hundredth part of a semitone, to quote Huygens directly: "We have here a system that in the same way includes all consonances and intervals for every pitch – every whole tone, every semitone, and every diësis going upwards or downwards." From this it follows that the scales are homogeneous.

> In this system the major semitone is 3/5ths of a whole tone and the minor semitone is the remaining 2/5ths. [The system] finally represents a complete harmonic circle in which the interval of the fifth or some other interval continually ascending and descending returns to the starting point after a certain number of revolutions.... I may mention here that the interval of the tritone is everywhere included in the relation of 7 to 5; only 1/12th of a comma is lacking... [and] I now discover that this interval 7 to 5, if examined attentively, has something harmonic to it [*etwas harmonisches hat*] – so at least I sense it with my ear; and [I feel] that whatever the composers who count it among the false tonal relations might say, one could count it among the consonances.[4]

With these concise and carefully weighed words from Huygens's *Nachlass*, written toward the end of the seventeenth century but not published until 1940, the advantages of the 31-tone scale were given decisive expression: uniform homogeneity and cyclical compactness (*Geschlossenheit*); distinction between major and minor semitone (their difference, the diësis, is the smallest interval of the 31-tone classification); pure thirds and harmonic sevenths. It was against this Huygenian background of mathematically reasoned guidelines that Fokker first undertook an in-depth investigation of the published works on music by the polymath Christiaan Huygens, and then sought to discover the contexts in which various writers and composers of the eighteenth and nineteenth centuries had written about the harmonic merits of the seventh and their reluctance to use the seventh because of being unable to find its proper place in music.[5]

---

[4] Fokker, *Neue Musik* (1966), pp. 12–13.

[5] Fokker, *Neue Musik* (1966), pp. 13–15.

With his original focus so selectively on the work of Huygens, Fokker nevertheless realized that much of what he had to say about just intonation was the more readily to be understood by consulting theorists before the time of Huygens and others who came later but who were not aware of Huygens's contributions to music. Fokker came to the study of these early music theorists during the war. "During the German occupation," wrote Fokker, "I was not able properly to accomplish my profession as a physicist. I turned to the study of problems in music theory. From the music library of the *Vereeniging voor Nederlandsche Musikgeschiedenis* in Amsterdam I was able to borrow various very rare books." Among the works that were examined with special benefit in relation to the works of Huygens was *Le istitutioni harmoniche* (1558) of Zarlino and the eighteenth-century writings of Rameau, Tartini, and Euler. They already have been earmarked elsewhere in these pages, but among these Euler merits special mention in this context because of the Teyler organ that was constructed on his model of his *Genera*.[6] The way in which that came about is explained by Fokker. Shortly before the war a foundation for the study of sound (*Geluidstichting*) was established in the Netherlands. Assigned mainly to the problem of noise abatement, the foundation also supported research in the solution of a number of other acoustical problems. A grant from the foundation was used to build a small organ with ten *Genera* (voices) of the third grade as described in Euler's *Tentamen novae theoriae musicae* of 1739. The organ was built by the brothers Van Leeuwen organ builders in Leiderscorp and installed in the Teyler Museum in Haarlem. It had ten wind chests corresponding to its ten *Generibus*; and each wind chest had 12 labial pipes corresponding the 12 equal white digital notes of the keyboard. By means of knobs each wind chest could be supplied with wind. Three of the *Genera* had only four tones; on the organ they were represented within the compass of three octaves. Six of the *Genera* had six tones and were represented within the compass of two octaves. The tenth *Genus* numbered eight tones and for them only one and a half octaves were available. The pipes were tuned so purely to the vibrational relationships of each *Genus* that the intervals sounded beat-free. All of such ventures, and others still to be examined in the above, were made possible with support from the Teyler Foundation.[7]

The circumstances of the establishment of the Teyler Foundation and the construction of the Teyler Museum are particularized by Fokker in *Neue Musik*. Pieter Teyler's arranged for the establishment of a cultural foundation and two working groups: a theological society (*Teyler's Godgeleerd Genootschap*) and a second society (*Teyler's Tweede Genootschap*) in which drawing, poetry, history, numismatics, physics, and biology were fostered. Both societies sponsored yearly competitive prize-winning essays that were published in the respective journals of

[6] Fokker, *Neue Musik* (1966), pp. 15–18.

[7] Fokker, *Neue Musik* (1966), pp. 17–18.

the societies. In 1943 the administrators of the second society proposed, as musical prize essay, short compositions for the Teyler Museum organ that would be based on the Euler *Genera*. The jury was composed of a professor of musicology at the university in Leiden, a specialist in music of the Middle Ages, the conductor of the *Concertgebouw* orchestra in Amsterdam, and Willem Pijper (1894–1947), the top-ranking composer in the Netherlands at the time. The resulting performances and demonstrations on the Euler organ at the Teyler Museum enticed Fokker, at the request of publishers Noorduyn en Zoon in Gorichem, to write a book about the physical and mathematical foundations of music theory. The prominent Dutch historian of science E. J. Dijksterhuis (1892–1965), who had been in charge of publishing the writings of Huygens and was the author of an important work on Simon Stevin, encouraged Fokker to set himself the task of writing a systematic account of early acoustic theory. Fokker accepted the challenge, received valuable counsel from the main teacher at the Royal Conservatory in the Hague, and followed composer Willem Pijper's advice to furnish an introduction of four historical sketches of Huygens's most important predecessors – Zarlino, Rameu, Tartini, and Euler. Fokker's final chapter also told the story of how, as a physicist, he came to write about music. The book, entitled *Rekenkundige Bespiegeling der Musiek* (*Arithmetic Reflections on Music*) was published in 1944 as a book of 226 pages in the Noorduijn scientific series. The title page indicated that its author was the curator of the physics laboratory at the Teyler Museum in Haarlem.[8]

Published in Dutch during the most difficult days of the war, *Rekenkundige Bespiegeling* reached a small and reserved audience of music theorists. It nevertheless elicited the interest of persons such as L. S. Lloyd and Alexander MacClure, who at the time were themselves involved in working on just intonation and the construction of enharmonic keyboards.[9] In musicological societies in the Netherlands Fokker encountered outright resistance. On suggesting that the society might want, at one of their meetings, to compare sixteenth- and seventeenth-century compositions for a harpsichord tuned according to contemporary methods with a harpsichord tuned according to mean-tone temperament, the suggestion was turned down with the response that the demonstration was too special and would not be of interest. In 1948 on the occasion of a festival in Hilversum, sponsored by the International Society for New Music, a radio performance of music by the seventeenth-century Dutch composer Sweelinck, when played on a harpsichord tuned to modern temperament and then on a harpsichord tuned to mean-tone temperament, produced a very different effect. "The huge difference between the tunings and the strange attraction of the older mean-tone tuning was obvious." Several of the participants at the 1948 festival, among them Alois Hába, came to

[8] The work is examined in the above on pages [***43–45***].

[9] Llewellyn S. Lloyd (1876–1956), British acoustician and enthusiastic promoter of Helmholtz's views in England, wrote on music and sound, the musical ear, the musical scale, and intonation problems connected with keyboard instruments. His article "Helmholtz and the Musical Ear," *Musical Quarterly*, 25 (1939) 161–175 is noteworthy.

Haarlem to inspect the organ at the Teyler Museum.[10] Hába showed great interest in Fokker's instrument and took the occasion to tell him that his quarter tones were not always 50 cents; there were both larger and smaller quarter tones depending on the circumstances. Fokker took Hába's remarks to signify that he was moving in the direction of the practicability of diësen tuning.[11]

With financial support coming from the Dutch society for the sciences in Haarlem, and from the ministry of culture and several other organizations, the time had come to undertake the construction of a large full-sized organ for the Teyler Museum. Fokker drew up plans for a 31-tone keyboard that would allow for a multi-voiced, rapid-functioning style of playing. It was attached on its side to a traditional keyboard of seven white keys and five black keys. By means of a guide track system, 12 of the pipes could be selected from among the 31 pipes and connected with a smaller attached console. By this means the 9 Euler *genera* with its 12 tones could be played on the traditional keyboard using customary fingering technique. In 1950 the organ was completed and installed in the Teyler Museum. it was provided with 648 pipes. The large console had two manuals with knobs for its guide track system.

A perceptive and critical review of Fokker's *Neue Musik* appeared in 1969 in *Die Musikforschung*, a journal of the *Gesellschaft der Musikforschung*, published in collaboration with the Prussian Institute of musical research in Berlin.[12] Its author, the Czech-born German composer and musicologist Erhard Karkoschka, professor in music theory and composition at the *Stuttgart Musikhochschule*, undoubtedly was the ideal person to review Fokker's programme simply because of the depth of his own institution-based involvement in furthering the feverishly pursued "new music" in Germany.[13] Karkoschka's primary research interest was to present musical time as a structural phenomenon that conveys the hearing and understanding of music directly to the listener. His dissertation was on the development of composition in the early works of Anton Webern, and although he never fully accepted the serial techniques of the avant-garde composers of the 1900s in permeated some of his compositions for chamber ensemble using strings, voice, brass, and electronic methods of instrumentation.

[10] The Czech composer Alois Hába (1898–1972), who was influenced by Schoenberg in his development of an athematic highly chromatic scale, taught microtonal compositions at the Prague conservatory and wrote choral and string music using quarter tones and sixth tones.

[11] Fokker, *Neue Musik* (1966), pp. 21–22.

[12] Erhard Karkoschka, "Adriaan D. Fokker: Neue Musik mit 31 Tönen." Düsseldorf, 1966, *Die Musikforschung*, 22 (1969), 115–117.

[13] Erhard Karkoschka (1923–2009) founded the *Ensemble Neue Musik* at the *Hochschule* in Stuttgart and became director of its *Studio für Elektronische Musik*. On the council of the *Institut für Neue Musik und Musikerzeihung* in Darmstadt, he later became president of the *Gesellschaftfür Neue Musik* which served as the Federal German Section of the *International Society for Contemporary Music*. Stefan Fricke, Erhard Karkoschka, *NG 13*, (2001) 318; Christian Martin Schmidt, Erhard Karkoschka, *MGG*, Personenteil *9* (2003), col. 1501–1503.

According to Karkoschka the proponents of the 31-tone system had not quite resolved the problem of how to treat the fifth. In that system the fifth deviates from the pure fifth by twice as much as the fifth deviates from the pure fifth in the 12-tone temperament. The 2 cents in that deviation actually can be detected. This is because the ear controls the fifths more readily than the thirds and the sevenths. Besides, it was seen to be questionable whether tonal music in the 31-tone system gains sufficiently in richness of expression to speak, as the 31-tonists had, of a "total revolution [*totaler Umwälzung*]." As Karkoschka pointed out, purity of intonation by no means depends solely on vibrational relationships. Other factors entered into the equation. There is first the matter of loudness level since an increase in loudness sounds to the ear as an increase in pitch. There is also the issue of variation in listening sensitivity, a factor that characterizes acoustically melodic and harmonic intervals even for the octave. Finally, in order to achieve correlation or togetherness (*Zusammenhang*) it is well known that all singers, and string and brass players, alter their intonations and deviate from purity by values that are much greater than the deviations resulting from impurity within the 12-tone system. How, asks Karkoschka, does the player of a keyboard instrument react to and achieve this togetherness? "Evidently with *Agogik* and differentiations in loudness in order to compensate for variations in intonation; since one always hears music as a whole."[14] Karkoschka had the opinion that the intonation of singers, and string and brass players, neither gains nor loses by using either the 31-tone or the 12-tone system. He maintained that it was much more important to be able to respond dynamically to the numerous demands placed on the musician. It therefore appeared to Karkoschka to be an illusory perspective to base intonation exclusively on 31 rigidly defined tonal steps.

According to Karkoschka the decisive objections to the 31-tone temperament system become apparent when one investigates what it is that constitutes the understanding of music (*Verständnis von Musik*). For the 31-tone system apologists it is, first and foremost, an understanding directed at primacy of tone. For the critic, however, tonal enjoyment (*klanglicher Genuss*) is the smallest part of music; the largest part is dynamic process. Thus it was that in the second half of the twentieth century microtones, and in fact all possible frequencies, became fertile ground for exploring new directions in music and not only in electronic music. The recognition of these new directions in music ensued from substantially different musical thinking and led to totally different manifestations. What Karkoschka had in mind in referring to a shift in musical thinking was that composer were drawing on tonal combinations that were chosen deliberately from among the infinite number of frequencies and intervals to produce certain effects without being unduly constrained to the use of acoustically pure sounds. "And so it turns out," Karkoschka remarked, "that the problem of intonation is exhibited more strongly than ever by the *desired* [emphasis added] tonal image (*Klangbild*), which in turn is

[14] An agogic accent is one that is created by duration rather than loudness or metrical position. The term was introduced by Hugo Riemann in his *dynamik* in 1884.

determined less often by the schematic model of the essentially pure than by the individual tone's color character (*Farbcharakter*). Impurities in inonation are, more or less, purposively composed (*gezielt komponiert*)." Karkoschka essentially was telling Fokker and the 31-tone system enthusiasts: there is no right way to listen to music.

Karkoschka suggested that with Fokker's concept so distinctly bound to the aesthetics of tonal music, it was reasonable to foresee that young composers of the day who throw themselves without hesitation into the most daring ventures would not latch on to the world of the 31-tone temperament. They already had more comprehensive possibilities to pursue at their disposal. Within tonal music, on the other hand, he believed that Fokker's ideas were no more than insignificant extensions of long worn out and abandoned paths. Karkoschka concluded his devastating critique of Fokker's approach to new music on a conciliatory note. "But perhaps tomorrow, or the day after tomorrow, the ideas of Fokker will acquire new significance when transposed into another context [*Zusammenhang*]. No such experiment, if meaningful and if at all possible, should be pushed aside. We need to keep under observation and respect each experiment."[15]

Reference

Fokker, Adriaan. 1966. *Neue Musik mit 31 Tönen.* Vertag der Gesellschaft fur Fürderung der systematischen Musikwissenschaft Düsseldorf.

[15] Erhard Karkoschka, "Review of *Neue Musik*", (1969), 116–117.

Appendix: Willem Pijper and the Efflorescence of Dutch Music

In order to allow for a balanced assessment of Fokker as a music theorist and as a composer, and to provide insights and critical commentary to his views on composing and performing music based on the 31-tone system that he and the other composers who employed that system used in their compositions, it is essential to explore both the composing styles and compositions of other Dutch composers of the early twentieth century and to examine the contextual musical environment in which they worked. We offer, first, some information about the state of Dutch music at the end of the nineteenth century and even earlier – this as background to an examination of the characteristic styles and modes of musical expression exhibited by several of the most important Dutch composers of the nineteenth century.[1]

One of the most striking characteristics of the history of music in the Netherlands is its general lack of continuity. "In this characteristic, which is shared only by England, lies the main difference with the three major cultures of Germany, France and Italy. It is a well-known fact that many of the leading figures in the world of music in the 15th and 16th centuries were Dutch."[2] With the death of Jan Pieterzoon Sweelingk (1562–1621), composer and organist at the *Oude Kerk* in Amsterdam, historians of music speak of a "withering" of musical activity in the Netherlands. Sweelingk, one of the leading composers of the seventeenth century, has been called the "master of German organists" by virtue of the fact that during the

[1] For short biographical sketches of composers who are considered to be Dutch citizens based on prolonged residency in the Netherlands or had a particular bond with the Netherlands: Jolande van der Klis (ed.), *The Essential Guide to Dutch Music. 100 Composers and Their Work*, Amsterdam, 2000.

[2] Marius Flothuis, "An unharmonious Figure in an Unharmonious Age", *Key Notes. Musical Life in the Netherlands, 3* (1976) 27–33. Quote on 27. Marius Flothuis (b. 1914) is a composer, chiefly self-taught, who studied at Amsterdam University and became artistic director of the Concertgebouw Orchestra.

DOI 10.1007/978-3-319-06602-8, © Springer International Publishing Switzerland 2014

eighteenth century the success of his Dutch pupils was transmitted directly to the North German school of composers that included J. S. Bach, Friedrich Händel, and Dietrich Buxtehude.

> However, for 250 years after Sweelinck's death no figures of international importance emerged: there was no Dutch Bach, Mozart or Berlioz, nor even a Scarlatti, Michael Hayden or Mendelssohn. The reasons for this are difficult to determine. A satisfactory answer has yet to be given to the question why in a country where, judging from the painting, music was often played and where many music publishers of international repute were established, so little creative talent came to the fore.[3]

For the Netherlands the year 1914 was a turning point in regard to the musical future of the country. Squeezed as it was between France and Germany as fighting powers, the Dutch managed to stay neutral – a neutrality that provided the Dutch with an opportunity to come to terms with itself at the level of its own intellectual and artistic past. Alexander Ringer has captured the essence of Dutch sentiments during the first World War:

> Despite the official neutrality of the Netherlands, the great war made the individual Dutchman suddenly quite conscious of his cultural allegiances. Intellectuals, artists, and students violently chose sides....Most of the younger composers followed suit, stirred undoubtedly by political sympathies and feelings of human justice, but also instinctively turning to the culture that alone held promise of their musical salvation.
>
> During the four years that saw their respective countries bleed one another to death, the artists and new compositions form both France and the Central Powers peacefully measured their strength before the Dutch public.

According to this interpretation it is unfair to infer that the Dutch abandoned their admiration of music for a stretch of two centuries after Sweelingk; rather it was not until the twentieth century, and more specifically not until after World War I (the motivating factor for the renewal of interest) that Dutch music flourished among Dutch composers and the Dutch people – this with an energetic enthusiasm that was coupled with decidedly nationalist sentiments and spearheaded by a group of young Dutch composers who were able to invent a characteristic mode of musical expression that elicited both Dutch and foreign acclaim.[4]

Our discussion of the contributions of post World War I composers and their role in the rejuvenation of Dutch music begins with Willem Pijper (1894–1947), who generally has been regarded by historians of music as the most significant Dutch composer since Sweelingk. Inasmuch as he set the pace for the renewal of Dutch music in the years between the two world wars his life, views, writings,

[3] Flothuis, "An Unharmonious Figure" (1976), 27.

[4] Alexander L. Ringer, Willem Pijper and the "Netherlands School" of the 20th Century, *The Musical Quarterly, 41* (1955) pp. 427–445. Quote on p. 432.

and compositions necessarily take on major significance for what follows.[5] Unlike Sweelingk, whose influence was directed primarily to North German organists rather than to his Dutch countrymen, Pijper's influence was felt most directly by Dutch musicians and the Dutch musical public. "His creative achievements were the vanguard of a Dutch musical efflorescence, and his acknowledged pedagogic skills stimulated a generation of Dutch taste and society."[6]

During the early years of the first half of the twentieth century, Gustav Mahler (1860–1911) and the French – typically Claude Debussy (1862–1918) – were the two most influential poles toward which young musicians in the Netherlands would feel drawn. Both signified a radical break with the past and to a large extent were mirrored in Dutch music of the twentieth century. As Wouters states the two alternatives: "There is on the one hand the influence of German music – in the case of Diepenbrock particularly that of Gustav Mahler, for whose work so much propaganda was made in the Netherlands by Willem Mengelberg – and on the other the influence of French music, particularly that of Debussy."[7]

---

[5] Willem Pijper, composer, was born in the village of Zeist outside of Utrecht in a working-class and rigid Calvinist milieu. His father was a deacon in the *Herformde Kerk*. After early interests that were divided equally between biology and music, he studied music composition at the Utrecht conservatory under the direction of the Utrecht composer and cathedral organist Johan Wagenaar (1862–1914). In 1918 he began teaching harmony, composition, and instrumentation at the Amsterdam conservatory, and from 1930 until his death he was director of the Rotterdam conservatory. Pijper wrote three symphonies, four string quartets, piano music, and concertos for piano and for piano and cello. Apart from the papers by Marius Flothuis and Alexander Ringer, already mentioned, the most useful writings on Pijper and the composers of his time have been Karel Mengelberg, Willem Pijper – 1894–1947, *Music Today, Journal of the International Society for Contemporary Music 1* (1949) 36042; Jos Wouters, Dutch Music in the 20th Century, *The Musical Quarterly 51* (1965) 97–110; Frank W. Hoogerwerf, Willem Pijper as Dutch Nationalist, *The Musical Quarterly 62* (1976) 358–373; Anton Haakman, Pijper on Pijper, *Key Notes. Musical Life in the Netherlands 23* (1986) 36–40; Harrison Ryker, Mosaics of Tone: Willem Pijper and his Music, *TLC. The Low Countries. Arts and Society in Flanders and the Netherlands* (2004) 136–141; Ryker, Willem Pijper, *NG 19* (2001) 740–43 and *MGG Personenteil 13* (2005) col. 577–580.

[6] Hoogerwerf, "Willem Pijper..." (1976), 358. "The most important source of information concerning Pijper's views and attitudes are the more than six hundred articles and reviews Pijper wrote in his capacity as essayist and critic. Contributing regularly to the *Utrechtsch Dagblad* (1917–23), his own monthly journal *De Muziek* (1926–33), and the weekly *Groene Amsterdammer* (1934–46), and irregularly to a host of other literary journals, Pijper's agile pen left a remarkable legacy of critical prose." *Ibid.*, 360.

[7] Wouters,"Dutch Music" (1965), 98. The Dutch composer Alphons Diepenbrock (1862–1921) was the first Dutch composer whose works would be judged by international standards. Willem Mengelberg (1871–1951) was a Dutch composer who studied in Utrecht and Cologne. In 1895 he became the conductor of the Amsterdam Concertgebouw Orchestra, which had been founded in 1888. He was noted as an outstanding interpreter of Mahler and Strauss as well as of the entire Romantic repertoire, but he exhibited a rather unsympathetic attitude to Dutch composers and especially shunned the compositions of Pijper. This became more and more apparent after 1920 – so that in 1927 while on tour in Germany and Switzerland with his orchestra, not a single Dutch work was programmed. After conducting in Germany during World War II he was banned for life from the Netherlands.

One of Pijper's students, Karel Mengelberg (1902–1984), a nephew of Willem Mengelberg, has remarked: "The performances of Mahler's works by Willem Mengelberg made a deep impression on [the young] Pijper, and their influence may be traced in his compositions, especially in his First Symphony (1917)."[8] Until the end of World War I most Dutch musicians still were under the influence of German music. Thereafter it swerved in favor of French and Italian music. The emphasis on melody, heaviness of sound, and the augmented triad gave way to soberness and the melodic and harmonic features characteristic of Latin music. When the war was over, Schönberg and the Italian Gian Francesco Malipiero (1888–1973) – the most original and inventive of Italian composers of his generation – became the chief object of Pijper's studies. "Though a great admirer of Schönberg's intellectual powers and sense of consequence, Schönberg's system as a whole seemed to Pijper lacking in imagination, in the touch of reality, to be 'insanely German.' Pijper never looked upon himself as a tonalist. . . .With all his new attainments, consistently followed up and forming a characteristic part of his very personal expression, there remained in Pijper an unshaken belief that in music melodic values stand first."[9] In regard to melody and the process of composition Pijper above all made a case for Beethoven. Haakman has maintained that "central to Pijper's consideration of history is Beethoven, against whom he measures every innovation. For Pijper there were two kinds of music, pure music and the subjective music of the 'disciples of Beethoven'. . . .Beethoven is the beacon in the turbulent waves of the time – the time being 1927, Beethoven's Anniversary year, which was Pijper's pretext for engaging in considerations of a general nature. . . .Beethoven. . . . was determined entirely by his defiance of traditional authority. . . .he did not, in my opinion, change the music of his time. He changed the process of composition."[10]

Apart from Pijper's opposition to extraneous romanticism and the German Romantic composers, who did not get much of a hearing from him, the composers with whom Pijper refused to sense any affinity and from whom he eventually distanced himself were Gustav Mahler (1860–1911), Richard Strauss (1864–1949), Arthur Honegger (1892–1955), and in the late 1920s Igor Stravinsky (1882–1971). The ambiguous position of Willem Mengelberg at the Concertgebouw Orchestra in Amsterdam, from the turn of the century until after World War II, especially in relation to Pijper and his Dutch fellow composers, is clarified by Hoogerwerf:

> While Mengelberg's positive attributes – his disciplined musicianship, his interpretive genius, his "artistic internationalism," as Pijper put it – had a beneficial effect on Dutch musical life, nevertheless Mengelberg displayed certain irritating shortcomings, which, as the years went by, became increasingly obvious. The first of these flaws was Mengelberg's essentially romantic orientation.
>
> Secondly, Mengelberg had a rather unsympathetic attitude toward Dutch composers, which became more and more apparent after 1920. As the progressive elements in Dutch

[8] Karel Mengelberg, "Willem Pijper" (1949), 36.

[9] Karel Mengelberg, "Willem Pijper" (1949), 37.

[10] Haakman,"Pijper on Pijper" (1986), 38.

> musical society were often nationalistic, a rather vocal opposition to Mengelberg's taste and influence developed, which found its most articulate representative in Willem Pijper.
>
> The musical situation in the Netherlands between 1920 and 1940 (the most productive period of Pijper's life) may be characterized by pointing out the existence of two factions: the more conservative "establishment" headed by Willem Mengelberg and his coterie, and the other, more progressive, idealistic faction, headed by Willem Pijper. The opposing sides moved further apart and became increasingly hostile as the two decades wore on.[11]

Mengelberg's response to the progressive idealistic faction headed by Pijper came to a head in 1922. As Ryker has reported: "In November Queen Wilhelmina, attending a Concertgebouw concert, plugged her ears with cotton-weel to keep out the sounds of modern music: namely Pijper's Second Symphony, which received a messy premiere under the composer's direction after Mengelberg reneged on a promise to conduct it."[12] At about the same time, in a festival of recent chamber music in Salzburg, Pijper's First Violin Sonata was performed (later repeated in London), participants took this as an opportunity to establish a London-based International Society for Contemporary Music (ISCM) for which Pijper served as Dutch representative. "Its festivals were a showcase for significant contemporary music until well after World War II." For its publication *A Dictionary of Modern Music and Musicians* (London, 1924), Pijper assembled and wrote the Dutch entries. In 1925 Pierre Monteux (1875–1964), American conductor of French birth, was engaged to share Mengelberg's duties with the Concertgebouw Orchestra. Pijper's third symphony, written at the behest of Monteux, was performed repeatedly in the Netherlands, Belgium, and France, and again in 1928 in Philadelphia and New York with the Philadelphia Orchestra. It continued to be performed in the U.S. for many years. "'Pijper's Third [Symphony],' due to its breezy style, touches of Blues, jazz and habañera rhythms, and abrupt outbursts of brilliant orchestration, became the work which would contribute most to the composer's international reputation."[13]

The music of Mahler that so singularly captivated the attention of the conductor of the Concertgebouw Orchestra not only was basically unfamiliar and even puzzling for audiences and critics in Holland but in America. Aaron Copland writing in the *New York Times* (April 2, 1925) remarked:

> The music critics of New York City are agreed upon at least one point – Gustav Mahler as a composer is hopeless. Year in and year out, the performance of one of Mahler's works is invariably accompanied by the same disparaging reviews. Yet no critic has been able to explain just what it is that [the conductor Willem] Mengelberg – and for that matter all Germany, Austria, and Holland – find so admirable in Mahler's music.

[11] Hoogerwerf, "Willem Pijper..." (1976), 361.

[12] Ryker, "Mosaics of Tone" (2004), 139.

[13] Ryker, "Mosaics of Tone" (2004), 139–40.

> If I write in defense of Mahler it is not merely for the pleasure of contradicting the critics. As a matter of fact, I also realize that Mahler has at times written music which is bombastic, longwinded, banal. What our critics say regarding his music is, as a rule, quite justified, but it is what they leave unsaid that seems to me unfair.
>
> If one discounts for the moment the banal themes, the old-fashioned romantico-philosophical conceptions so dear to Mahler – if one looks at the music *qua* music – then it is undeniable that Mahler is a composer of today. . . .
>
> That Mahler has on occasion been grandiloquent is indeniable. . . .[yet most critics] are so prone to discussing Mahler's music in generalities that any one unfamiliar with that composition would be led to suppose that it, too, was full of sound and fury signifying nothing.[14]

According to Marius Flothuis (b. 1914), Holland's outstanding chamber music composer, the work of Pijper falls into three distinguishable periods. During the earliest period, from 1914 to 1920, while searching among composers for musical ideas and style, Pijper wrote his first Mahler-influenced symphony, his first cello sonata, and his first violin sonata. The secondary period, one of "secondary break-through," and characterized as a time of conflict between the heritage of Mahler and French composers, led to Pijper's second and third symphonies and the bulk of his chamber music, and his choral works and theatre music. In the third period, 1933–1947, Pijper's creative output seems to have declined as he returned to more tonal elements, more simplified compositions and older formal principles.[15]

More than anyone else it was Willem Pijper whose compositions, writings, and crusading spirit began to reverse the neglect of Dutch music in the years following World War I. The singular and characteristic earmarks that contributed to that crusading spirit are readily identified in his early compositions of the 1920s based largely on the four-tone germ-cell principle. Pijper originally studied biology and had a great love of plants and animals and the study of flora and fauna for which the term "germ-cell" is invoked to represent the idea that each plant and each animal organism evolves from a single cell, and develops, but never returns to a previous stage. In the organic world everything develops from a single cell and there are no repetitions. In music this idea led to the abandonment of literal repetition and meant that the structure of Pijper's music was essentially different from much of the music written in his day and earlier. It has been reported that "[o]ne of Pijper's mottoes, 'a composer only composes what he cannot express or communicate in other ways,' fits his own works. They say much, in a vital and unique way; they are of high intrinsic value. His is some of the most inventive and compelling music created in Western Europe during the inter-War years, and it has found its place in the concert repertoire."[16]

---

[14] Boston Symphony Orchestra, *Program Notes for the 128th Season 2008/9*, 29–30.

[15] The two most comprehensive book-length studies of Piijper as composer are Harrison Ryker, *The Symphonic Music of Willem Pijper* (Ph.D. Diss., Univ. of Washington, 1971); and Frank Hoogerwerf, *The Chamber Music of Willem Pijper* (Ph.D. Diss., Univ. of Michigan, 1974).

[16] Ryker, *The Symphonic Music* (2004), 141.

The significance of the germ-cell principle and the derivative Pijper scale are explained and rationalized by Ryker as follows:

> Pijper's germ cell was like a random piece of stone mosaic, running throughout the musical texture with little regard to key or metre. The approach was uncompromising. No Dutch composer in several centuries had written anything so forthrightly experimental.
>
> The innovations which made Pijper's music unique were not drastic like those of Schönberg, but incremental. The 1920s germ-cell pieces were notable for their ever-changing structures, as these small motivic fragments did not meld with inherited musical forms.
>
> The build-up of chord layers in contrasting keys seems to be his own discovery. Thereafter he stumbled on a remarkable octatonic (eight-note) scale, neither major nor minor, of alternating tones and semitones. This scale retained several familiar properties from the tonal past but could also be used polytonally. Though employed extensively by Stravinsky, it became known as the 'Pijper Scale' due to its widespread popularity in the Netherlands. The polytonal, octatonic and polymetric properties, taken as a whole, make his music immediately recognizable to the ear.[17]

In regard to the principle of repetition that traditionally has been an element of all kinds of musical technique, and that Pijper sought to avoid, Flothuis offers the following clarification:

> Literal repetition was without doubt used much less frequently in early 20th-century music than, for instance, in the music written in the days of Haydn and Mozart. But Pijper had another reason for wishing to avoid repetition. The way in which he developed certain harmonic phenomena in the music of Mahler and Debussy gave rise to combinations which we call "polytonal." Pijper was convinced that the harmonic structure was closely related to the formal structure, and indeed the repetition found in "classical" forms cannot be viewed in isolation from the principles of harmony. So it followed that he had to modify not one of the two, but both at the same time. And of course the same applied to melody and rhythm. He was more consistent in this respect that Schoenberg, whom he accused – rightly, to my mind – of trying to graft new compositional procedures onto old formal principles in some of his works.[18]

Pijper occupied a position at the center of Dutch music throughout his life. As a Francophile who berated the Austro-German hegemony in music and the teutonization of Dutch culture, Pijper became the spokesman and self-appointed figurehead of the younger generation of avant-garde composers. His efforts to arouse the Dutch public to realizations of a creative talent among their own composers was a lifelong crusade. His interest in biology and emphasis on experimentation in music was coupled with fascination for all sorts of technologies that bordered on affinity for a futuristic outlook. He attracted a large number of students and rose to prominence in the Netherlands as composer, polemicist, and above all as a teacher. Widely acknowledged is the "fact that the majority of Dutch composers whose works form part and parcel of the [20th-century] Dutch repertory have been pupils of Willem

[17] Ryker, *The Symphonic Music* (2004), 138.

[18] Flothuis, "An unharmonious figure" (1976), 29.

Pijper. It was in fact primarily Pijper who determined the nature of Dutch music in the years between the two World Wars and who at the same time succeeded in giving it a place in the musical life of Europe."[19] As a teacher at the Amsterdam and Rotterdam Conservatories for more than two decades, he trained a large number of the generation of composers of the first decades of the twentieth century.

> In his teaching, along with his fiery idealism there was the strictness of a highly logical way of thinking and clarity of expression. His teaching was however always directed towards making his pupils think musically in an independent and personal way, and although many of them willingly conformed to his principles of composition, he considered it absolutely necessary to leave each of them free in processing the musical material. Pijper's starting point as a teacher was the discovery of one's own personality and the development of one's own particular manner of musical expression.[20]

Others portrayed Pijper's teaching style as a game in which the dormant musical potentialities of Dutch composers were on a par with those of any other nation.

> Pijper never imposed his own views upon others with this one exception: he insisted with an almost fanatical perseverance that the Dutch could produce good music like any other nation if only they explored their own possibilities to the Germans, he once said, music may be a religion; the Dutch will always consider it a game, but a serious one.[21]

As a teacher Pijper formed an entire generation of composers.[22]

In 1930 Paul F. Sanders, one of Pijper's close friends, wrote: "The music of Willem Pijper is of a richness and variety nowhere to be found in the history of the music of our time. . . .With Pijper, for the first time in centuries Dutch music is once again part of the European musical scene."[23] In 1986 the composer and writer Leo Samama remarked that alongside Matthijs Vermeulen, Henk Badings, Rudolf Escher, and Ton de Leeuw, he counted Pijper among the five greatest composers of Dutch music between 1915 and 1970.[24] After the second World War Pijper's compositions became more widely disseminated than they were during his lifetime, and the music of his students and their students in turn followed suit. Pijper's music continued to be performed and recorded 40 years after his death. In 1987 the publisher Donemus, in cooperation with the Rotterdam Philharmonic, issued a

[19] Wouters, "Dutch Music" (1965), 97.

[20] Wouters, "Dutch Music" (1965), 101.

[21] Ringer, "Pijper and the 'Netherlands School'" (1955), 443.

[22] Karel Mengelberg lists the following: Henriette Bosmans, Hans Henkemans, Piet Ketting, Guillaume Landré, Bertus van Lier, Iet Stants, Wolfgang Wijdeveld, Kees van Baaren, Rudolf Escher, Johan Franco, Henk Badings and Karel Mengelberg. Mengelberg, "Willem Pijper..." (1949), 42.

[23] Paul F. Sanders, *Moderne Nederlandsche* Componisten, 's Gravenhage (1930), 105–107.

[24] Leo Samama, *Zeventig jaar Nederlandse muziek 1915–1985: Voorspel tot een nieuwe dag*, Amsterdam, 1986, Chap. IV.

series of his orchestral works.[25] In *De Quintencirkel* (1929), a work in which Pijper appealed to the new national consciousness and to Dutch intelligence and statesmanship, he laid out his vision of the twentieth-century renaissance in Dutch music:

> [W]e lack faith in our own strength, in honest appreciation of our own importance. Dutch musicality is not an emulsion of German profundity and French *savoir faire*; Dutch musicality is the consequence of five centuries of cultural unity. We have repeatedly lost sight of that unity. . .but after periods of oblivion it has repeatedly surfaced again. At the present moment we are standing on the threshold of a new national consciousness; our people will have to wake up to the fact that a powerful new Dutch music exists. We no longer need to turn to other countries for advice and instruction; our years of vassalage are gone for good. Dutch musicality is not inferior to Dutch intelligence, to Dutch statesmanship. The new Dutch music has arisen from a conscious rediscovery of the truths which our great men of the seventeenth century knew.[26]

References

de Beer, Roland. 1986. A Photographic Portrait of Willem Pijper. *Key Notes* 23: 34–34.

Flothuis, Marius. 1976. An unharmonious figure in an unharmonious age. *Key Notes. Musical Life in the Netherlands* 3: 27–33.

Haakman, Anton. 1986. Pijper on Pijper. *Key Notes. Musical Life in the Netherlands* 23: 36–40.

Hoogerwerf, Frank W. 1976. Willem Pijper as Dutch Nationalist. *The Musical Quarterly* 62: 358–373.

Ringer, Alexander L. 1955. Willem Pijper and the "Netherlands School" of the 20th century. *The Musical Quarterly* 41: 427–445.

Wouters, Jos. 1965. Dutch music in the 20th century. *The Musical Quarterly* 51: 97–110.

[25] Roland de Beer, A Photographic Portrait of Willem Pijper, *Key Notes 23* (1986), 34–34. Pijper's works were issued six times on semi-commercial labels by companies such as Philips and CBS.

[26] Hoogerwerf, "Willem Pijper" (1976), 365.

MIX
Papier aus verantwortungsvollen Quellen
Paper from responsible sources
FSC® C105338

If you have any concerns about our products,
you can contact us on
ProductSafety@springernature.com

In case Publisher is established outside the EU,
the EU authorized representative is:
Springer Nature Customer Service Center GmbH
Europaplatz 3, 69115 Heidelberg, Germany

Printed by Libri Plureos GmbH
in Hamburg, Germany